Integrated Researches in Lactic Acid Bacteria

Edited by **Sean Layman**

R Callisto
Reference

New York

Published by Callisto Reference,
106 Park Avenue, Suite 200,
New York, NY 10016, USA
www.callistoreference.com

Integrated Researches in Lactic Acid Bacteria
Edited by Sean Layman

International Standard Book Number: 978-1-63239-432-3 (Hardback)

Printed in the United States of America.

Contents

Preface

This book has been a concerted effort by a group of academicians, researchers and scientists, who have contributed their research works for the realization of the book. This book has materialized in the wake of emerging advancements and innovations in this field. Therefore, the need of the hour was to compile all the required researches and disseminate the knowledge to a broad spectrum of people comprising of students, researchers and specialists of the field.

This book collates the researches on Lactic Acid Bacteria (LAB). Continuing scientific research in several parts of the globe on the proteomics, genomics and genetic engineering of lactic acid bacteria is enhancing our knowledge of their physiology, pushing further the boundaries for their potential applications. This book represents a collection of the excellent scientific research activities regarding the future applications of LAB. The book has been organized under four sections: Dairy Food Products, Meat Products, Vegetable & Cereal Products, and Health Applications Purposes. It is intended for a great spectrum of readers including researchers, students, corporate R&D and academicians interested in this subject.

At the end of the preface, I would like to thank the authors for their brilliant chapters and the publisher for guiding us all-through the making of the book till its final stage. Also, I would like to thank my family for providing the support and encouragement throughout my academic career and research projects.

Editor

Dairy Food Products

Lactic Acid Bacteria Resistance to Bacteriophage and Prevention Techniques to Lower Phage Contamination in Dairy Fermentation

A.K. Szczepankowska, R.K. Górecki,
P. Kołakowski and J.K. Bardowski

Additional information is available at the end of the chapter

1. Introduction

The first negative effect of bacteriophages on dairy fermentation was reported in the mid 30s of the XX century [1]. Regardless of sanitary precautions, starter strain rotations and constant development of new phage-resistant bacterial strains, phages remain one of the main and economically most serious sources of fermentation failures. Due to their natural presence in the milk environment, bacteriophages cause problems in industrial dairy fermentations world-wide. Their short latent period, relatively large burst size and/or resistance to pasteurization makes them difficult to eliminate [2]. Phage-induced bacterial cell lysis leads to failed or slow fermentation, decrease in acid production and reduction of milk product quality (e.g. nutritive value, taste, texture, etc.), which in effect cause profound economical losses [3]. An intriguing high number of bacteriophages of *Lactococcus* and *Streptococcus* bacteria reflects the biotechnological interest and engagement of the dairy industry in research on biology of these phages [4].

Since *Lactococcus lactis* strains are widely used as starter cultures for milk fermentation during manufacturing of many types of cheeses, sour cream and buttermilk, bacteriophages virulent against these strains appear commonly in the fermentation environment. It is estimated that 60 – 70% of technological problems in production of cottage and hard cheeses are caused by bacteriophage infection of bacteria from the *Lactococcus* genus [5]. The raise of interest in lactococcal phages due to economical aspects has subsequently led to a more global research on the biology of lactococcal phages, ways of their appearance in dairy environments and means of their elimination as well as characterization of phage resistance mechanisms encoded by bacteria exploited by the industry.

2. Lactic acid bacteria used in dairy industry

Lactic acid bacteria (LAB) comprise different groups of microorganisms, such as *Carnobacterium, Enterococcus, Tetragenococcus, Vagococcus, Weissella* as well as species of genera which constitute the "industrial" core of LAB, like *Lactococcus, Lactobacillus, Streptococcus, Pediococcus* and *Leuconostoc* [6]. LAB reside in different natural habitats, including healthy and decaying plants, milk and dairy products, oral cavity and gastrointestinal tract of humans and animals. In addition, lactic acid bacteria can grow on meat and wine. These features are used in the production of fermented sausages (*Lactobacillus, Pediococcus*) and to improve the organoleptic characteristics of wine (*Oenococcus oeni*) [6].

The genus *Lactococcus* is the best characterized food-related LAB. As lactococcal strains are able to grow in milk and transform lactose to lactic acid, they are commonly used as starter cultures in industrial fermentations for cheese production. The ability of LAB to transform raw milk into other products suitable for consumption has been used by man for millenniums. Such long history record of interactions of man with lactic acid bacteria and present knowledge led to assigning these bacteria the GRAS status (generally recognized as safe) [7]. Dairy products and the respective LAB species are gathered in Table 1 based on specifications and recommendations released by the main culture suppliers.

A typical lactococcal mixed starter culture consists of 2-3 well defined strains, which specific properties have significant impact on the texture and flavor of the end product. Nowadays, large dairy plants process up to 10^6 liters of milk per day, producing annually approximately 10^7 tons of cheese [8]. Therefore, technological problems in production of cottage and hard cheeses caused by bacteriophage infections have serious economical consequences.

3. Lactic acid bacteria phages – history background, morphology, classification

The history of discovery of bacteriophages originates in the research of Felix d'Herelle and Frederick Twort in the beginning of the XX century and further development of phage biology studies spans the fourth quarter of the last century. Bacteriophages (phages) are defined as viruses that exert their activity against prokaryotic cells – both bacterial as well as archeal.

The name "bacteriophage" derives from the Greek word "phagein", meaning "to eat", which points to their destructive action. Bacteriophages exist in two states – extra- and intracellular – which place them half-way between live organisms and non-viable forms. As obligate intracellular parasites their survival is dependent on host organisms. Phage "life functions", such as genome replication and synthesis of capsid components, are restricted to occur within infected cells. Outside of the host phages are regarded as metabolically inert, unable to carry out neither biosynthetic nor respiratory functions.

Phages intrigue by their simplistic organization and submicroscopic sizes. These infectious particles consist of a single- or double-stranded nucleic acid genome (DNA or RNA), enveloped in a protein structure (capsid). Current taxonomy and classification of bacteriophages rely on the type of nucleic acid genome and phage morphology, physiology (temperate and virulent life cycles) and genomics. Taxonomy of viruses is supervised by the International Committee for Taxonomy of Viruses (ICTV) that imposes rules for names and writings.

Product	LAB species
Yoghurt	*Streptococcus thermophilus, Lactobacillus delbrueckii* subsp. *bulgaricus*
Cottage cheese, Cheddar, Pasta Filata	*Lactococcus lactis* subsp. *lactis, Lactococcus lactis* subsp. *cremoris* *Streptococcus thermophilus**
Tvarog, blue cheese	*Lactococcus lactis* subsp. *lactis, Lactococcus lactis* subsp. *cremoris* *Lactococcus lactis* subsp. *lactis* var. *diacetylactis,* *Leuconostoc mesenteroides* subsp. *cremoris* *Leuconostoc mesenteroides* subsp. *mesenteroides*
Butter milk, fermented cream, butter	*Lactococcus lactis* subsp. *lactis, Lactococcus lactis* subsp. *cremoris* *Lactococcus lactis* subsp. *lactis* var. *diacetylactis*
Ryazanka	*Streptococcus thermophilus***
Cheddar, Feta	*Lactococcus lactis* subsp. *lactis, Lactococcus lactis* subsp. *cremoris*
Mozzarella, Pizza cheese	*Streptococcus thermophilus, Lactobacillus delbrueckii subsp. bulgaricus*
Masdamer, Gouda, Edam, Tilsitter, soft mould ripened cheese, quark, fermented milk beverages	*Lactococcus lactis* subsp. *lactis, Lactococcus lactis* subsp. *cremoris* *Lactococcus lactis* subsp. *lactis* var. *diacetylactis* *Leuconostoc mesenteroides* subsp. *cremoris*
Mozzarella, Swiss, stabilized soft mould ripened cheese	*Streptococcus thermophilus*
Swiss, Grana	*Lactobacillus helveticus, Lactobacillus delbrueckii* subsp. *lactis*
Fermented cream, fermented milk beverages	*Lactobacillus acidophilus, Streptococcus thermophilus*
Actimel®-like products	*Lactobacillus casei, Lactobacillus paracasei* subsp. *paracasei* *Lactobacillus rhamnosus, Lactobacillus acidophilus*
Swiss, Italian	*Lactococcus lactis* subsp. *lactis, Lactococcus lactis* subsp. *cremoris* *Lactococcus lactis* subsp. *lactis* var. *diacetylactis* *Lactobacillus helveticus, Lactobacillus delbrueckii* subsp. *lactis*

* seldom applied in cottage cheese, ** texturizing strains

Table 1. Various dairy products and LAB species applied in their production.

The majority of known viruses are bacteriophages, which infect cells of Eubacteria and Archaea. It is also accepted that most phages (96%) isolated so far belong to one taxonomic order of *Caudovirales* [9]. Bacteriophages within this order contain tails and a linear dsDNA genome. They are further classified into three phylogenetically linked families of: *Myoviridae, Siphoviridae, Podoviridae* [9]. *Myoviridae* phages contain a long and contractile tail, while *Siphoviridae* and *Podoviridae* are equipped with a non-contractile tail, long and short, respectively [10]. Isometric heads are dominating (85%) in the morphology of phages from all three families [11]. It is worth to mention that 61% of known phages are classified into the *Siphoviridae* family, of which most of them infect strains of enterobacteria (906 phages), *Lactococcus* (700), *Bacillus* (380) and *Streptococcus* (290) [9]. Apart from the tailed *Caudovirales* phages, there are others demonstrating filamentous, pleomorphic or polyhedral morphology.

Bacteriophages, although simple in organization, are the most diverse life forms in the biosphere. Their apparent heterogeneity is reflected by various features – both morphological as genetic, and their persistence on Earth, estimated as high as 10^{31}, outnumbers by far their bacterial hosts [12]. Phages inhabit various niches, like oceans [13], thermal waters [14], gastrointestinal tract [15] and superficial ecosystems created by man, including fermentation tanks in dairy industry [16]. Hence, their impact on the microbial world cannot be underestimated.

Bacteriophage genome structure, indicating linear and double-stranded characteristics of the DNA molecule, containing or not cohesive ends and sometimes presenting terminal redundancy and circular permutation, describes the general feature of LAB phage genomes.

4. Molecular mechanisms of phage infection of LAB

To enter the host, phages firstly come in contact and adsorb to the bacterial cell wall. The adsorption process has been well studied in Gram-negative bacteria, where it was found that two components are involved in the phage-host interaction. One of them is a receptor located in the bacterial cell envelope (membrane or wall), whereas the second component, called the receptor binding protein (RBP), is presented on the phage surface. RBP is responsible for recognition and binding of the phage particle to the bacterial receptor [17]. In the first stage of phage infection, the RBP protein recognizes and binds to a suitable sugar receptor. However, such binding is reversible and thus, the initial phage-bacteria interaction does not ensure commencement of a successful infection event. In contrast to this, in the second stage, a stable phage attachment to the bacterial cell occurs due to an irreversible binding between proteins located on bacterial and phage surfaces [18]. Both stages of adsorption are observed in Gram-positive bacteria: phages that attack *Lactococcus lactis* cells bind to specific receptors, mainly sugars, located in the cell wall. It is widely known that rhamnose, glucose, galactose, and galactosamine are compounds with which the phage RBP interacts at the initial stage of adsorption [19]. In the case of *Lactococcus* c2-type phages, effective infection requires interaction between phage and the bacterial protein Pip (phage infection protein) [20]. The Pip protein of *L. lactis* is an integral membrane protein [21] and its interaction is crucial both for establishing the reversible and irreversible contact between the phage and the host. In contrast to c2-type phages, phages representing P335 and 936 groups bind to other various bacterial membrane proteins and have been examined in a lesser extent [22]. After establishing a tight connection, they inject their genetic material inside the host cytoplasm, while the capsid remains outside the cell. Then, subsequent steps of phage infection are effectuated which follows either the lytic or lysogenic life cycle. Phages entering the lytic mode immediately redirect the host replication machinery and metabolic functions to replicate its own genetic material and synthesize phage encoded proteins. In effect, abundant amounts of progeny particles are produced. Phages executing only the lytic cycle are designated as virulent and their infection implicates cell death. Yet, certain phages termed as temperate can lead an alternating existence between a dormant state inside the bacterial cell and lytic growth. These phages can exist in the cell in a latent form for generations, replicating in synchrony with the bacterial chromosome. A dormant form of the phage is called a prophage and leads a lysogenic life cycle in a bacterial host strain, which is regarded as a lysogen. Conversion from the lysogenic life cycle to the lytic often occurs spontaneously or can be induced by various mutagens (UV, mitomycin).

5. Phage sensitivity of LAB starters used in dairy industry

Virulent phages of *Lactococcus lactis* spp. are the most frequently encountered phages in milk plants during cheese and dairy beverages production. Additionally, phages attacking *Streptococcus thermophilus* are often observed in cheese and less distinct in yoghurt manufacturing. Phages against *Lactobacillus* spp. and *Leuconostoc* spp. starter cultures represent a minor problem [23]. Currently, in production of dairy beverages functional *Lactococcus* and *Streptococcus thermophilus* texturizing strains with ability to produce exopolysaccharides (EPS) are commonly used. In nature it is very difficult to find strains with similar rheological properties differing in resistance to phages. Thus, phage contamination of texturizing strains can lead to serious problems in ensuring quality dairy products.

6. Defense mechanisms of lactic acid bacteria

It is well documented that lactic acid bacteria evolved defense systems against bacteriophages, which allow them to survive in an environment full of their predators. These anti-phage systems have been organized into five groups depending on the manner by which they operate: (i) inhibition of phage adsorption, (ii) blocking of phage DNA injection, (iii) restriction modification systems, (iv) phage abortive infection systems, and finally, the most recently described, (v) CRISPR/*cas* systems. The knowledge about natural phage resistance mechanisms together with a set of genetic tools were applied to develop also (vi) engineered defense systems that confer higher levels of resistance and/or broader phage specificity.

6.1. Inhibition of phage adsorption

Basic mechanisms of inhibition of phage adsorption to the bacterial cell are associated either with physical masking of the receptor or with changes in its structure, or even with its absence in the cell envelopes [24]. Lack of a functional receptor might be due to spontaneous mutations in the genetic material, leading in turn to bacteriophage insensitive mutants (BIM). A good illustration of the BIM phenomenon is a lactococcal mutant in the chromosomally-encoded *pip* gene. The resultant strain is unable to interact with phages of the c2 group, revealing high level of c2-specific resistance [24] (for further details on BIMs see section 12.2.).

Mechanisms preventing phage adsorption are not only mediated by the bacterial chromosome, but also by acquired plasmids. The best documented plasmid-encoded mechanisms of inhibition of phage adsorption rely on either direct synthesis of cell surface antigens or the production of extracellular carbohydrates. Of the two modes of action, the former reveals phage specificity, whereas the latter seems to restrict access to the bacterial cell for various harmful factors, including bacteriophages [25]. Studies carried by Tuncer and Akcelic demonstrated that a 28.5-kb plasmid, isolated from *L. lactis* subsp. *lactis* MPL56, causes complete inhibition of four lactococcal phages due to the production of a 55.4-kDa protein [25]. The protein exhibits similarity to lectins, a group of proteins that adsorb to specific monosaccharide components of polysaccharides in the cell wall, hence, impairing specific recognition of the phage receptor sites by these four phages. Thus, this plasmid-encoded 55.4-kDa protein shields specifically the galactose-containing receptor rather than interacts with the phage, in other words, the bacterial lectin and the phage RBP compete for the receptor [25]. Another example of physical masking of the receptor is the plasmid-mediated production of extracellular carbohydrates, called exopolysaccharides (EPS) [26]. Such EPS envelope coats the cell surface giving bacteria extra protection, not only against bacteriophages, but also against desiccation. There is some evidence that EPSs contain sugar residues that are similar or even identical to initial phage receptors. Therefore, phage insensitivity of LAB strains that carry EPS-encoding plasmids, for instance, pCI658, might be due to phage immobilization by binding to EPS [26]. On the other hand, polysaccharides have an impact on the properties of dairy products, like: texture, viscosity and smoothness of mouthfeel. Thereby, application of EPS-producing phage-resistant strains might be limited to a narrow range of dairy products [25-26].

6.2. Blocking of phage DNA injection

After phage binding to the receptor, phage DNA is introduced into the bacterial cell. In the cytoplasm, phage genetic information is amplified and consequently progeny particles are produced. However, studies of Watanabe on the interaction between phage PL-1 and a *Lactobacillus casei* strain showed no bacterial lysis, despite phage adsorption to cell envelopes [27]. An electron microscopy image indicated that the phage DNA remains intact in the capsid. In contrast to this, a significant increase in the number of empty capsids was observed on the surface of the sensitive strain. In the light of this evidence, it is obvious that phage DNA injection might be interrupted, although the adsorption of phages to the cell surface occurred. Intensive attempts to elucidate the injection blocking phenomenon have allowed identifying different Sie (superinfection exclusion) or Sie-like systems. On the other hand, only few of them have been well characterized [28]. Therefore, the mechanism preventing entry of phage DNA to the cell is still poorly understood, both in LAB and other microorganisms. Surprisingly, it was discovered that most *sie* genes are located within the prophage regions of the bacterial chromosome [28]. However, the first lactococcal injection blocking system was identified on the pNP40 plasmid, which blocks DNA penetration specifically for φc2 phage of the lactococcal c2 phage group [29]. As it was described in the previous section, the membrane Pip protein is essential for c2 adsorption to *Lactococcus lactis*. It was speculated that the pNP40-encoded protein product might have an impact on the activity, production, or membrane insertion of Pip, thereby affect its biological function and prevent phage DNA entry [29]. The first description of a *sie* system of *Lactococcus* was published in 2002 and referred to the P335-type temperate lactococcal bacteriophage Tuc2009 [30]. After integration of the bacteriophage Tuc2009 genome into the lactococcal chromosome, the prophage protein Sie_{2009} is produced and blocks superinfecting phage DNA entry into the cell. The blocking mechanism has not been fully elucidated; nevertheless, it has been proposed that Sie_{2009} interacts with factor(s) responsible for initiating the phage DNA release from the capsid. Alternatively, the Sie_{2009} protein might interact with cell membrane proteins that are essential for DNA translocation. The effect of Sie_{2009} seems to be analogous to the effect of the lysogenic phage repressor (CI) preventing re-infection. In contrast, the presence of the sie_{2009} gene determines resistance to various phages, also to phages from other species [28,30]. Similarly to lactococci, in lactobacilli prophages are also a common phenomenon [31]. Comparative genomics of lactobacilli revealed the presence of genes coding for putative proteins with a close sequence match to a surface-exposed lipoprotein encoded by bacteriophage TP-J34 of *Streptococcus thermophilus*, another bacterial species used in industrial milk fermentation processes. The TP-J34 prophage carries a Sie-like system consisting of the *ltp* (lipoprotein of temperate phage) gene, encoding a surface-exposed lipoprotein of biologically proven phage-resistance functions. In view of the fact that the *sie* genes of lactic acid bacteria are located on lysogeny modules of prophages and confer infection exclusion, they have been termed phage-derived phage resistance systems [32].

6.3. Restriction modification systems

Following successful injection of DNA, phage infection might be completed or hindered by the presence of restriction modification systems (RM). RM systems comprise two activities represented by the following enzymes: endonuclease (restriction) and methyltransferase (modification) [33]. Simultaneously, both activities are specific to the same target sequences. The endonucleolytic activity is responsible for degradation of invading foreign DNA, including phage DNA, which lack a unique methylation pattern, while the methyltransferase activity protects the host DNA against degradation by introducing a methyl group into a specific nucleotide of the target site [34]. In detail, phage DNA usually reveal different methylation patterns than those recognized by innate RM systems. Unmethylated target sequences are significantly susceptible to endonucleolytic attack, resulting in DNA degradation [35]. Such mode of action guarantees that the presence of RM systems limits phage proliferation in the cytoplasm, causing no harm to the cell. RM systems are classified into four groups, based on their molecular structure, co-factor requirements, sequence recognition and cleavage position [34-36].

6.3.1. Type I RM

Type I is the most complex RM system in terms of genetic organization and biochemical activity. It is composed of three different *hsd* (host specificity determinant) genes coding for the following subunits: HsdR - responsible for restriction, HsdM - involved in modification and HsdS - responsible for specific sequence recognition. None of them reveals any activity as a single protein [36]. In order for the modification activity to occur, a combination of one HsdS and two HsdM subunits is required. The M_2S_1 multifunctional enzyme acts as protective methyltransferase, which modifies DNA through the transfer of the methyl group from S-Adenosyl-methionine (AdoMet) to the specific adenines in the recognition site [36,38]. For restriction activity, all subunits are absolutely required in a stoichiometric ratio of $R_2M_2S_1$. This holoenzyme exhibits both endonucleolytic and helicase activities, and is active only in presence of Mg^{2+}, AdoMet and ATP [36].

Besides the complex structure of this multifunctional enzyme, also structure of the recognized sequences and cleavage position are the distinguishing features of type I RM systems. Type I RM enzymes specifically recognize asymmetric and bipartite sequences. These non-palindromic DNA sequences consist of two specific components, one of 3-4 bp and the other of 4-5 bp, separated by a 6-8 bp non-specific sequence [34,36-37]. The innate methylation state of the target sequence determines the activity of the multifunctional $R_2M_2S_1$ enzyme. When the target sequence is methylated or semi-methylated (e.g. just after replication), the enzyme will exhibit activity of a methyltransferase, which completes DNA modification. In contrast, if the holoenzyme binds to an unmethylated recognition site, DNA translocation past the DNA-enzyme complex occurs in an ATP-dependent manner [35,38]. In spite of DNA translocation, the enzyme remains bound to the target site. DNA is cleaved at a position, where either collision with another translocating complex has appeared or translocation is halted due to the topology of the DNA substrate. Consequently, type I restriction enzymes cleave DNA randomly at a nonspecific site, far from the recognition sequence [38].

Interaction between subunits, leading to formation of multifunctional enzymes as well as interaction of resultant enzyme molecules with DNA, are determined by the structure of the HsdS subunits. HsdS subunits consist of regions, which amino acid sequences are conserved within an enzyme family, and two independent target recognition domains (TRD) that share low level of amino acid identity [34,39]. TRDs are involved in target sequence recognition, each TRD recognizes one-half of the split target site and is responsible for DNA binding. Since TRDs are highly variable, they recognize multiple target sequences, and thus, provide a variety of phage resistance types [34,36,39]. The central domain, located between two TRDs, is responsible for interaction with one HsdM subunit. Other conserved regions located at N and C termini have been proposed to form a split domain, which makes contact with a second HsdM subunit [35,37].

Type I systems have been further classified into four families based on genetic and biochemical criteria, such as: gene order, identity at amino acid level, complementation assay and enzymatic properties. RM systems belonging to type IA, IB, and ID are only chromosomally-encoded, while most complete type IC systems are either chromosomal or carried on large conjugative plasmids [36]. Additionally, numerous small plasmids carry the *hsdS* gene alone [34,40]. While all subunits belong to the same subtype, a plasmid-encoded HsdS protein is able to form a multifunctional enzyme with chromosomally-encoded HsdM and HsdR subunits [41]. Thus, acquisition of a new *hsdS*, revealing new sequence specificity, leads to the increase of phage resistance.

Among LAB, type IC systems seem to be most widespread. Type IC RM loci of both *L. lactis* IL1403 and *L. cremoris* MG1363 consist of three genes: *hsdR, hsdM, hsdS*, and two promoters, one for transcription of *hsdR* and the other for transcription of both *hsdM* and *hsdS* [17,42]. Nevertheless, there is no clear evidence for transcription regulation of type I RM enzymes [42]. Under these circumstances, an unmodified chromosome is exposed to endonucleolytic digestion after acquisition of either a new system or just the subunit specificity genes. It was observed that a delay in the appearance of restriction activity, which ensures the survival of recipient cells in the absence of complete modification of chromosomal target sites, depends on host function [36,43]. Chromosomally-encoded energy-dependent proteases ClpP and ClpX, co-operating in a complex, are implicated in the regulation of restriction activity [36]. The ClpXP complex is responsible for restriction alleviation through proteolytic degradation of HsdR subunits. Based on results of Janscak and colleagues concerning the EcoR124I endonuclease, an alternative mechanism of delay in restriction alleviation has been proposed. As each of the two HsdR subunits interacts differently with HsdM, it has been postulated that the control of restriction activity is implemented at the level of subunit assembly [38]. Formation of a weak $R_2M_2S_1$ restriction complex will be suspended, unless accumulation of HsdR molecules occurs. Excess of HsdR over HsdM is observed in the late stage of establishing of the RM system in a recipient cell; hence, the unmodified chromosome is protected against premature restriction activity [38].

6.3.2. Type II RM

In contrast to type I, type II RM systems are structurally the simplest of all restriction modification systems. They are generally encoded by two genes, but the key defining feature of this RM type is the independent activity of restriction and modification enzymes [33]. Methyltransferase is active as an asymmetric monomer, requires only AdoMet, and recognizes the same target sequences as the cognate endonuclease. In contrast, restriction endonuclease is a homodimer and requires divalent Mg^{2+} cations for proper activity. Endonucleases generally recognize a palindromic 4-8 bp DNA sequence and cleave within or in a fixed distance of the recognition site. In contrast to type I, ATP has no effect on the cleavage activity of type II endonucleases [44].

As this RM type is more heterogeneous in respect to endonucleolytic activity than originally thought based on their structural simplicity, the described mode of action refers mainly to typical (orthodox) type II endonucleases [45].

Apart from the orthodox type (called IIP), type II restriction enzymes have been categorized into the following subclasses: IIA, IIB, IIC, IIE, IIF, IIG, IIH, IIM, IIS and IIT. Endonucleases of these subclasses differ in structure of the recognized sequence (asymmetric or symmetric), cleavage positions and cofactor requirements. Type IIA endonucleases behave similarly to the orthodox class, but recognize asymmetric sequences [45]. The unique feature of subclass IIB refers to the cleavage position. These endonucleases cut DNA from both sides, which results in complete extraction of the target sequence from the DNA molecule [46]. Subclasses IIC and IIE have both modification and restriction domains present in one polypeptide. Additionally, class IIE endonucleases interact with two copies of their recognition site, one copy being the target for cleavage, the other serving as an allosteric effector [47]. Similarly to subclass IIE, class IIF restriction enzymes interact with two copies of their recognition sequences, but cleavage occurs at both sequences. Type IIG restriction enzymes seem to combine properties of both IIB and IIC subclasses. The methyltransferase activity of class IIG, like IIB, is stimulated by AdoMet. The main similarity between IIG and IIC is that they both have restriction and modification activities located on one polypeptide chain [45,47]. Subclass IIH, represented by the AhdI system, appears to be a novel RM system due to its genetic organization resembling that of type I. As in type II systems, the AhdI endonuclease is encoded by a single gene; on the other hand, similarly to type I, its cognate methyltransferase forms a complex consisting of two modification and two specificity subunits [44,48]. Subclass IIM is at the opposite extreme from other type II subclasses as it recognizes and cleavages methylated target sequences. The key distinguishing feature of type IIS is the cleavage position outside of the recognition sequence at a defined distance [49]. Subclass IIT is an example of a variation in the typical genetic organization of type II RM systems, as the endonuclease is composed of two different subunits. Moreover, some IIT endonucleases function not only as heterodimers, but also as heterotetramers [44-45].

As enzymes belonging to type II systems are the most abundant and mainly encoded on plasmids, they can be acquired by the bacterial cell through plasmid transfer events. Therefore, a question arises as how to protect the host cell against an incoming endonuclease. In many cases, each gene of the type II RM system has its own promoter. Thus, a delay in appearance of the endonuclease activity is regulated at the transcriptional level. The lactococcal LlaDII RM system is a good example which illustrates this type of regulation [50]. At the initial stage of establishing in the host cell, the LlaDII methyltransferase is overexpressed, whereas the restriction enzyme is produced in small amounts due to the weak constitutive expression of its gene. On the other hand, a permanently high concentration of methylases is an unfavorable circumstance due to possible methylation and therefore protection of the invading phage DNA. The LlaDII methyltransferase contains HTH motifs, which were shown to be engaged in direct interaction with its promoter sequence, causing silencing of its own gene expression [50].

6.3.3. Type III RM

Unlike types I and II, type III systems are less spread among lactic acid bacteria. The *Lla*FI system identified on the lactococcal pND801 plasmid is the first type III RM system described not just in LAB, but generally in Gram-positive bacteria [51]. Based on computational analyses of genome sequences, type III systems were observed to occur also in lactobacilli (for instance *Lactobacillus johnsonii* and *Lb. rhamnosus*) [52]. On the one hand, type III resemble type II systems in their structural and genetic organization. Type III, like type II systems, consists of two genes, one encoding a methyltransferase (Mod) and the other - an endonuclease (Res). Mod is responsible for binding and methylating the recognition sequences, regardless of the presence of Res. On the other hand, type III systems are similar to type I, in respect to endonuclease activity, as the Res subunit is only active in a complex with Mod. Another basic similarity to type I systems is the fact that they both comprise the helicase domain and require both AdoMet and ATP for full restriction activity. The distinctive features characterizing type III systems concern recognition sequences and cleavage sites. The Mod subunit recognizes asymmetric, opposite-oriented sequences and methylation takes place only on one strand of the DNA [53]. The Res endonuclease cuts both strands of the DNA at the distance of 24-27 nucleotides downstream of the unmethylated specific sites [53].

Lactococci have been found to possess three types of RM systems: type I, II and III. Based on genomic sequence data, it is evident that RM genes are both chromosomally- and plasmid-encoded. However, a variety of RM determinants is generally associated with plasmids [17]. In contrast, very few phage defense mechanisms have been described for *S. thermophilus*. In 2001, Solow and Somkuti reported on the discovery of a complete type I RM system encoded on a streptococcal plasmid pER35 [54]. Further progress in genome sequencing led to finding complete type I and III RM systems in chromosomes of *S. thermophilus* strains. Genome sequence analyses revealed that lactobacilli, like lactococci and streptococci, possess in their chromosomes three types (I-III) of RM systems [55].

6.3.4. Type IV RM

To date, no type IV RM systems has been distinguished in lactic acid bacteria. It is highly likely that in the future members of this class will be discovered in LAB. For that reason as well as from the evolutionary point of view, the type IV RM system is worth mentioning. A fusion of genes coding for Mod and Res subunits of type III systems was a key step for evolution of type IV RM [56]. The resulting endonuclease (revealing also methyltransferase activity) has an asymmetrical recognition sequence and cleavage occurs at a fixed distance from the recognition site, like for the type IIS enzymes. On the other hand, this endonuclease requires AdoMet, which distinguishes it from type II endonuclease activity. Therefore, taking into account the enzymatic features of model type IV *Eco*57I and *Bse*MII endonucleases, it has been hypothesized that type IV endonucleases are an intermediate between type III and type IIS enzymes.

In summary, it has been well documented that phage restriction-modification systems are widely spread among lactic acid bacteria. Nevertheless, comparative genomics of LAB demonstrated that bacteria representing different niches vary in the presence of restriction-modification genes. The lack of RM systems is a common feature for LAB isolated from the gut, whereas the presence of RM genes is a typical feature for dairy species. Therefore, it was proposed that genes constituting the restriction-modification systems, together with certain genes of sugar metabolism and the proteolytic system, constitute "a barcode" of genes, which can indicate the ability of the microorganism to occupy either dairy or gut niches [57].

6.4. Phage abortive infection systems

When the RM systems fail in protecting the bacterium against invading phage DNA, initiation of the phage propagation cycle occurs. However, proliferation of progeny particles might be dramatically limited due to systems that abort the infection at various points of the phage cycle. Abortive infection mechanisms (Abi) have different targets in the cell. They are able to interrupt phage DNA replication, transcription, protein synthesis, phage particle assembly or induce premature cell lysis [17,58]. The Abi mechanisms have been found in many bacterial species, including *Escherichia coli, Bacillus subtilis, Streptococcus pyogenes, Vibrio cholerae* and *Lactococcus lactis* [58]. The most known Abis have been found in the latter species. To date, 22 lactococcal Abi mechanisms have been identified and designated into various groups distinguished by a subsequent letter of the alphabet [58-60]. Most of Abi systems are plasmid-encoded and only three are located on chromosomal DNA (*abi*H, *abi*N, *abi*V) [60]. For instance, *abi*N is located in a prophage region of the *L. lactis* subsp. *cremoris* MG1364 genome and exhibits significant similarity to a corresponding region of the lactococcal temperate phage rlt [61]. Abi systems present simple genetic organization. The Abi phenotype is most frequently encoded by a single gene; however, more complex structures have been identified in six systems. AbiE, AbiG, AbiL, AbiT and AbiU are encoded by two genes, whereas AbiR is the only system identified until now that is encoded by three separate genes [58, 62-63]. Proteins encoded by *abi* genes are cytoplasm-located, where they reveal their activity. In contrast, the AbiP system is represented by a membrane-anchored protein [64].

Abi systems reveal a variety of modes of action. However, in many cases, mechanisms of action of the individual systems were not fully elucidated. Some Abis, like AbiA, AbiD1, AbiF, AbiK, AbiP and AbiT, have been found to interfere with DNA replication, whereas AbiB, AbiG and AbiU arrest mRNA synthesis or have a negative impact on stabilization of transcripts. Haaber and colleagues presented that the AbiV system strongly affects translation of both early and late phage proteins, shortly after infection. Based on this observation, it was concluded that the AbiV system arrests the bacterial translation apparatus [60]. AbiE, AbiI, AbiQ and AbiZ systems affect maturation of phage particles [59,65]. The AbiZ system, identified in 2007 by Durmaz and Klaenhammer, induces premature lysis of phage-infected cells, resulting in the release of the developing phage particles before completion of the maturation process. The timing of phage lysis is controlled by the phage holin protein; thus, AbiZ might interact cooperatively with the phage holin or with a holin inhibitor to make it active prematurely [59].

While the mechanism of cell death in the AbiZ system is self-explanatory, in case of other Abi systems is poorly elucidated. The most likely explanation for this phenomenon is that Abi proteins interfere with processes essential not only for phage, but also for bacterial development; therefore, death of individual bacterial cells is always observed following activation of the Abi systems [17,58-59]. As a consequence, release of progeny particles is limited and the bacterial population survives. Hence, the Abi systems constitute a barrier against bacteriophage proliferation, in which "altruistic suicide" of infected bacterial cells provides protection of the whole uninfected population [17,58].

6.5. CRISPR/cas systems

Another naturally-occurring distinct phage defense system recently described in Prokaryotes is CRISPR/cas. Besides RM mechanisms, this system is also directly engaged in protecting bacterial cells against invading genetic elements, such as phages or plasmids [66]. In brief, CRISPR-conferred phage resistance relies on incorporation of short phage-derived sequences within specific loci of the bacterial genome. In effect, the bacterial cell becomes immune to phages which carry homologous sequences.

CRISPR/cas systems are composed of two specific determinants: (i) clustered regularly interspaced short palindromic regions (CRISPR array) and (ii) regions encoding CRISPR-associated (Cas) proteins. The CRISPR arrays consist of non-coding sequences composed of unique phage-derived spacers (21-72 bp) separated by short direct repeated sequences (21-48 bp) of bacterial origin. The length of spacers and repeats within a single array is always the same, while their number may vary from 2-375, depending on the species. On the other hand, Cas proteins constitute a heterologous group of proteins, which contain various functional domains, e.g. typical for nucleases, helicases, nucleic acid binding proteins, etc. [66]. The specific role of individual Cas proteins vary as they were shown to be engaged at various stages of CRISPR-conferred resistance. Interestingly, cas genes were detected only in CRISPR-containing genomes, suggesting their tight association. The number of cas genes within a CRISPR locus varies from 4 to 20 [67]. Their position can be either upstream or downstream of repeat-spacer units, but always from the same side for a given CRISPR locus type. The CRISPR array and Cas-encoding genes are separated by an A-T rich leader region, suggested to be the promoter region of CRISPR transcription; yet, mechanisms regulating expression still remain to be elucidated [68]. Together these two elements, CRISPR spacer-repeat array and Cas proteins, provide "immunity" to the bacterial cell against invading foreign DNA molecules, including phages (for detailed review see: [67-69]). CRISPR arrays are widely distributed within the Prokaryotic world and are detected in the genomes of 40% of Bacteria and 90% of Archea [70]. Depending on the species, a single genome can carry up to 18 CRISPR loci, which are suggested to confer resistance to various phages [66].

The mechanism of CRISPR/*cas* conferred protection of bacterial cells against phage infection is rather complex and can be divided into three main stages: (i) adaptation, (ii) CRISPR expression and (iii) CRISPR-mediated interference. The first stage relies on incorporation into the bacterial genome within the CRISPR locus of short phage-derived fragments (proto-spacers). Despite the fact that the exact mechanism of spacer acquisition is not known, it is not accidental. Recognition of specific phage sequences for integration is suggested to be linked with sequences termed PAMs (proto-spacer adjacent motifs), located up- or downstream of the proto-spacer. Integration of new spacers occurs from the end of the leader region, between the palindromic repeats and involves certain Cas proteins. Stage 2 is CRISPR expression, which involves transcription of the whole CRISPR spacer-repeat array (pre-mRNA). The presence of palindromic repeat sequences within the transcript, leads to formation of secondary hair-pin like structures. These are subsequently processed into short CRISPR RNAs (crRNAs) by endonucleolytic digestion at a cleavage site located downstream from the last nucleotide forming the hairpin. Finally, the last stage of CRISPR/*cas* activity is based on interaction of mature crRNAs with invading foreign DNA elements (phages), which leads to silencing/degradation of the latter by a certain group of Cas proteins. By this activity, CRISPR/*cas*-carrying hosts are protected from invasion by phages carrying sequences homologous to the integrated spacers. Application of the CRISPR/*cas* system for developing novel phage resistant dairy starter strains may be an attractive alternative, which will be discussed in further parts of this chapter (see section: 12.4.).

6.6. Engineered defense systems

Besides the naturally-occurring defense mechanisms against recurrent phage infections (discussed above), new methods involving molecular techniques are designed to combat phages. The constantly growing knowledge on phage development and their genome sequences allows currently to develop engineered defense systems, which are otherwise not encountered in nature (for review see also: [71]). The idea of such systems relies on engineering bacterial strains in a way which impairs genes vital for phage development, e.g. phage replication proteins or other replication factors. Moreover, identification of homologues of these crucial genes within multiple phage genomes allows creating broad-range phage defense systems. As presented below, numerous studies deliver clear evidence that such engineered systems provide efficient protection against phage infections. The following parts of this chapter will delineate each of these systems in more details. Studies on developing engineered systems for lactic acid bacteria were performed in most part in *Streptococcus thermophilus* and *Lactococcus lactis* as strains from both species find wide applications in dairy fermentation processes.

6.6.1. Antisense RNA-based phage defense systems

Bacterial-engineered expression of antisense RNA directed against phage transcripts has been described as one of the most efficient phage defense systems. The mode of action of such RNAs is hybridization to phage sense strand RNAs upon infection. By these means the system interferes with the phage life cycle, inhibiting translation of essential phage genes or degradation of their mRNAs [72].

An example are systems developed in *Streptococcus thermophilus*, which were shown to provide protection against Sfi21-type phages, including κ3 [73-74]. These systems are based on expression of antisense RNAs against genes from the replication module of the Sfi21-type phage κ3 genome, e.g. putative primase (*pri3.1*) or helicase (*hel3.1*) genes. Strong conservation of the whole replication module among the already sequenced Sfi21-type phages makes it a good target for inhibiting phage development [73]. Moreover, hybridization studies revealed that the Sfi21-type replication module is commonly encountered in majority of industrially isolated phages. This reinforced the choice to use it as a phage defense element [73,75].

To test the efficiency of the Sfi21-type module antisense RNA system, constructs expressing antisense RNA cassettes of different length were introduced into *S. thermophilus* strains, which were then challenged with phage infection. The most effective were constructs expressing antisense RNA covering the whole region of target (primase or helicase gene). Also shorter RNAs provided sufficient phage resistance, which was speculated to be due to the presence of specific structural or potential regulatory domains within these fragments. Furthermore, in case of constructs harboring antisense RNAs of similar length, more efficient were usually those comprising the RBS (ribosome-binding sequence) sites. Such effect was believed to be due to the fact that the antisense RBS sequences prevented gene translation by impeding efficient loading of ribosomes onto phage mRNAs [72]. Overall, expression of phage antisense RNAs in *S. thermophilus* was shown to interfere/delay the intracellular phage DNA replication, decrease phage plaque formation (EOP, efficiency of plating), lower the abundance of phage sense mRNA transcripts and reduce phage progeny particles released from infected cells [73,75].

Similar systems were also developed in *Lactococcus lactis* by expressing anti-sense RNAs directed against various phage genes (e.g. P335-type *gp18C* and *gp24C*, *gp15C* alone, or putative replication genes, 936-type phage F4-1 major coat protein (*mcp*) gene) [76-79]. In these cases, similarly as for *S. thermophilus* systems, the most efficient antisense RNAs in inhibiting phage development were those comprising the RBS site.

Current data allow to conclude that the most effective antiRNA-based phage defense systems, apart from some exceptions, are those which target: (i) genes vital for phage development (e.g. involved in synthesis of phage DNA), (ii) preferably early-expressed phage genes, (iii) genes expressed at low levels, (iv) genes which respective transcripts are unstable [73,79]. Sequencing of novel phage genomes and development of comparative genomics allows identification of other conserved phage genome regions that could serve as potential targets of antiRNAs.

6.6.2. Origin-derived phage-encoded resistance

Defense systems that employ elements derived from lytic phage genomes are termed phage-encoded resistance (PER). One type of engineered PER systems is based on the origin (*ori*) of phage replication [71]. The principle of such systems relies on presenting *in trans* false targets (in this case, phage-derived *oris*), which titrate phage replication factors and make them inaccessible for the phage. In result, phage development is inhibited due to arrested replication of its DNA. These engineered systems resemble the naturally-occurring abortive infection mechanisms as they exploit the same principle (for details see: 6.4. Phage abortive infection systems).

One of the first phage origin-derived systems developed was for *Lactococcus lactis* and employed the *ori* of replication of an industrial phage Φ50 (*ori50*) [80]. Introduction of the Φ50 *ori* region on a high copy number plasmid into the *L. lactis* NCK203 strain provided resistance to not only to phage Φ50 itself, but also to other small isometric phage isolates from industrial environments [80-81]. It was suggested that all of these sensitive phages are part of the same family and most probably exhibit significant homologies within their *ori* regions. Additionally, replication of the *ori50⁺* plasmid was shown to be stimulated by Φ50 infection, implying that phage factors are engaged in the process [80]. Further studies determined that the system affects neither adsorption nor phage DNA injection, which suggested that this defense mechanism acts at a later stage of phage development, i.e. DNA replication. It was also clear from the study that the origin-derived phage-encoded resistance phenotype was strongly dependent on the plasmid copy number. Most probably, low copy number plasmids are insufficient in providing enough phage *ori* sites that could efficiently titrate and attract phage replication factors. Yet, on the other hand, when the copy number of *ori⁺* plasmids exceeded a certain level, resistant phage mutants were observed as a side-effect. Characterization of these mutants by DNA restriction analysis revealed mutations within the *ori* region, which enabled them to escape the phage defense system.

More recently, a similar origin-derived phage-encoded resistance system was developed for *S. thermophilus* strain Sfi1 based on the *ori* of phage Sfi21 [82]. The presence of this non-encoding phage DNA fragment rendered the Sfi1 host strain resistant to the concomitant phage infection by Sfi21 and 17 other *S. thermophilus* phages. Interestingly, all of them were found to exhibit homology within the *ori* region. However, resistant phages that could overcome this defense mechanism were also detected. They, on the other hand, exhibited differences in the *ori* sequence compared to the wild-type Sfi21-like *ori*. Examination of other *S. thermophilus* phage genomes (~ 30) allowed identifying other distinct replication *oris* and to divide them into separate groups: replication group I, IIA and IIB [83]. Plasmid constructs harboring these three phage *ori* types increased phage resistance in certain host backgrounds. However, in some strains this origin-derived phage-encoded resistance was not observed. It is therefore speculated whether the efficiency of these systems could be also dependent on some still undetermined host factors.

Development of analogous systems for other lactic acid bacteria involves identification and functional characterization of *ori* regions of their respective phages. This approach can be especially useful for phage-sensitive strains for which other plasmid-encoded defense systems have not yet been determined.

6.6.3. Superinfection immunity and exclusion

During the lysogenic life cycle of temperate phages, the lytic module is inactive due to the activity of the CI repressor. However, certain prophage genes - the superinfection-immunity (CI-like repressor) gene itself and the superinfection-exclusion gene, are actively expressed. Both functions were determined to provide protection to the lysogenic host against phage superinfection. Application of these genes to create engineered phage defense systems is yet another strategy of protecting bacterial cells from incoming infections. Multiple bacterial genomes carry prophage-derived sequences, which can count up to 10% of the total genomic content of the cell. Therefore, despite the fact that phage-related sequences are a burden for bacterial cells, they are also believed to provide some advantage to the host by increasing its fitness.

Genomic studies in *S. thermophilus* led to the identification of superinfection-immunity (*orf127*) and superinfection-exclusion (*orf203*) genes from the lysogeny module of the Sfi21 prophage [84-85]. Expression of *S. thermophilus* phage Sfi21 *orf127* gene from a plasmid vector conferred the phage resistance phenotype against homologous phage, but was ineffective against other heterologous phages [86]. Analysis of the respective ORF127 product revealed its structural homology with phage λ CI repressor and amino acid homology (15% identity) to a potential CI-like repressor of the lactococcal phage Tuc2009. Gel shift experiments allowed determining the ability of the Sfi21 CI-like repressor to bind to two operator sites identified in the genome of the superinfecting homologous phage Sfi21. Superinfection immunity genes (CI repressors) were also identified in phages of other lactic acid bacteria species (e.g. for *Lactococcus* phage TP901-1 and *Lactobacillus* phages A2 and Φadh) [87-89]. Their expression *in trans* was also reported to provide immunity against homologous phage infection.

In contrast, superinfection exclusion genes are not engaged in maintaining the lysogenic state, yet are also active during the lysogenic cycle. Experiments based on expression of the *S. thermophilus* phage Sfi21 *orf203* gene *in trans* in high copies determined that it confers resistance to superinfection of a range of heterologous lytic streptococcal phages [85]. Contrarily to the Sfi-21-derived superinfection immunity, in this case resistance to the Sfi21 phage itself was not observed. Moreover, the mechanistic background of the *orf203*-dependent resistance phenotype was shown to involve inhibition of phage DNA injection.

A superinfection exclusion system was also developed in *Lactococcus lactis* based on the *sie2009* gene from the temperate phage Tuc2009 [30]. When cloned *in trans*, *sie2009* provided resistance only to some 936-type phages used in the study. Moreover, neither c2- nor P335-type phages were affected. It was determined that Sie2009 is a cell membrane-associated protein interfering at the stage of phage DNA injection. However, the exact mechanism by which Sie2009 acts was not yet established. The ability of the designed system to confer resistance only to certain 936-type phages might indicate different mechanisms of DNA injection exhibited by various phages. A similar membrane protein was detected for *S. thermophilus* phage TP-J34, Ltp. Expression of the *ltp* gene provided protection against TP-J34 in *S. thermophilus* and, interestingly in *L. lactis* against a 936-type phage, P008 [32]. This, quite surprising observation of *sie*-encoded cross-resistance was argued to be due to a recent genetic transfer event between the two species. In both cases, it was noted that phage adsorption was not impaired, but there was significant inhibition of phage DNA accumulation within the host cell. Based on these observations it was proposed that the *ltp* gene product acts at the stage of phage DNA injection by either impairing insertion of the phage tail into the cytoplasmic membrane or by obscuring the host membrane protein responsible for inducing the release of phage DNA from the capsid.

Putative superinfection exclusion genes seem to be widespread among prophage-containing lactococcal and streptococcal strains and localized in the same genomic region limited by repressor and integrase gene from each side. Although *sie* genes lack significant homology, all currently identified Sie proteins are small with hydrophobic N' tail and at least one transmembrane domain. Various studies of lactococcal Sie proteins allowed grouping them into several phylogenetic groups, depending on the subset of 936-type phages they target (sk1/jj50, bIL170/p008 or 712 group) [28]. At present, it is argued that all *sie* systems identified for lactococcal lysogenic strains interact or mask cell membrane associated factors engaged in phage DNA injection or come into direct contact with structural proteins of the infecting phage. The most probable theory is that Sie function is aimed against the tail tape measure protein (function implicated in phage DNA injection process), as its encoding region is among the few divergent genomic regions between the different subsets of 936-type phages.

Superinfection exclusion and immunity genes in natural conditions can also be provided by defective prophages. The nature of defective phages is that they cannot be efficiently induced by environmental factors; hence, cured from the host strain. Such lysogenic lactic acid bacterial strains (particularly *Lactobacillus* species), which exhibit no threat to the fermentation processes due to uncontrolled prophage induction, are of special interest to the dairy industry as naturally-resistant strains to superinfection events.

6.6.4. Phage-triggered suicide systems

Phage-triggered suicide systems rely on expression of toxic elements under the strict control of phage-inducible promoters. Such specifically engineered systems most closely resemble the naturally-occurring abortive infection systems, which trap the phage within infected cells and lead to programmed cell death. Upon phage infection, host cells are lysed, disabling at the same time phage propagation and the concomitant spread of the phage. In effect, the uninfected bacterial population is saved (for details see: 6.4. Phage abortive infection systems). Suicide systems are based on three genetic components: (i) a lethal gene cassette, (ii) a phage promoter induced only after phage infection, and (iii) an appropriate vector, providing sufficient amount of the lethal gene product.

Such system, based on an inducible plasmid strategy, was created for *L. lactis* to control phage infections [90]. The system comprises a lethal three gene cassette, *llaIR⁺*, encoding a restriction endonuclease of the *L. lactis* LlaI R/M system, cloned under the tight control of the phage Φ31 middle-expressed promoter (Φ31*p*) that is active at a significant level solely after Φ31 infection [91]. Expression of this plasmid-encoded suicide system was designed to restrict unmethylated host and phage DNA upon infection. During infection, induction of the *llaIR*⁺ cassette caused a significant drop of phage Φ31 EOP. Only a small fraction of infected cells produced progeny phage particles. The system provided also protection against other Φ31-like phages. Yet, despite the observed inhibition of phage development, some phages were found to escape this defense system. Phage mutants that emerged during the assays were all found to be altered in the sequence encoding the transcriptional activator of the Φ31*p* promoter [92]. Thus, these mutants escaped the system due to lack of efficient transcription of the *llaIR*⁺ cassette. The drawback of this system is the fact that it is only active against phages that can trigger the phage-derived promoter; in this case, against Φ31 and its closely related phages. Another disadvantage is the fact that the *llaIR*⁺ cassette is not expressed immediately, but after a time necessary for the infecting phage to synthesize transcription factors activating the middle-expressed promoter. This, in effect, allows for replication of a low number of phage particles that escape restriction. It would seem more appropriate to use early phage promoters; yet, these are usually host-controlled. Improvement of the efficiency of the already existing suicide systems should involve cautious selection of effectively controlled and adequately strong phage promoters and more proficient restriction endonucleases. This is best illustrated by the described earlier observation that application of a stronger promoter, despite a more efficient reduction of phage EOP, can have a negative effect on bacterial cell growth. As a method to enhance the efficacy of the suicide system, it was proposed to include within the suicide cassette another gene - *llaIC*, encoding a regulator protein [93]. The presence of this regulator protein was suggested to significantly increase the anti-phage restriction activity of LlaIR.

6.6.5. Subunit poisoning

The subunit poisoning system is an engineered phage defense strategy that relies on expression *in trans* of truncated/mutated proteins which impair (poison) the function of their wild type variants. To achieve this, mutant proteins should be expressed at levels higher than their wild-type counterparts. Moreover, despite alterations in their amino acid sequences, they must have an intact structural form in order to titrate sites or substrates, or other protein components, away from the wild-type proteins.

An example of such system is based on the CI-like repressor of lytic *Lactococcus lactis* phage Φ31. The general idea of this strategy resembles very much the superinfection immunity approach, where *S. thermophilus* bacteria expressing the phage Sfi21 CI repressor were protected from closely-related phages (for details see: *6.6.3. Superinfection immunity and exclusion*). However, in this case, the exact mechanism is somewhat different [94]. Studies of the wild type CI repressor of phage Φ31 showed that it is non-functional and, when expressed in the *L. lactis* host, does not provide protection against superinfecting phages nor represses the transcription of phage lytic genes. Yet, when this wild-type Φ31 CI protein or its truncated variants were expressed *in trans*, they could efficiently inhibit infection of Φ31 and other lytic P335-type phages. Expression of Φ31-derived *cI* mutant genes from a high-copy number vector was shown to inhibit growth of Φ31 (to EOP 10⁻⁶ or lower, depending on the mutation) and of other lytic P335-type phages. The observed effect was determined to be due to the competitive binding of the non-functional Φ31 CI and phage-expressed CI repressors to two of the three wild-type operators identified within the genetic switch region. It was suggested that the truncated variants of the Φ31 CI repressor exhibit a higher affinity for these sites than the phage-encoded CI protein. In effect of Φ31 CI binding, expression of lytic phage functions was repressed, impairing phage DNA replication.

During the study resistant phages were also detected. Sequence analysis studies within the genetic switch regions revealed alterations in their operator sites, which impaired binding of the Φ31-derived CI repressor.

Another example of subunit poisoning phage defense is a system developed in *Streptococcus thermophilus*. The strategy was based on mutating the primase-encoding gene, an essential component of the replication module of the *S. thermophilus* Sfi-21-like κ3 phage [95]. Mutation of sequences within the highly conserved domains of the primase gene resulted in obtaining dysfunctional protein variants. Expression at high level of such primase derivatives *in trans* resulted in reduction of EOP of κ3 and several other Sfi-21-type phages due to inhibition of phage DNA replication. The mutated primase was implied to titrate replication factors and/or the origin of replication, making them inaccessible for the wild-type primase. This suggestion seems to be credible as introduction of the STOP codon upstream of the initial gene mutations restored the phage sensitivity phenotype. Such alteration lead to the synthesis of a truncated protein, which, most probably, could no longer mimic the structure of the native primase. A great advantage of this primase-based subunit poisoning system is lack of phage mutants resistant to the mutated primase proteins under study.

Overall, subunit poisoning is an approach that is believed to constitute a broad phage defense system, as it was shown to be effective against more than one lytic P335-type phage. In this aspect, it differs from the earlier described superinfection immunity systems, where expression of phage repressor genes from phages of various lactic acid bacterial species (e.g. *Streptococcus* phage Sfi21, *Lactococcus* phage TP901-1 or *Lactobacillus* A2 or Φadh phages) provided immunity against the respective single phage only [85,87-89].

6.6.6. Host-factor elimination

Eliminating a genetic element from the genome of starter bacteria to obtain phage-resistant strains is yet another strategy of engineering a phage defense system. This approach can target different stages of the phage life cycle, which are often host-dependent, e.g. phage injection dependent on host membrane proteins, host factors necessary for phage DNA replication.

Among methods identifying such host-encoded factors is random mutagenesis using the pGhost::IS*S1* mutation vector. This approach allowed to identify genes necessary for phage development in the genome of the *S. thermophilus* Sfi strain [96]. The *orf394* gene, encoding a putative transmembrane protein, was one of the host loci determined to confer phage resistance to Sfi19 as well as more than 10 other heterologous *S. thermophilus* phages. After infection by phage Sfi19, no phage DNA synthesis was detected for such mutant strain. Based on this observation, it was suggested that this transmembrane protein may be implicated in the stage of phage DNA injection, analogously to the Pip (phage infection protein) of *Lactococcus lactis* [97]. Among advantages of the host-factor elimination system is the fact that it is food-grade and that the engineered phage resistant strains can be successfully used in production processes. Yet, it must be noted that before its application in the industry, the strain should be assayed for its phage resistance phenotype during several rotation rounds of culturing. There is also no data on the phage mutants that can evolve due to the continuous use of such Pip⁻ mutant strains.

Other host factors that were suggested to be efficient targets for developing phage defense systems are auxotrophic genes. Pedersen et al. developed a strategy of impairing phage replication in an industrial *Lactococcus lactis* strain by deleting the thymidylate synthase (*thyA*) gene from its genome [98]. This patented strategy is based on the process of phage DNA replication [99]. Upon infection, phages take advantage of the DNA replication machinery of the host to amplify its own genetic material. However, when the host is lacking one of the main DNA building factors, formation of novel replicated DNA molecules is inhibited. The *thyA* gene is responsible for *de novo* synthesis of dTTP in the cell. Strains devoid of *thyA* cannot synthesize dTTP in the medium that lacks thymidine, such as milk. Under such conditions, the Δ*thyA* mutant was resistant to infection by selected P335- and 936-type phages, which efficiently infected the parental wild type strain, and what is important its acidifying properties remained undisturbed. Addition of thymidine to the milk medium restored phage sensitivity of the strain. Among the drawbacks of this system is the fact that the mutant strain lacking thymidylate synthase is impaired for growth. Therefore, in industrial conditions it must be inoculated into milk tanks at higher concentrations than the parental wild type strain in order to meet the technological criteria. A solution to this problem could be addition of limiting amounts of thymidine to promote growth of such starter strain. Yet, thymidine as an additive in the cheese industry is not allowed. Among various options proposed by Pedersen to obtain a phage-resistant *thy* mutant for industrial use could be construction of a thermosensitive mutant, in which expression of the *thy* gene is inhibited at temperatures at which technological processes are carried out [98]. The greatest concern when selecting for phage-resistant strains is their ability to prevail a broad range of phages over long periods of time. In this case, it seems that the host-factor elimination system is the most universal among the presented systems as it acts against all phage types. Moreover, at this point it seems that the probability of occurrence of phage mutants overcoming this resistance mechanism is low. However, it should also be noted that some phages are known to encode own *thy* genes or utilize nucleotides of the host by hydrolyzing its DNA, which can be a weak point [100-101].

7. Problem of phage contamination in dairy industry

There are no commercial LAB cultures available which would be completely insensitive to all phages. Even when a starter culture that is launched on the market appears to be phage resistant, phages are detected usually after a certain period of use.

Phage contaminations in dairy plants can cause 3 main serious drawbacks:

- problems in obtaining expected technological parameters and product quality consistence
- staff stress, decrease of motivation and engagement, irregular working hours, staff economical consequences, job resignation
- financial losses (failed production, non-standard product, lower unit price, delayed deliveries, customer losses).

8. Phage detection in dairy industry

8.1. Simple tools for phage detection at the dairy plant level

A simple test assessing acidification activity of currently used starter cultures on a daily basis can be used successfully to monitor phage contaminations in dairy plants. Briefly, a cheese whey sample from the last production vat of the shift is collected and, before use, sterilized by filtration (0.45 µm filter-pore size). In the case of dairy beverages, a sample of the final product, before its filtration, is clarified with addition of lactic acid and centrifuged. Processed pasteurized milk or sterilized milk reconstituted from powder is inoculated in duplicate with starter cultures (including a phage alternative culture) at a standard dosage. One sample of each culture is inoculated with a whey filtrate (usually 1-2%) and the second one - with a temperature sterilized whey filtrate. After incubation (the temperature and time depend on the culture and process), the pH of the milk is measured. When the pH of the milk containing the filtrate is 0.2 units higher in comparison to the sample containing the sterilized filtrate, it indicates that phage contamination is rather high and phage-unrelated culture rotation as well as disinfection with higher concentrations of active substances should be recommended.

To avoid direct measurements of pH, bromocresol purple (100 µg ml[-1]) as a pH indicator may also be used. The test lasts around 6 h, for mesophilic starters, and 4 h, for thermophilic cultures. When pH of the milk drops below 5.4, the indicator turns from purple to yellow. If, at the same time, the color of the sample containing the non-sterilized filtrate becomes green or purple, it means, with high probability, that phages are present and may adversely influence the fermentation process [102].

Another approach of phage detection is continuous monitoring of pH during fermentation processes conducted in vats or tanks with short time intervals and plotting the data on a graph. Even in the case when delay of the fermentation process is not observed, but the graph shows an irregular shape not related to temperature deviation, phage contamination is suspected (Fig.1). However, in this method a delay in acidification can also result from other inhibitors than phages (e.g. antibiotics, detergents) present in the sample.

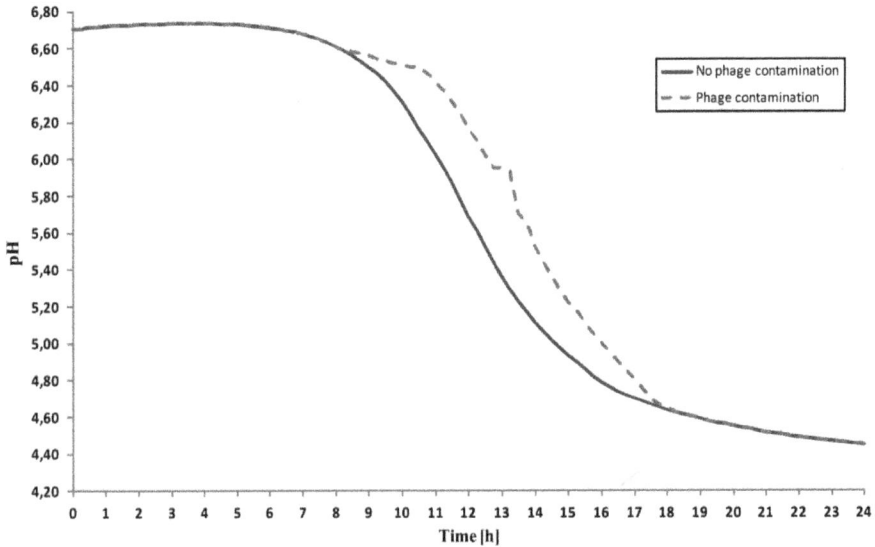

Figure 1. Example of pH curve during milk fermentation in the presence of virulent phages incubated with the multistrain and multispecies culture.

8.2. Routine service at culture supplier level

The most common and most useful method of phage enumeration is the plaque assay. The method is quite old and was first described by d'Herelle shortly after the discovery of bacteriophages. Currently it is used in many labs with some modifications, but its principle has not changed [103]. The most common, practical, cheap, without using large numbers of plates and sufficiently accurate method in the dairy industry is the semi-quantitative spot test method. Using this approach, results are available after 24-48 h. The method is well suitable for detection of phages of pure lactic acid bacterial strains at relatively low levels (< 100 phages ml⁻¹). Plague assays allow detecting the presence of phages as well as determining the number of phages in dairy samples against all individual strains present in the applied defined cultures. In case of phage contamination in a dairy plant, the method is a good tool for selecting the best phage-resistant alternative cultures. The method can also be used for hygiene monitoring by enumeration of phages in samples collected from critical places if the plant. For dairy culture producers, permanent phage monitoring can identify strains which are most sensitive in defined cultures. These strains can be systematically replaced with more phage resistant strains. Semi-solid medium supporting bacterial growth is used for multiplication of strains in form of a smooth opaque layer or lawn on the medium surface using standard Petri dishes. Serial dilutions of phage solution previously sterilized with a filter are placed (5-20 µl) on the surface of the opaque layer. When a single phage particle develops on a recipient bacterial lawn, it forms a plaque (clear spot, no bacterial lawn) visible to the naked eye. This plaque results from the destruction of bacterial cells by the phage progeny. Growth of the plaque is limited by slow diffusion of the phage in the semi-solid medium and bacterial cell growth stops, so phage growth is also inhibited due to the fact that host cells support phage growth. No visible plaques on the plate mean that the sample is not contaminated by phages. Large clear zones (no separate plaques) on the plate indicate with high probability that the level of phages is rather high and further dilutions of the sample are required to precisely determine the phage titer. The presence of a plaque means that: i) the tested sample contains phages; ii) the phage is virulent against the tested strain; iii) the strain is sensitive to the phage. Each phage particle that gives rise to a plaque is called a plaque-forming unit (PFU). One plaque corresponds to a single phage particle and phages can easily be counted. In result, the number of PFUs corresponds to the viable phage concentration in a given sample volume.

8.3. Sensitive methods (including ELISA and molecular DNA techniques) at the level of academic or innovation labs

Plaque assays and acidification tests are microbiological methods that are economically accessible and sensitive enough for detection of phages in the dairy industry. These techniques are time consuming, but provide many practical data for both dairy plants and starter producers. The polymerase chain reaction (PCR), ELISA and flow cytometry-based methods have been designed for detecting phages and are often used to complement microbiology tests. However, they have still many drawbacks to be applied for routine analyses in the dairy industry [104].

PCR-based methods detect virulent and non-virulent phages; thus, microbial methods should be used in parallel to precisely distinguish the virulent phages. PCR-based methods can also be too expensive and too specific (only phages targeted by specifically-designed primers are detected) for routine experiments. However, PCR is a fast method able to confirm the presence of bacteriophages within 30 minutes and can be applied to determine the potential utility and quality of big batches of milk. At the same time, the method could be handy in finding niches of phage accumulation, in order to reduce their impact in dairy fermentations [105-108].

ELISA techniques use for phage detection antibodies which are highly specific against structural proteins of phage capsids. Due to the wide phage diversity in the dairy environment, development of several antibodies detecting various groups of phages was required. ELISA is regarded as a highly useful method for monitoring specific phages in the dairy environment, but a single assay cannot be used to detect phages with different structural proteins. For this reason, the sensitivity of an ELISA method to detect phages in dairy a sample is rather low.

Flow-cytometry can also be used for detection of phages in dairy samples by discriminating the phage-infected cells from non-infected based on cell morphological changes leading to lysis. Running on the flow-cytometry of samples containing phages gives a broad distribution of cell mass (wide peak), which demonstrates the presence of both lysed and live cells, while non-infected samples give narrow peaks. Flow-cytometry allows detection of phages in real time, but expensive equipment and well-trained staff needed to perform the assays limits application of this technique in the dairy industry [104].

9. Sources of phage contamination

In dairy plants phages can originate from a variety of sources. The prime importance is to identify the potential sources of phage contamination and limit their entry to the fermentation process.

9.1. Raw milk

The most probable source of virulent phages is raw milk. LAB phages occur naturally in raw milk at low titers (between 10^1-10^3 PFU ml^{-1}) and constitute a continuous supply of bacteriophages in dairy plants [109-110]. Phage concentrations in raw milk also depend on conditions of collecting, handling and storing of milk by the supplier (farm), on transport to the plant and, finally, handling of the milk in the plant itself. For example, reverse osmosis used to concentrate raw milk at a farm can impact the level of phages detected in milk. Almost 10% of 900 milk samples examined from various geographical areas in Spain contained *Lactococcus lactis* phages [110]. Using a multiplex PCR method *Streptococcus thermophilus* phages have been detected in more than one third of milk samples used for yoghurt production in Spain [106]. Phage biodiversity is increased by combining milk collected from different farms and these numbers can be even higher in processed milk.

9.2. Milk powder and whey protein concentrates

Reconstituted milk from powder is used in many countries for yoghurt, fresh cheese (tvarog and quark) and even maturated cheese production. Also whey proteins are used to standardize milk before the fermentation process or to improve the taste and texture as well as the nutrient value of the final product. Recently, the modern technology of milk powder and whey protein concentrate production applies often lower temperatures of treatment than during traditional technologies. Both milk powder and whey protein concentrates can be sources of high temperature-resistant phages and can influence the quality of the final product [111-112]. For separating whey proteins, ultrafiltration or/and microfiltration are more frequently employed. Applied separation processes result in higher concentrations of phages in the permeate or the retentate. Depending on which fraction is used in subsequent processes, different concentrations of phages in whey protein samples can be detected.

9.3. Starter cultures

The starter culture itself can be a source of phages, when strains contain temperate phages. Temperate phages are incorporated into the bacterial chromosome and their genome replicates in synchrony with the bacterial genome. Prophages are carried in many LAB strains. The analysis of bacterial genomes revealed that prophages are more widespread than previously considered [113-114]. Phages may be induced from lysogenic to lytic form by the manufacturing conditions. Serial subculturing of temperate phages in milk may result in their replacement by a virulent mutant. Prophage induction from multiple lysogenic starter culture strains has the potential to influence fermentation. Induction can occur under stress conditions, such as heat, salts, acidity, bacteriocins, starvation or UV [115-116], and can also occur naturally with a frequency of even up to 9% [117]. Starter culture producers make huge efforts to eliminate strains containing prophages using a screening assay for strain lysogeny. Usually, easily lysogenized strains are difficult to find in defined strain cultures. The main source of lysogenic strains are undefined cultures, which are still commonly used (for example, kefir grains). This is due to two main reasons: i) the exact strain composition of these starters is unknown; ii) elimination of lysogenic strains from undefined culture is very difficult.

9.4. Equipment/air

The one of the most probable sources of virulent phages is the dairy plant environment. Phages are commonly present on working surfaces. For propagation, phages need the presence of their bacterial hosts, in this case lactic acid bacteria. Due to this fact, they are usually found in places where conditions for LAB development are favorable. The most common sources of phage contaminations are valves, crevices and "dead ends" (difficult cleaning and disinfection places) of production lines. Also, the formation of biofilms on dairy equipment can lead to serious phage problems. Moreover, phages were detected at high levels on various equipments and objects found in cheese plants, such as walls, pipes, door handles, floors, office tables and even on cleaning materials [118]. Raw milk handling, cheese milk processed in open vats and whey handling can lead to spreading of phages in the air. Phage aerolization can occur during air displacements around contaminated places (fluids or surfaces) or by liquid splashes. Virulent phages can circulate through the air far away from their aerosolization source due to the ability to bind to small particles (< 2.1 µm) [118]. Taking into account high levels of phages detected in the air, it is hard to precisely determine whether phage propagation already took place or if it is likely to occur. Concentrations of up to 10^8 PFU per m³ of air have been detected in a cheese manufacturing plant in Germany; however, mainly in specific areas of the fermentation line [119-120].

10. Phage problem frequencies and consequences depend on product portfolio

Fermentation problems in the dairy plant can be related with: low starter activity, fermentation conditions (e.g. temperature fluctuations), milk composition (year, season, occurrence of mastitis, mineral levels, lactation period, microbial and enzymatic composition), presence of inhibitors in milk (antibiotics, detergents) and phage infections.

However, phages are the primary source of fermentation problems in the dairy industry. Bacteriophages can cause great economic losses due to fermentation failure in dairy plants. About one third of the annual world production of around 500 million tons is converted into fermented products. Two thirds of all processed milk is fermented by *Lactococcus lactis* and *Leuconostoc* spp. Thermophilic *Lactobacillus* and *Streptococcus thermophilus* spp. account for fermentation of the remaining major part of the milk. According to estimations, from 0.1% to 10% of all milk fermentations are negatively affected by virulent phages [102]. Phage contaminations can slow down or even halt the milk fermentation process. Consequences of the phage presence include: alteration of the product quality, such as taste, flavor, texture, and its microbiota composition. Phage contaminations due to the delay in lactic acid production can also lead to development of undesired microbiota during the fermentation process. In the worst cases, the inoculated milk must be discarded. The frequencies of phage contaminations and their consequences depend on the type of milk product produced. Phages can also sometimes turn a dairy staff life into a 'nightmare'.

10.1. Fermented milk beverages

Among dairy products, the least phage affected are fermented milk beverages (yoghurt, kefir, butter-milk, Actimel®-like products, etc.). There are many reasons behind this phenomenon. Milk for beverage production usually undergoes treatment at temperatures much higher than in cheese manufacturing. Moreover, some drinking yoghurts are produced from UHT milk. Beverages are made in relatively aseptic conditions, including more and more aseptic inoculation systems, where the fermented product is minimally exposed to the factory environment. In spite of that, phage contamination is sufficiently frequent and has become the primary source of fermentation problems in milk beverage production. Phage contaminations in this particular case lead to fermentation delays or inhibition, product alterations in taste and flavor as well as texture properties.

10.2. Ripened cheese

In cheese production the risk of phage infection is very high. A large cheese plant can process more than 500 tons of milk per day, very often in many vats, lasting more than one shift. Pasteurized milk (very often low temperature-treated milk or even raw milk) is used in cheese fermentation and many phages as well as microorganisms remain viable after pasteurization. Contamination, also by phages, increases during curd handling and whey separation in open vats. The consequences of phage infection in cheese production can be: delay or halt in milk acidification, cheese contamination with foreign microbiota, including pathogens, preferential growth of post-pasteurization microbiota, problems in whey separation (syneresis), higher water and lactose content in the final cheese product, abnormal or irregular holes (eyes), or no eyes, and alterations of flavor and texture [5]. To conclude, phage contamination may result in lower quality of cheese or cheese quality suitable only for processed cheese production and, in some extreme cases, complete loss of product.

10.3. Fresh cheese (cottage cheese, quark, tvarog)

Cottage cheese and traditional tvarog productions are the most sensitive processes to phages infection. Fermentation delays in production of cottage cheese lead often to complete loss of the final product. However, symptoms of phage contamination are most visible in production of traditional tvarog, where curd quality depends on the activity of lactic acid bacteria alone (rennet is not used). It is estimated that more than 70% of technological disruptions during tvarog manufacture is related to phage contaminations, which usually lead to the following consequences: delay or halt in milk acidification, curd lamination or its drop to the bottom of the tank or vat (which, in effect, causes problem with curd handling), prolonged process of whey separation due to the loss of the curd syneresis, low tvarog yield, contamination with foreign microbiota, including pathogens, intensive growth of post-pasteurization microbiota, off-flavor and texture alterations of the tvarog [121].

11. Phage control strategy

As previously stated, phages represent a constant threat of serious economic losses in the dairy industry. Dairy microbiologists have attempted for almost 80 years to eliminate or, at least, bring under better control, bacteriophages that interfere with the manufacture of fermented milk products. Phages rapidly disseminate in dairy environment and are difficult to eliminate. The important procedures for phage control are: adapted factory design, design of starters, cleaning and disinfection, and air control [102].

11.1. Cleaning and disinfection

The classical operations of cleaning and disinfection are an essential part of milk processing. Cleaning-in-place (CIP) procedures are usually applied in milk processing lines. The basic procedure consists of the following sequence operations: i) pre-rinse with cold water to remove gross residues; ii) circulation of alkali detergent to remove the remaining minor residues (from time to time acidic detergent is incorporated to remove precipitated minerals and milkstone deposits in the following sequence: alkali detergent, water rinse, acidic detergent); iii) rinse with cold water to flush out the detergent; iv) circulation of disinfectant to inactivate residual microorganisms and phages (still in many dairies this stage is not performed in each cleaning cycle); v) final rinse with cold water to flush out the detergent and cooling line [122]. The cleaning process can remove 90% or more of microorganisms associated with the surface, but cannot kill all of them. One of the drawbacks of the cleaning process is that residual live bacteria can redeposit and, in longer periods of time, can form a biofilm. The presence of LAB among the residual microorganisms increases phage risk contamination. The main role of disinfection is to kill microorganisms that survive the cleaning procedures.

Disinfectant	Supplier/ Producer	Main active substances	Conditions recommended by supplier		
			Concentration (%)	Temp. (°C)	Time* (min.)
Deptil PA 5	Hypred	Hydrogen peroxide, Peracetic acid, Acetic acid	0.1- antiseptic 2.5-fungicidal	< 30	20
Divosan Hypochlorite VT3	Johnson Diversey	Sodium hypochlorite	0.1 – 3.0	cool	10 – 20
Oxidan special 150	Novadan	Hydrogen peroxide, Peracetic acid, Acetic acid	0.1 – 0.35	5 – 40	5 – 60
Hypochlor DES	Novadan	Sodium hypochlorite, Sodium hydride	0.25 – 1.0	20	15
Desinfect CL	Novadan	Sodium hypochlorite	0.20 – 1.0	5 – 40	10 – 15
P-3 Oxonia	Ecolab	Hydrogen peroxide	0.5 – 1.0	ca. 10	5 – 30
P-3 Oxonia active 150	Ecolab	Hydrogen peroxide, Peracetic acid, Acetic acid	0.1 – 0.2	ca. 10	5 – 30
P3 – Oxysan ZS	Ecolab	Hydrogen peroxide, Peracetic acid, Acetic acid, Peroxyoctanoic acid	0.10	ca. 10 max. 40	5 – 30
P-3 Hypochloran	Ecolab	Sodium hypochlorite, Sodium hydride	0.2 – 0.5	20 – 60	15
P-3 Horolith CD	Ecolab	Nitric acid, Phosphoric acid, Polyhexamethylene biguanide hydrochloride	0.5 – 1.5	50 – 70	10
Clarin spezial	Clarin	Peracetic acid, Hydrogen peroxide	0.2 – 0.5	20	5 – 20

*exposure time

Table 2. Characteristics of CIP disinfectants used in the dairy industry.

Disinfection is becoming more and more important in the current strategies used by the dairy industry to limit bacteriophage infections. The virucidal efficacy of disinfectants against bacteria, yeasts, moulds, including pathogens, is well-documented in supplier specifications, but very seldom the information on the efficacy against phages is available. It is wrong to consider that disinfectants active against bacteria will also inactivate bacteriophages [123]. The virucidal activity of commercially available disinfectants is unknown or known only against lab reference phages proposed by the established in 1989

Disinfectant	Supplier/ Producer	Main active substances	Conditions recommended by supplier		
			Concentration (%)	Temp*. (°C)	Time** (min.)
Deptil Mycocide S	Hypred	Propan-2-ol Didecyldimonium chloride	0.3 – 2.5	RT	5
Deptil HDS	Hypred	Ethanol Sorbic acid	undiluted	RT	5
Deptil BFC	Hypred	Laurylamine dipropylenediamine	1.0	20 - 90	5 - 15
Tego 2000 VT 25	Johnson Diversey	Amphoteric surfactants (amines, N-C10-C16- alkyl trimethylenedi, reaction products with chloroacetic acid)	0.5 – 1.0	TR max 50	15 - 60
Divodes FG VT 29	Johnson Diversey	Propan-1-ol Propan-2-ol	50 – 100	RT	5 - 15
Divosan Extra VT 55	Johnson Diversey	Benzalkonium chloride (CAS No 8001-54-5)	0.4 – 0.8	RT	60 - 240
Suredis VT1	Johnson Diversey	Cationic surfactants (N-(3-aminopropyl)-N-dodecylpropane-1,3-diamine CAS: 2372-82-9 Sodium carbonate Disodium tetraborate decahydrate	0.5 – 2.0	RT . max 50	5 - 30
Tego Hygiene 2001	Johnson Diversey	Trisodium nitrilotriacetate (CAS:5064-31-3) N-Dodecylpropane-1,3-diamine (CAS: 5538-95-4) 2-methoxymethylethoxy propanol (CAS: 34590-94-8) reaction product of alkylamino acetic acid and alkyl diazapentane (CAS: 139734-65-9)	1.0 – 2.0	RT max 50	15- 60
Virocid	CID Lines	Benzalkonium chloride Dimetylodidecyloammonium chloride Glutaraldehyde Propan-2-ol	0.5 – 1.0	RT	60

Disinfectant	Supplier/ Producer	Main active substances	Conditions recommended by supplier		
			Concentration (%)	Temp*. (°C)	Time** (min.)
Eko Javel	PUT Ekoserwis	Sodium hypochlorite Sodium hydride	0.5 – 1.5	RT	15
P-3 Topax 91	Ecolab	Benzalkonium chloride (CAS No 8001-54-5)	0.50 - 1	RT	10 – 20
P-3 Topax 99	Ecolab	Alkyl ammonium acetate Acetic acid	1.0 - static method 2.0 - foam method	RT	10 – 20
P-3 topactive DES	Ecolab	Hydrogen peroxide Acetic acid Amino-oxide	1.0-3.0	RT	10-30
P-3 Monodes	Ecolab	Benzyl alcohol Propanol-2-ol Ethanol	undiluted	RT	0.5
Anthium Dioxide 5% active chlorine	GSG	Chlorine dioxide Activator – citric acid	0.01 – 0.05	RT	10

* RT – room temperature, ** exposure time

Table 3. Characteristics of the disinfectants for surfaces, equipment, shoe baths and hands used in dairy industry.

CEN committee for harmonizing the method of evaluating the efficacy of disinfectants [124]. Factors influencing the efficiency of a given disinfectant are: concentration, temperature and exposure time. Among them, the most important is the concentration of active substances. Most of disinfectants are less effective against phages in the presence of interfering proteins (milk or whey) or hard water. The virucidal activity of most disinfectants is improved by increasing the temperature and is usually the lowest in cold water. Therefore, at low temperatures and/or in the presence of proteins, disinfectant concentration and/or contact time should be increased. It is always advisable to combine biocides and heat rather than use them separately at extreme conditions [125]. However, no disinfectant will be fully effective when sanitized surfaces are not cleaned and proteins or biofilm-living cells are present [126]. Under certain conditions phage particles may exist as aggregates, which may also impair complete inactivation. Peracetic acid and sodium hypochlorite are the most efficient biocides of the CIP system in the dairy industry; however, literature data indicate that some LAB phages may be resistant to sodium hypochlorite [125,127-130]. Nonetheless, the most recently available disinfectants are a combination of several biocides. Table 2 presents the chemical content of CIP disinfectants and conditions of their use in the dairy industry as recommended by the suppliers.

Disinfectants recommended mainly for surfaces, equipment, hands and shoe sanitization are listed in Table 3. Disinfectants are in liquid, foam or aerosol form, depending on their application. The efficacy of such disinfectants for phage inactivation, especially those based on alcohols, are lower in comparison to CIP disinfectants. Among biocides, particularly ineffective in phage inactivation is isopropanol [125]. However, taking into account a lower number of phages in an environment, it can be sufficient for their elimination.

11.2. Design of starter cultures rotation system based on phage contamination control

Starter cultures are a key factor influencing the diversity of phage population in a dairy plant. Application of undefined multispecies and multistrain cultures was the main strategy to overcome production problems related to phages in many factories (Flora Danica - Chr. Hansen, Probat 505 - Danisco) in the past. One complex culture (e.g. Flora Danica) allowed producing many products: maturated cheese, fresh cheese (tvarog and quark), butter, butter-milk and other mesophilic fermented beverages. Complex multispecies and multistrain cultures are relatively phage tolerant and even upon high phage contamination give products with small deviations that are accepted for marketing. In the past, when dairy plants produced a wide range of products, mainly for the local market, complex undefined cultures fulfilled the expectations of the dairy business.

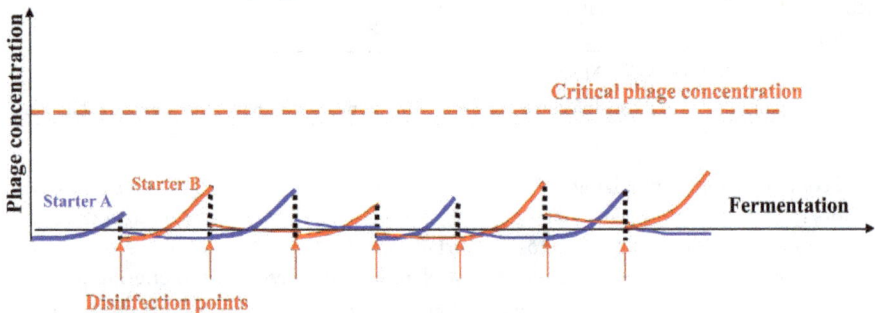

Figure 2. Example of well-designed culture rotation and disinfection frequency strategy for phage control in dairy plant.

Modern industrial fermentations increasingly rely on well-defined, direct vat inoculated (DVI), high concentrated (> 10^{10} cfu g^{-1}) and product-optimized starters, containing from two to five phage-unrelated strains [131-132]. Market share of bulk starters (semi-direct inoculation) diminished very fast in the last two decades and does not exceeded 20% for dairy beverages and 60% for cheese of the total global processed milk. The defined cultures have been widely adapted in large-scale production facilities due to the significant degree of control over fermentation processes and complementary fermentation properties, such as rapid acidification, gas formation, texturization, and development of flavor and aroma compounds. Each defined culture is designed in two or three phase-unrelated options, which can consistently enable producers to obtain high quality standard products. Rotation of defined phage-unrelated cultures is an efficient phage control method. Usually the rotation strategy in big dairy plants is elaborated in tight collaboration with culture suppliers based on individual phage monitoring programs. Ideally, sterilized products or whey samples are delivered on a routine basis at agreed intervals to the phage lab of the culture supplier. In longer perspective, successful cooperation of culture suppliers and users in monitoring different culture rotation strategies allows designing sequences of culture rotation and safe intervals between rotations as well as elaborate the cleaning and disinfection strategy adapted to specific dairy environments (Fig.2).

Rotation strategy of defined multiple strain cultures demands selection of strains resistant to a wide range of phages, which could replace infected strains. This aspect can be a drawback when considering continuous and effective use of this method. Moreover, continual rotation of multiple strains during fermentation processes has an effect on phage co-evolution and was shown to increase phage diversity and their abundance in the dairy environment [133]. It also requires constant selection of starter strains with specific fermentative properties. An alternative is the use of a single, highly specialized phage-resistant strain and its variants carrying phage resistance plasmids obtained from naturally resistant strains. This strategy was termed by Sing and Klaenhammer as the phage defense rotation strategy (PDRS) [134]. The success of designed rotations systems of phage-resistant single strain derivatives is assessed by the Heap-Lawrence starter culture activity test (SAT) performed usually in phage-contaminated milk or whey from earlier cycles [135]. Continuous rotation in repeated cycles of single starter lactococcal strain derivatives, where each carries a different type or a combination of various phage defense systems (e.g. R/M or Abi), has been recognized as an effective method of limiting phages during industrial processes [134,136]. Sing and Klaenhammer have shown that the rotation system of three *Lactococcus lactis* derivative strains encoding different phage defense mechanisms provided resistance to the culture during nine rotation cycles against 10^6 PFU ml^{-1} of whey composition containing as many as 160 phage isolates [134]. The strategy was then shown to demand precise determination of the type of defense systems to be used as well as the rotation order of the strains. Expression of several phage defense systems relying on different mechanisms conferred complementary defense against phage infection of single strain-derived cultures. Even if one defense system has been overcome, the phage can be inactivated by another. In the study of Durmaz and Klaenhammer (1995) three single starter *Lactococcus lactis* subsp. *lactis* derivatives, containing different plasmid-encoded phage defense mechanisms, were subjected to a 9-day rotation process challenged by two isometric phages (ul36 and Φ31) or a combination of 10 industrial phages at high titer [136]. Moreover, in most cases examined, an additive effect of different phage R/M and Abi defense systems was observed [136]. As assessed by SAT, the culture persisted incoming infections and only one Φ31-derived mutant phage was detected, but did not disturb culture growth during 17 rotation rounds. Based on these observations, it seems that continuous rotation of at least three derivatives of a single starter strain, where each carries a different phage defense system, is an attractive method to overcome phages as well as all types of resulting phage mutants. Moreover, the use of a limited number of strains, in this case one strain and its variants, limits the phage number as well as the occurrence of novel phages in fermentation plants [135,137]. A great advantage for the industry is also the use of only one indicator strain to monitor phage occurrence. Application of PDRS by construction of novel strains carrying newly identified phage-resistance mechanisms makes this strategy broad range with unlimited variants.

11.3. Production organization

An important element reducing the spread of phages in the dairy plant is the organization of production. The control of phage risk in dairy plants relies on development and implementation of a variety of procedures. To keep phages under control one should [5,102,123]:

- perform daily tests for phage detection
- avoid crossing paths for raw milk, pasteurized milk and whey
- reduce the diversity of products made on a given day in one production hall
- rotate manufacturing processes
- directly inoculate milk with high concentrated cultures
- rotate starter cultures
- use anti-phage media for bulk starter (BS) propagation
- perform aseptic inoculation where possible
- use air filtration (HEPA) and positive pressure in production facilities
- use positive pressure in fermentation tanks where possible
- use steam sterilization of production lines where possible, especially when phage contamination is high
- dispose stagnant zones of water, whey, milk and foam from production hall or other liquid pools containing live cultures
- clean and disinfect lines, floors, walls, bins and drains used immediately after the process completion
- redisinfect lines after longer production break (e.g. weekend, bank holiday, breakdown)
- disinfect of small equipment used in milk processing after each use (pH-electrode, temperature sensors, etc.)
- use footbaths with disinfecting agents at the entry of production facilities
- avoid using the same equipment for raw milk and whey transportation and treatment
- separate fermentation and packaging areas
- limit personnel path movements (staff in contact with raw milk has no admission to the production facilities)

Plant staff should be aware of the importance of phage control risk, well acquainted with procedures and follow them.

12. Selection of phage tolerant strains

12.1. Classical methods (isolation and selection of phage tolerant strains against the most aggressive phages from the dairy environment)

In order to isolate phage-resistant mutants, a secondary culture method can be used [138], in which sensitive strains undergo selective pressure of their specific phages. Sensitive strains are inoculated in liquid medium and subsequently infected with suspensions of a selected lytic phage at specific titer. Liquid cultures exhibiting complete lysis are incubated for 24-48 hours (secondary growth). After incubation, bacteria are streaked on adequate solid medium. The grown colonies are consecutively cultured in liquid medium with the same selected phage during at least three rounds. Resultant isolates that are able to grow normally in the presence of the specific phage are considered as true phage-resistant mutants [139].

Another means of natural selection of phage-resistant strains was developed by Viscardi and colleagues [140]. The approach is based on flow-cytometry technique that senses and selects bacterial cells to which phage particles that have been added to the medium did not adsorb. Two detection methods have been designed, which rely on recognition of either specifically labeled anti-phage antibodies or fluorochrome-stained phages. The presented method is an attractive alternative to other means of isolating phage-resistant strains (described earlier). In the study, several different *Streptococcus thermophilus* strains were analyzed for their potential to develop spontaneous phage resistance that could be detected by flow-cytometry technique. The designed selection methods proved quite sensitive, as phage-resistant cells could be detected after only one selection round. Nonetheless, a two-round selection based on selection with anti-phage antibodies or labeled phages and then with unlabeled phage alone was more efficient in obtaining stable and proper phage-resistant mutants. Phage adsorption assays determined that majority of the isolated mutants resisted phage infection at the level of phage adsorption. Moreover, several selection rounds using different labeled phages lead to isolating multi-phage resistant cells.

The great advantage of the method is its high sensitivity (detection of 2 out 10^7 cells) and high analysis rates (10^3 cells per second). As the occurrence of spontaneous phage-resistant cells is rather low in nature, the method allows increasing the level of detection of such mutants. Furthermore, the selected *S. thermophilus* mutants were resistant to phage attack throughout multiple generations, indicating the stability of this property. The novelty of the method is the short amount (several days) of time necessary for obtaining phage-resistant mutants. This creates a possibility of fast selection of new resistant starter strains in the presence of novel phages, which constantly break away from the current defense systems.

12.2. BIM system - exposure of sensitive strains to lytic phages (spontaneous mutation in chromosomal or plasmidic genes)

Selection of BIMs (bacteriophage insensitive mutants) is a way to obtain phage-resistant strains without genetic manipulations. The idea of obtaining such cells is to infect a starter strain culture and select for mutants which have sustained phage attack.

This approach has its drawbacks, as it is based solely on the occurrence of random potential mutations in genes coding for receptor materials. The lack of a functional initial receptor for 936- and P335-type phages, such as a polysaccharide, is associated with mutations in genes involved in its synthesis or transport. It is well documented that phage insensitivity of *L. lactis* strains is correlated with loss of the galactose-associated receptor in the cell wall. This disturbs the synthesis of wall components and, as a consequence, insensitive strains often lose their industrial properties, such as the ability to produce acids, and reveal weaker growth in comparison to wild type strains.

Apart from altering cell growth, other two features, such as narrow phage specificity and spontaneous reversion to sensitive phenotype, limit exploitation of BIM mutants in industrial applications [17]. However, mutations in the *pip* gene, encoding a specific receptor for c2-type phages only (for further details on Pip function see: 4 and 6.1-2.), have no significant impact on vitality of lactococcal cells and resultant mutants are stably maintained [17,24]. Genetic engineering methods, which possess a huge potential for developing protection against phages, based on specific point mutations, and construction of stable mutants, might be the solution to this problem. However, at present methods utilizing recombinant DNA approaches restrict the industrial use of genetically modified strains. Mills and colleagues presented a simple 3-step approach, devoid of genetic engineering methods, for generating BIMs of *S. thermophilus* [141]. In the first step, sensitive bacteria were completely lysed in soft top agar plates by adding a selected industrial phage at a MOI > 1 (multiplicity of infection above 1). Subsequently, plates were incubated up to 48 hours after which appearance of resistant colonies was observed. In the next step, all colonies were collected and used to inoculate fresh liquid medium. Harvested bacteria from step 2 were used for conducting a continuous culture in milk with 20–25 passages in the presence of phage at a high concentration (MOI = 10). In order to obtain BIM colonies, the last passage was poured on solid agar from which phage-resistant BIMs were selected after overnight growth. Resistance to another phage could be generated by repeating the whole process on the resultant BIM strain. The insensitive phenotype was initially attributed to nonspecific mutations in receptor genes. However, further studies revealed that phage insensitivity is due to alteration of the CRISPR (clustered regularly interspaced short palindromic repeats) locus, not associated with the previously thought mutations [142] (for further details on CRISPRs see section 6.5 and 12.4).

12.3. Plasmid concept

Among the acknowledged and widely applied methods of obtaining starter strains resistant to phage infections is conjugational transfer of plasmids conferring phage resistance determinants [143-144]. In lactococci, there is a range of bacteriophage defense systems occurring naturally on plasmids (**natural, plasmid-encoded phage-resistance systems**). Among the plasmid-encoded phage resistance are such defense mechanisms as restriction/modification (R/M) or abortive infection (Hsp+ or other Abi+) (for more details see sections: 6.3. and 6.4.). First studies, which linked the presence of phage resistance mechanisms to plasmid molecules, were simple assays based on isolation of plasmids from resistant strains and their reintroduction into susceptible cells to obtain cells immune to attack by a particular phage. The later discovery of phage resistance determinants encoded on conjugational plasmids attracted great interest of the food production industry. Most of the data on conjugative plasmids conferring phage resistance comes from studies in *Lactococcus lactis*. In this species many various conjugal plasmids conferring phage-resistance have been identified, including: pTN20, pNP40 and pCI1750, carrying both conjugal transfer (Tra⁺) and abortive infection (Abi⁺) determinants, or pAJ1106, exhibiting Tra⁺ and Hsp⁺ phenotype [145-149]. Extensive studies of various research groups showed that indeed construction of phage-resistant strains via simple conjugational transfer is an effective means of generating phage resistant starter strains, some of which found application in the dairy industry [143,150].

Among the first conjugal plasmids discovered in *Lactococcus lactis* was pTR2030 isolated from strain ME2. It was characterized to encode heat-sensitive phage resistance (Hsp⁺), restriction-modification (LlaIR/M) as well as conjugal transfer (Tra⁺) genes [151]. Its introduction via conjugation into other lactococcal strains, including *Lactococcus lactis* subsp. *cremoris*, resulted in phage-resistance phenotypes [152]. Application of these genetic elements was hence proclaimed as an attractive and acceptable alternative for generating resistant strains, in contrast to strain construction using genetic engineering. The study of Sanders et al. (1986) described the successful attempt of introducing the pTR2030 plasmid via conjugation from a *L. lactis* donor into several industrial recipient strains, from both *lactis* and *cremoris* subspecies [143]. Resulting transconjugants proved resistant to homologous phage infection. Curing of pTR2030 from transconjugants restored phage-sensitive phenotypes, proving visibly that phage resistance is conferred by the plasmid. Noteworthy is the fact that selection of phage-resistant transconjugants was performed in an antibiotic-free background, which is most appropriate for manipulations with strains intended for food production. Another important advantage of this approach was the fact that transconjugant strains maintained their acid-producing properties. This aspect is quite important as it shows that conjugative plasmid manipulations do not alter the industrially attractive features of starter bacteria. The pTR2030 plasmid was maintained throughout multiple generations, indicating that phage resistance will be a stable feature during prolonged use of the transconjugant in industrial applications. Resistance mechanisms identified on conjugative plasmids were also applied in developing engineered bacterial phage defense systems, e.g. the LlaIR/M function encoded on the pTR2030 plasmid was used in constructing phage-triggered suicide systems (see section: 6.6.4.).

The plasmid-concept of generating phage-resistant strains has also its limitations. First of all, it should be taken into account that many industrially-applied strains are hard to transform. Furthermore, there is a chance that introduction of new plasmids might destabilize industrially attractive strain properties that are also plasmid-encoded (issue of plasmid incompatibility). Introduction of plasmids transferring phage resistance into the bacterial chromosome could be a way of stabilizing this feature; yet, on the other hand, will demand approval of appropriate authorities. Furthermore, some industrially-exploited lactic acid bacteria species, e.g. *S. thermophilus*, carry few plasmids (including conjugal plasmids). This can be an obstacle in generating novel phage-resistant strains via conjugational events [153]. Yet, studies performed by Burrus et al. (2001) revealed the presence of an integrative conjugative element ICR*St1* in *S. thermophilus* strain CNRZ368, shown to encode a II-type R/M system that provided resistance to phage φST84 infection [154]. Identification of a phage defense system on an integrative element suggests that also such genetic elements as transposons can be responsible for the spread of phage-resistance mechanisms within bacterial populations.

12.4. CRISPR/*cas* defense in LAB

The CRISPR/*cas* defense system was first described in the 1980s for *E. coli*, but only recently recognized for lactic acid bacteria (2007), including such genera as *Lactobacillus*, *Bifidobacterium*, *Symbiobacterium*, *Enterococcus* and *Streptococcus*. Examination of more than 100 genomes of various LAB species allowed identifying over 60 different CRISPR loci, which were grouped into eight distinct families [155]. This indicates the highly diverse nature of LAB CRISPR loci. Additionally, it was observed that clustering of LAB CRISPRs was not in accordance with the classical phylogenetic correlations observed between the LAB phyla. This strongly implies that dissemination of CRISPR loci within the Prokaryotic world into separate lineages occurred by horizontal gene transfer events and their further evolution was imposed by the selective pressure due to phage infections. In general, CRISPR loci were determined to be located on the chromosome, except for one *E. faecium* strain found to carry the CRISPR array on a plasmid. Most LAB species harbor more than one CRISPR locus; yet, despite the common occurrence of CRISPR/*cas* systems, they have still not been identified for such species as *Lactococcus*, *Leuconostoc*, *Carnobacterium*, *Pediococcus*, and *Oenococcus*. This surprising absence of CRISPR loci was implied to be connected with an insufficient amount of sequencing data for these species in public databases. Examination of other strains of these species, involving genome sequencing, should be performed in order to fully resolve the issue on the existence of CRISPR/*cas* systems in these LABs. The identified various CRISPR arrays were determined to contain in total 100 different spacer sequences, including sequences of phage (26%) or prophage (47%) origin.

As CRISPR/*cas* systems confer phage resistance to host cells, they are quite of interest for the dairy industry where microbial production plays a significant role. Application of CRISPR/*cas* systems for construction of new LAB strain variants with differentiated resistance to phage infections is a novel alternative approach [67,142,156]. Moreover, such strains are regarded as safer for industrial applications, as the possibility for them to incorporate or disseminate foreign mobile genetic elements of unknown impact is low. Natural methods of selecting CRISPR-containing BIM cells (see section: 12.2.) of industrially applied bacteria could be an interesting solution for obtaining resistant strains, without deliberate genetic modifications. The first report on isolating CRISPR-containing lactic acid bacteria came from Barrangou et al. (2007) [67], who described the an approach of obtaining spontaneous *S. thermophilus* BIM cells by providing selection pressure due to phage infection. Protocols of isolating such strains have been later developed for dairy *S. thermophilus*, applied in the manufacturing of cheese and yoghurts [141]. The strategy is based on exposition of bacterial starter culture to high phage titers. Several rounds of growth in milk media under the constant selection pressure due to the phage presence resulted in obtaining phage-resistant mutants able to efficiently grow under industrial conditions. The great advantage of such approach is the fact that the presence of naturally acquired spacer sequences renders the strain resistant to phage infections, while preserving the industrially-attractive features of the initial starter cultures. Another strategy of constructing phage-resistant strains could be deliberate integration of synthetic spacers homologous to conserved sequences of industrial phage isolates into the CRISPR array of starter bacteria. However, this approach would involve certain molecular manipulations at the DNA level. Nonetheless, controlled modification of phage resistance of LAB strains using the CRISPR/*cas* regions is not considered by the food industry as a genetic modification method within the meaning of the existing rules in this area.

Author details

A.K. Szczepankowska, R.K. Górecki and J.K. Bardowski
Institute of Biochemistry and Biophysics of Polish Academy of Sciences, Warsaw, Poland

P. Kołakowski *
Danisco Biolacta, Innovation, Olsztyn, Poland

13. References

[1] Whitehead HR, Cox GA. The occurrence of bacteriophage in lactic streptococci. N. Z. J. Dairy Sci. Technol. 1935;16 319–320.

[2] Daly C, Fitzgerald GF, Davis R. Biotechnology of lactic acid bacteria with special reference to bacteriophage resistance. Antonie Van Leeuwenhoek 1996;70(2-4) 99-110.

[3] Lawrence R C. Action of bacteriophages on lactic acid bacteria: consequences and protection. N. Z. J. Dairy Sci. Technol. 1978;13 129-136.

[4] Brüssow H, Hendrix RW. Phage genomics: small is beautiful. Cell 2002;108(1) 13-6.

[5] Kołakowski P, Rybka J. Causes of disorders of fermentation failures (in Polish). BIBIT (Information Bulletin Rhodia Food Biolacta) 2001;2(24) 7-14.

[6] Klaenhammer T, Altermann E, Arigoni F, Bolotin A, Breidt F, Broadbent J, Cano R, Chaillou S, Deutscher J, Gasson M, van de Guchte M, Guzzo J, Hartke A, Hawkins T, Hols P, Hutkins R, Kleerebezem M, Kok J, Kuipers O, Lubbers M, Maguin E, McKay L, Mills D, Nauta A, Overbeek R, Pel H, Pridmore D, Saier M, van Sinderen D, Sorokin A, Steele J, O'Sullivan D, de Vos W, Weimer B, Zagorec M, Siezen R. Discovering lactic acid bacteria by genomics. Antonie Van Leeuwenhoek 2002;82(1-4) 29-58.

* Corresponding Author

[7]   Klaenhammer TR, Barrangou R, Buck BL, Azcarate-Peril MA, Altermann E. Genomic features of lactic acid bacteria effecting bioprocessing and health. FEMS Microbiol Rev. 2005;29(3) 393-409.

[8]   Fox PF, McSweeney PLH. Cheese: An Overview. Cheese: Chemistry, Physics and Microbiology 2004;1 1-18.

[9]   Ackermann HW, Kropinski AM. Curated list of prokaryote viruses with fully sequenced genomes. Res Microbiol. 2007;158(7) 555-566.

[10] Ackermann HW. Bacteriophage observations and evolution. Res Microbiol. 2003;154(4) 245-251.

[11] Ackermann HW. Frequency of morphological phage descriptions in the year 2000. Brief review. Arch Virol. 2001;146(5) 843-857.

[12] Hendrix RW. Bacteriophage genomics. Current Opinion in Microbiology 2003;6 506–511.

[13] Breitbart M, Salamon P, Andresen B, Mahaffy JM, Segall AM, Mead D, Azam F, Rohwer F. Genomic analysis of uncultured marine viral communities. Proc Natl Acad Sci USA 2002;99(22) 14250-5.

[14] Yu MX, Slater MR, Ackermann HW. Isolation and characterization of Thermus bacteriophages. Arch Virol. 2006;151(4) 663-79.

[15] Breitbart M, Hewson I, Felts B, Mahaffy JM, Nulton J, Salamon P, Rohwer F. Metagenomic analyses of an uncultured viral community from human feces. J Bacteriol. 2003;185(20) 6220-6223.

[16] Mc Grath S, Fitzgerald GF, van Sinderen D. Bacteriophages in dairy products: pros and cons. Biotechnology Journal 2007;2(4) 450-455.

[17] Forde A, Fitzgerald GF. Bacteriophage defence systems in lactic acid bacteria. Antonie Van Leeuwenhoek 1999;76(1-4) 89-113.

[18] Heller K, Braun V. Polymannose O-antigens of *Escherichia coli*, the binding sites for the reversible adsorption of bacteriophage T5+ via the L-shaped tail fibers. J. Virol. 1982;41(1) 222-7.

[19] Dupont K, Janzen T, Vogensen FK, Josephsen J, Stuer-Lauridsen B. Identification of *Lactococcus lactis* genes required for bacteriophage adsorption. Appl Environ Microbiol. 2004;70(10) 5825-32.

[20] Babu KS, Spence WS, Monteville MR, Geller BL. Characterization of a cloned gene (*pip*) from *Lactococcus lactis* required for phage infection. Dev Biol Stand. 1995;85 569-75.

[21] Mooney DT, Jann M, Geller BL. Subcellular location of phage infection protein (Pip) in *Lactococcus lactis*. Can J Microbiol. 2006;52(7) 664-72.

[22] Kraus J, Geller BL. Membrane receptor for prolate phages is not required for infection of *Lactococcus lactis* by small or large isometric phages. J Dairy Sci. 1998;81(9) 2329-2335.

[23] Hoier E, Janzen T, Rattray F, Sorensen K, Borsting MW, Brockmann E. The production, application and action of lactic cheese starter cultures. In: Law BA, Tamime AY. (eds.) Technology of Cheese Making. Oxford UK: Wiley-Blackwell; 2010. p166-192.

[24] Coffey A, Ross RP. Bacteriophage-resistance systems in dairy starter strains: molecular analysis to application. Antonie Van Leeuwenhoek 2002;82(1-4) 303-21.

[25] Tuncer Y, Akçelik M. A protein which masks galactose receptor mediated phage susceptibility in *Lactococcus lactis* subsp. *lactis* MPL56. International Journal of Food Science & Technology 2002;37(2) 139–144.

[26] Forde A, Fitzgerald GF. Molecular organization of exopolysaccharide (EPS) encoding genes on the lactococcal bacteriophage adsorption blocking plasmid, pCI658. Plasmid 2003;49(2) 130-42.

[27] Watanabe K, Ishibashi K, Nakashima Y, Sakurai T. A Phage-resistant mutant of *Lactobacillus casei* which permits phage adsorption but not genome injection. J Gen Virol. 1984;65 981 - 986.

[28] Mahony J, McGrath S, Fitzgerald GF, van Sinderen D. Identification and characterization of lactococcal-prophage-carried superinfection exclusion genes. Appl Environ Microbiol. 2008;74(20) 6206-15.

[29] Garvey P, Hill C, Fitzgerald GF. The lactococcal plasmid pNP40 encodes a third bacteriophage resistance mechanism, one which affects phage DNA penetration. Appl Environ Microbiol. 1996;62(2) 676-679.

[30] McGrath S, Fitzgerald GF, van Sinderen D. Identification and characterization of phage-resistance genes in temperate lactococcal bacteriophages. Mol Microbiol. 2002;43(2) 509-20.

[31] Ventura M, Canchaya C, Pridmore RD, Brüssow H. The prophages of *Lactobacillus johnsonii* NCC 533: comparative genomics and transcription analysis. Virology 2004;320(2) 229-242.

[32] Sun X, Göhler A, Heller KJ, Neve H. The ltp gene of temperate *Streptococcus thermophilus* phage TP-J34 confers superinfection exclusion to *Streptococcus thermophilus* and *Lactococcus lactis*. Virology 2006;350(1) 146-57.

[33] Blumenthal RM, Cheng X. Restriction-modification systems. In: Streips UN, Yasbin RE. (eds.) Modern microbial genetics, 2nd ed. New York: Wiley; 2002. p177–226

[34] Seegers JF, van Sinderen D, Fitzgerald GF. Molecular characterization of the lactococcal plasmid pCIS3: natural stacking of specificity subunits of a type I restriction/modification system in a single lactococcal strain. Microbiology 2000;146(2) 435-443.

[35] Smith MA, Read CM, Kneale GG. Domain structure and subunit interactions in the type I DNA methyltransferase M.EcoR124I. J. Mol. Biol. 2001;314(1) 41-50.

[36] Murray NE. Type I restriction systems: sophisticated molecular machines (a legacy of Bertani and Weigle). Microbiol. Mol. Biol. Rev. 2000;64(2) 412-434.

[37] Janscak P, Bickle TA. The DNA recognition subunit of the type IB restriction-modification enzyme EcoAI tolerates circular permutions of its polypeptide chain. J Mol Biol. 1998;284(4) 937-948.

[38] Janscak P, Dryden DT, Firman K. Analysis of the subunit assembly of the type IC restriction-modification enzyme EcoR124I. Nucleic Acids Res. 1998;26(19) 4439-45.

[39] O'Sullivan D, Twomey DP, Coffey A, Hill C, Fitzgerald GF, Ross RP. Novel type I restriction specificities through domain shuffling of HsdS subunits in *Lactococcus lactis*. *Mol. Microbiol.* 2000;36(4) 866-75.

[40] Gorecki RK, Koryszewska-Bagińska A, Gołębiewski M, Żylińska J, Grynberg M, Bardowski J. Adaptative potential of the *Lactococcus lactis* IL594 strain encoded in its 7 plasmids. PLoS ONE 2011;6(7) e22238.

[41] Adamczyk-Popławska M, Kondrzycka A, Urbanek K, Piekarowicz A. Tetra-amino-acid tandem repeats are involved in HsdS complementation in type IC restriction–modification systems. Microbiology 2003;149(11) 3311–3319.

[42] [42]Holubova I, Vejsadova S, Firman K, Weiserova M. Cellular localization of Type I restriction-modification enzymes is family dependent. Biochem Biophys Res Commun 2004;319(2) 375–380.

[43] Keatch SA, Su TJ, Dryden DT. Alleviation of restriction by DNA condensation and non-specific DNA binding ligands Nucleic Acids Res. 2004;32(19) 325841–5850.

[44] Pingoud A, Fuxreiter M, Pingoud V, Wende W. Type II restriction endonucleases: structure and mechanism. Cell Mol. Life Sci. 2005;62(6) 685-707.

[45] Pingoud A, Jeltsch A. Structure and function of type II restriction endonucleases. Nucleic Acids Res. 2001;29(18) 3705-27.

[46] Marshall JJ, Gowers DM, Halford SE. Restriction endonucleases that bridge and excise two recognition sites from DNA. J Mol Biol. 2007;367(2) 419-431.

[47] Gowers DM, Bellamy SRW, Halford SE. One recognition sequence, seven restriction enzymes, five reaction mechanisms. Nucl. Acids Res. 2004;32(11) 3469–3479.

[48] Marks P, McGeehan J, Wilson G, Errington N, Kneale G. Purification and characterisation of a novel DNA methyltransferase, M.AhdI. Nucleic Acids Res. 2003;31(11) 2803–2810.

[49] Szybalski W, Kim SC, Hasan N, Podhajska AJ. Class-IIS restriction enzymes–a review. Gene 1991;100 13–26.

[50] Christensen LL, Josephsen J. The methyltransferase from the LlaDII restriction modification system influences the level of expression of its own gene. J Bacteriol. 2004;186(2) 287-295.

[51] Su P, Im H, Hsich H, Kang AS, Dunn NW. LlaFI, a type III restriction and modification system in *Lactococcus lactis*. Appl Environ Microbiol. 1999;65(2) 686–693.

[52] Guinane CM, Kent RM, Norberg S, Hill C, Fitzgerald GF, Stanton C, Ross RP. Host specific diversity in *Lactobacillus johnsonii* as evidenced by a major chromosomal inversion and phage resistance mechanisms. Plos One 2011;6 e18740.

[53] Su P, Im H, Hsieh H, Kang AS, Dunn NW. LlaFI, a type III restriction and modification system in *Lactococcus lactis*. Appl. Environ. Microbiol 1999;65(2) 686-93

[54] Solow BT, Somkuti GA. Molecular properties of *Streptococcus thermophilus* plasmid pER35 encoding a restriction modification system. Curr Microbiol 2001;42(2) 122–128.

[55] Hao P, Zheng H, Yu Y, Ding G, Gu W, Chen S, Yu Z, Ren S, Oda M, Konno T, Wang S, Li X, Ji ZS, Zhao G. Complete sequencing and pan-genomic analysis of *Lactobacillus*

delbrueckii subsp. *bulgaricus* reveal its genetic basis for industrial yogurt production. PLoS ONE 2011;6(1) e15964.

[56] Janulaitis A, Petrusyte M, Maneliene Z, Klimasauskas S, Butkus V. Purification and properties of the *Eco*57I restriction endonuclease and methylase—prototypes of a new class (type IV). Nucleic Acids Res. 1992;20(22) 6043–6049.

[57] O'Sullivan O, O'Callaghan J, Sangrador-Vegas A, McAuliffe O, Slattery L, Kaleta P, Callanan M, Fitzgerald GF, Ross RP, Beresford T. Comparative genomics of lactic acid bacteria reveals a niche-specific gene set. BMC Microbiol. 2009;5 9–50.

[58] Chopin MC, Chopin A, Bidnenko E. Phage abortive infection in lactococci: variations on a theme. Curr. Opin. Microbiol. 2005;8(4) 473–479.

[59] Durmaz E, Klaenhammer TR. Abortive phage-resistance mechanism AbiZ speeds the lysis clock to cause premature lysis of phage-infected *Lactococcus lactis*. J Bacteriol. 2007;189(4) 1417–1425.

[60] Haaber J, Fortier LC, Moineau S, Hammer K. AbiV, a novel abortive phage infection mechanism on the chromosome of *Lactococcus lactis* subsp. *cremoris* MG1363. Appl. Environ. Microbiol. 2008;74(21) 6528–6537.

[61] Prevots F, Tolou S, Delpech B, Kaghad M, Daloyau M. Nucleotide sequence and analysis of the new chromosomal abortive infection gene *abiN* of *Lactococcus lactis* subsp. *cremoris* S114. FEMS Microbiol Lett. 1998;159(2) 331–336.

[62] O'Connor L, Tangney M, Fitzgerald GF. Expression, regulation, and mode of action of the AbiG abortive infection system of *Lactococcus lactis* subsp. *cremoris* UC653. Appl. Environ. Microbiol. 1999;65(1) 330–335.

[63] Twomey DP, De Urraza PJ, McKay LL, O'Sullivan DJ. Characterization of AbiR, a novel multicomponent abortive infection mechanism encoded by plasmid pKR223 of *Lactococcus lactis* subsp. *lactis* KR2. Appl Environ Microbiol. 2000;66(6) 2647–2651.

[64] Domingues S, McGovern S, Plochocka D, Santos MA, Ehrlich SD, Polard P, Chopin MC. The lactococcal abortive infection protein AbiP is membrane-anchored and binds nucleic acids. Virology 2008;373(1) 14–24.

[65] Emond E, Dion E, Walker SA, Vedamuthu ER, Kondo JK, Moineau S. AbiQ, an abortive infection mechanism from *Lactococcus lactis*. Appl Environ Microbiol. 1998;64(12) 4748–4756.

[66] Horvath P, Barrangou P. CRISPR/Cas, the immune system of Bacteria and Archea. Science 2010;327(5962) 167-170.

[67] Barrangou R, Fremaux C, Deveau H, Richards M, Boyaval P, Moineau S, Romero DA, Horvath P. CRISPR provides acquired resistance against viruses in prokaryotes. Science 2007;315(5819) 1709–1712.

[68] Lillestøl RK, Shah SA, Brügger K, Redder P, Phan H, Christiansen J, Garrett RA. CRISPR families of crenarcheal genus Sulfolobus: bidirectional transcription and dynamic properties. Mol Microbiol 2006;72(1) 259-272.

[69] Szczepankowska AK. Role of CRISPR/cas system in the development of bacteriophage resistance. In: Łobocka M, Szybalski WT. (eds.) Advances in Virus Research 82 (part A). USA: Academic Press (ELSEVIER); 2012. p289-338.

[70] Rousseau C, Nicolas J, Gonnet, M. CRISPI: A CRISPR Interactive database. Bioinformatics 2009;25(24) 3317–3318.

[71] Sturino JM, Klaenhammer TR. Engineered bacteriophage-defense systems in bioprocessing. Nat. Rev. Microbiol. 2006;4(5) 395-404.

[72] Inouye M. Antisense RNA: its functions and applications in gene regulation - a review. Gene 1988;72(1-2) 25–34.

[73] Sturino JM, Klaenhammer TR. Expression of antisense RNA targeted against *Streptococcus thermophilus* bacteriophages. Appl. Environ. Microbiol. 2002;68(2) 588–596.

[74] Sturino JM, Klaenhammer TR. Antisense RNA targeting of primase interferes with bacteriophage replication in *Streptococcus thermophilus.* Appl. Environ. Microbiol. 2004;70(3) 1735-1743.

[75] Brüssow H, Probst A, Frémont M, Sidoti J. Distinct *Streptococcus thermophilus* bacteriophages share an extremely conserved DNA fragment. Virology 1994;200(2) 854–857.

[76] Kim SG, Bor YC, Batt CA. Bacteriophage resistance in *Lactococcus lactis* subsp. *lactis* using antisense ribonucleic acid. J. Dairy Sci. 1992;75(7) 1761–1767.

[77] Chung DK, Chung SK, Batt CA. Antisense RNA directed against the major capsid protein of *Lactococcus lactis* subsp. *cremoris* bacteriophage F4-1 confers partial resistance to the host. Appl. Microbiol. Biotechnol. 1992;37(1) 79–83.

[78] Kim SG, Batt CA. Antisense mRNA-mediated bacteriophage resistance in *Lactococcus lactis* subsp. *lactis.* Appl. Environ. Microbiol. 1991;57(4) 1109–1113.

[79] McGrath S, Fitzgerald GF, van Sinderen D. Improvement and optimization of two engineered phage resistance mechanisms in *Lactococcus lactis.* Appl. Environ. Microbiol. 2001;67(2) 608–616.

[80] Hill C, Miller LA, Klaenhammer TR. Cloning, expression, and sequence determination of a bacteriophage fragment encoding bacteriophage resistance in *Lactococcus lactis.* J. Bacteriol. 1990;172(11) 6419–6426.

[81] O'Sullivan DJ, Hill C, Klaenhammer TR. Effect of increasing the copy number of bacteriophage origins of replication, *in trans*, on incoming-phage proliferation. Appl. Environ. Microbiol. 1993; 59(8) 2449–2456.

[82] Foley S, Lucchini S, Zwahlen MC, Brüssow H. A short noncoding viral DNA element showing characteristics of a replication origin confers bacteriophage resistance to *Streptococcus thermophilus.* Virology 1998;250(2) 377–387.

[83] Stanley E, Walsh L, van der Zwet A, Fitzgerald GF, van Sinderen D. Identification of four loci isolated from two *Streptococcus thermophilus* phage genomes responsible for mediating bacteriophage resistance. FEMS Microbiol. Lett. 2000;182(2) 271-277.

[84] Bruttin A, Desiere F, Lucchini S, Foley S, Brüssow H. Characterization of the lysogeny DNA module from the temperate *Streptococcus thermophilus* bacteriophage Sfi21. Virology 1997;233(1) 136–148.

[85] Bruttin A, Foley S, Brüssow H. DNA-binding activity of the *Streptococcus thermophilus* phage Sfi21 repressor. Virology 2002;303(1) 100–109.

[86] Bruttin A, Foley S, Brüssow, H. The site-specific integration system of the temperate *Streptococcus thermophilus* bacteriophage Sfi21. Virology 1997;237(1) 148-158.

[87] Engel G, Altermann E, Klein JR, Henrich B. Structure of a genome region of the *Lactobacillus gasseri* temperate phage phi adh covering a repressor gene and cognate promoters. Gene 1998;210(1) 61–70.

[88] Alvarez MA, Rodriguez A, Suárez JE. Stable expression of the *Lactobacillus casei* bacteriophage A2 repressor blocks phage propagation during milk fermentation. J. Appl. Microbiol. 1999;86(5) 812–816.

[89] Madsen PL, Johansen AH, Hammer K, Brøndsted L. The genetic switch regulating activity of early promoters of the temperate lactococcal bacteriophage TP901–1. J. Bacteriol. 1999;181(24) 7430–7438.

[90] Djordjevic GM, O'Sullivan DJ, Walker SA, Conkling MA, Klaenhammer TR. A triggered-suicide system designed as a defense against bacteriophages. J. Bacteriol. 1997;179(21) 6741–6748.

[91] O'Sullivan DJ, Walker SA, West SG, Klaenhammer TR. Development of an expression strategy using a lytic phage to trigger explosive plasmid amplification and gene expression. Biotechnology 1996;14(1) 82–87.

[92] Djordjevic GM, Klaenhammer TR. Bacteriophage-triggered defense systems: phage adaptation and design improvements. Appl. Environ. Microbiol. 1997;63(11) 4370-4376.

[93] O'Sullivan DJ, Klaenhammer TR. C *LlaI* is a bifunctional regulatory protein of the *llaI* restriction modification operon from *Lactococcus lactis*. Dev. Biol. Stand. 1995; 85: 591–595.

[94] Durmaz E, Madsen SM, Israelsen H, Klaenhammer TR. *Lactococcus lactis* lytic bacteriophages of the P335 group are inhibited by overexpression of a truncated CI repressor. J. Bacteriol. 2002;184(23) 6532–6544.

[95] Sturino JM, Klaenhammer TR. Inhibition of bacteriophage replication in *Streptococcus thermophilus* by subunit poisoning of primase. Microbiology 2007;153(10) 3295–3302.

[96] Lucchini S, Sidoti J, Brüssow H. Broad-range bacteriophage resistance in *Streptococcus thermophilus* by insertional mutagenesis. Virology 2000;275(2) 267-277.

[97] Garbutt KC, Kraus J, Geller BL. Bacteriophage resistance in *Lactococcus lactis* engineered by replacement of a gene for a bacteriophage receptor. J. Dairy Sci. 1997;80 1512–1519.

[98] Pedersen MB, Jensen PR, Janzen T, Nilsson D. Bacteriophage resistance of a Δ*thyA* mutant of *Lactococcus lactis* blocked in DNA replication. Appl. Environ. Microbiol. 2002;68(6) 3010–3023.

[99] Nilsson D, Janzen T. Method of preventing bacteriophage infection of bacterial cultures. International patent application no. PCT/DK99/00382 (1998).

[100] Kenny E, Atkinson T, Hartley BS. Nucleotide sequence of the thymidylate synthase gene (*thyP3*) from the *Bacillus subtilis* phage Φ3T. Gene 1985;34(2-3) 335–342.

[101] Powell IB, Tulloch DL, Hillier AJ, Davidson BE. Phage DNA synthesis and host DNA degradation in the life cycle of *Lactococcus lactis* bacteriophage c6A. J. Gen. Microbiol. 1992;138(5) 945–950.

[102] Moineau S, Levesque C. Control of bacteriophages in industrial fermentations. In: Kutter, Sulakvelidze A. (Eds.) Bacteriophages: Biology and Applications. Boca Raton, USA: CRC Press; 2005. p285-296.

[103] Carlson K. Appendix: Working with bacteriophages common techniques and methodological approaches. In: Kutter, Sulakvelidze A. (Eds.) Bacteriophages: Biology and Applications. Boca Raton, USA: CRC Press; 2005. p435-492.

[104] Garneau JE, Moineau S. Bacteriophages of lactic acid bacteria and their impact on milk fermentations. Microbial Cell Factories 2011;10(Suppl 1) S20, 1-10. http://www.microbialcellfactories.com/content/10/S1/S20.

[105] Labrie S, Moineau S. Multiplex PCR for detection and identification of lactococcal bacteriophages. Applied and Environmental Microbiology 2000;66(3) 987-994.

[106] Del Rio B, Binetti AG, Martin MC, Fernandez M, Magadan AH, Alvarez MA. Multiplex PCR for the detection and identification of dairy bacteriophages in milk. Food Microbiology 2007;24(1) 75-81.

[107] Martin MC, del Rio B, Martinez N, Magadan AH, Alvarez MA. Fast real-time polymerase chain reaction for quantitative detection of *Lactobacillus delbrueckii* bacteriophages in milk. Food Microbiology 2008;25(8) 978-982.

[108] Binetti AG, De Rio B, Martin MC, Alvarez MA. Detection and characterization of *Streptococcus thermophilus* bacteriophages by use of the antireceptor gene sequence. Appl. Environ. Microbiol. 2005;71(10) 6096-6103.

[109] Kleppen HP, Bang T, Nes IF, Holo H. Bacteriophages in milk fermentations: diversity fluctuations of normal and failed fermentations. International Dairy Journal 2011;21(9) 592-600.

[110] Madera C, Monjardin C, Suarez JE. Milk contamination and resistance to processing conditions determine the fate of *Lactococcus lactis* bacteriophages in dairies. Applied and Environmental Microbiology 2004;70(12) 7365-7371.

[111] Atamer Z, Ali Y, Neve H, Heller KJ, Hinrichs J. Thermal resistance of bacteriophages attacking flavor-producing dairy *Leuconostoc* starter cultures. International Dairy Journal, 2011;21(5) 327-334.

[112] Chopin MC. Resistance of 17 mesophilic lactic *Streptococcus* bacteriophages to pasteurization and spray-drying. Journal of Dairy Research 1980;47(1) 131-139.

[113] Canchaya C, Proux C, Fournous G, Bruttin A, Brüssov H. Prophage genomics. Microbiology and Molecular Biology Reviews 2003;67(2) 238-276.

[114] Mercanti DJ, Carminati D, Reinheimer JA, Quiberoni A. Widely distributed lysogeny in probiotic lactobacilli represents a potentially high risk for the fermentative dairy industry. International Journal of Food Microbiology 2011; 144(3) 503-510.

[115] Lunde M, Aastveit AH, Blatny JM, Nes IF. Effects of diverse environmental conditions on φLC3 prophage stability in *Lactococcus lactis*. Applied Environmental Microbiology 2005;71(2) 721-727.

[116] Madera C, Garcia P, Rodriguez A, Suarez JE, Martinez B. Prophage induction in *Lactococcus lactis* by the bacteriocin Lactococcin 972. International Journal of Food Microbiology 2009;129(1) 99-102.

[117] Lunde M, Blatny JM, Lillehaug D, Aastveit AH, Nes IF. Use of real-time quantitative PCR for the analysis of φLC3 prophage stability in lactococci. Applied Environmental Microbiology 2003;69(1) 41-48.

[118] Verreault D, Gendron L, Rousseau GM, Veillette M, Masse D, Lindsley WG, Moineau S, Duchaine C. Detection of airborne lactococcal bacteriophages in cheese manufacturing plants. Applied and Environmental Microbiology 2011;77(2) 491-497.

[119] Neve H, Kemper U, Geis A, Heller KJ. Monitoring and characterization of lactococcal bacteriophages in a dairy plant. Kieler Milchw. Forsh. 1994;46 167-178

[120] Neve H, Laborius A, Heller KJ. Testing of the applicability of battery-powered portable microbial air samplers for detection and enumeration of airborne *Lactococcus lactis* dairy bacteriophages. Kieler Milchw. Forsh. 2003;55(4) 301-315.

[121] Kołakowski P, Rybka J, Fetlinski A. Bacteriophages in milk industry (in Polish). BIBIT (Information Bulletin Rhodia Food Biolacta) 2001;3(25) 8-11.

[122] Simoes M, Simoes LC, Vieira MJ. A review of current and emergent biofilm control strategies. LWT-Food Science and Technology 2010;43(4) 573-583.

[123] Primrose SB. Controlling bacteriophage infections in industrial bioprocesses. Advances in Biochemical Engineering Biotechnology 1990;43 1-10.

[124] Deutsches Institut Fur Normung E.V. DIN EN 13610:2002. Chemical disinfectants-Quantitative suspension test for the evaluation of virucidal activity against bacteriophages of chemical disinfectants used in food and industrial areas- Test method and requirements (phase2, step1); 2003.

[125] Guglielmotti DM, Mercanti DJ, Reinheimer JA, Quiberoni AL. Rewiev: efficiency of physical and chemical treatments on the inactivation of dairy bacteriophages. Frontiers in Microbiology 2011;2(282) 282-297

[126] Simoes M, Simoes LC, Machado I, Pereira MO, Vieira MJ. Control of flow generated biofilms using surfactants – evidence of resistance and recovery. Food and Bioproducts Processing 2006;84(4) 338-345.

[127] Briggiler MM, De Antoni GL, Reinheimer JA, Quiberoni A. Thermal, chemical, and photocatalytic inactivation of *Lactobacillus plantarum* bacteriophages. Journal of Food Protection 2009;72(5) 1012-1019.

[128] Buzrul S, Ozturk P, Alpas H, Akcelik M. Thermal and chemical inactivation of lactococcal bacteriophages. LWT Food Science Technology 2007;40(10) 1671-1677.

[129] Capra ML, Quiberoni A, Reinheimer JA. Thermal and chemical resistance of *Lactobacillus casei* and *Lactobacillus paracasei* bacteriophages. Letters Applied Microbiology 204;38(6) 499-504.

[130] Suarez VB, Reinheimer JA. Effectiveness of thermal treatments and biocides in the inactivation of Argentinian *Lactococcus lactis* phages. Journal Food Protection 2002;65(11) 1756-1759.

[131] Thunell RK, Sandine WE. Types of starter cultures. In: Gilland SE. (ed.) Bacterial starter cultures for foods. Boca Raton USA: CRC Press; 1985. p127–144.

[132] Cogan TM, Peitersen N, Sellars RL. Starter systems. In: Bulletin of the International Dairy Federation, no. 263/1991. Practical phage control. International Dairy Federation, Brussels. 1991; p16–23.

[133] Heap HA, Lawrence RC. The contribution of starter strains to the level of phage infection in a commercial cheese factor. N. Z. J. Dairy Sci. Technol. 1977;12(4) 213.

[134] Sing WD, Klaenhammer TR. A strategy for rotation of different bacteriophage defenses in a lactococcal single-strain starter culture system. Appl. Environ. Microbiol. 1993;59(2) 365–372.

[135] Heap HA, Lawrence RC. The selection of starter strains for cheesemaking. N. Z. J. Dairy Sci. Technol. 1976;11(1) 16–20.

[136] Durmaz E, Klaenhammer TR. A starter culture rotation strategy incorporating paired restriction/modification and abortive infection bacteriophage defenses in a single *Lactococcus lactis* strain. Appl. Environ. Microbiol. 1995; 61(4) 1266-1273.

[137] Thunell RK, Sandine WE, Bodyfelt FW. Phage-insensitive, multiple-strain starter approach to cheddar cheesemaking. J. Dairy Sci. 1981;64(11) 2270-2277.

[138] Quiberoni A, Reinheimer JA, Tailliez P. Characterization of *Lactobacillus helveticus* spontaneous phage resistant mutants by RAPD–PCR fingerprints and phenotypic parameters. Food Research International 1998;31(8) 537–542.

[139] Guglielmotti DM, Reinheimer JA, Binetti AG, Giraffa G, Carminati D, Quiberoni A. Characterization of spontaneous phage-resistant derivatives of *Lactobacillus delbrueckii* commercial strains. Int J Food Microbiol. 2006;111(2) 126-33.

[140] Viscardi M, Capparelli R, Di Matteo R, Carminati D, Giraffa G, Iannelli D. Selection of bacteriophage-resistant mutants of *Streptococcus thermophilus*. J. Microbiol. Met. 2003;55(1) 109–119.

[141] Mills S, Coffey A, McAuliffe OE, Meijer WC, Hafkamp B, Ross RP. Efficient method for generation of bacteriophage insensitive mutants of *Streptococcus thermophilus* yoghurt and mozzarella strains. J Microbiol Methods 2007;70(1) 159-64.

[142] Mills S, Griffin C, Coffey A, Meijer WC, Hafkamp B, Ross RP. CRISPR analysis of bacteriophage-insensitive mutants (BIMs) of industrial *Streptococcus thermophilus* — implications for starter design. J Appl Microbiol. 2010;108(3) 945-55.

[143] Sanders ME, Leonhard PJ, Sing WD, Klaenhammer TR. Conjugal strategy for construction of fast acid-producing, bacteriophage-resistant lactic streptococci for use in dairy fermentations. Appl. Environ. Microbiol. 1986;52(2) 1001-1007.

[144] Pillidge CJ, Collins LJ, Ward LJH, Cantillon BM, Shaw BD, Timmins MJ, Heap HA, Polzin KM. Efficacy of four lactococcal phage resistance plasmids against phage in commercial *Lactococcus lactis* subsp. *cremoris* cheese starter strains. Int. Dairy J. 2000;10(9) 617– 625.

[145] McKay LL, Baldwin KA. Conjugative 40-megadalton plasmid in *Streptococcus lactis* subsp. *diacetylactis* DRC3 is associated with resistance to nisin and bacteriophage. Appl. Environ. Microbiol. 1983;47(1) 68-74.

[146] Higgins DL, Sanozky-Dawes RB, Klaenhammer TR. Restriction and modification activities from *Streptococcus lactis* ME2 are encoded by a self-transmissible plasmid, pTN20, that forms cointegrates during mobilization of lactose-fermenting ability. J. Bacteriol. 1988;170(8) 3435-3442.

[147] Coffey A, Fitzgerald GF, Daly C. Identification and characterization of a plasmid encoding phage abortive infection from *Lactococcus lactis* ssp. *lactis* UC811. Netherlands Milk and Dairy Journal 1989;43(3) 229-244.

[148] Jarvis AW, Heap HA, Limsowtin GKY. Resistance against industrial bacteriophages conferred on lactococci by plasmid pAJ1106 and related plasmids. Appl. Environ. Microbiol. 1989;55(6) 1537-1543.

[149] Harrington A, Hill C. Construction of a bacteriophage-resistant derivative of *Lactococcus lactis* subsp. *lactis* 425A by using the conjugal plasmid pNP40. Appl. Environ. Microbiol. 1991;57(12) 3405-3409.

[150] Alatossava T, Klaenhammer TR. Molecular characterization of three small isometric-headed bacteriophages which vary in their sensitivity to the lactococcal phage resistance plasmid pTR2030. Appl. Environ. Microbiol. 1991;57(5) 1346–1353.

[151] Klaenhammer TR, Sanozky RB. Conjugal transfer from *Streptococcus lactis* ME2 of plasmids encoding phage resistance, nisin resistance and lactose-fermenting ability: evidence for a high-frequency conjugative plasmid responsible for abortive infection of virulent bacteriophage. J. Gen. Microbiol. 1985;131(6) 1531-1541.

[152] Steenson LR, Klaenhammer TR. *Streptococcus cremoris* M12R transconjugants carrying the conjugal plasmid pTR2030 are insensitive to attack by lytic bacteriophages. Appl. Environ. Microbiol. 1985;50(4) 851-858.

[153] Mercenier A. Molecular genetics of *Streptococcus thermophilus*. FEMS Microbiol. Rev. 1990;87 61–78.

[154] Burrus V, Bontemps C, Decaris B, Guédon G. Characterization of a novel type II restriction-modification system, *Sth368I*, encoded by the integrative element ICE*St1* of *Streptococcus thermophilus* CNRZ368. Appl. Environ. Microbiol. 2001;67(4) 1522–1528.

[155] Horvath P, Coûté-Monvoisin AC, Romero DA, Boyaval P, Fremaux C, Barrangou R. Comparative analysis of CRISPR loci in lactic acid bacteria genomes. Int. J. Food Microbiol. 2009;131(1) 62–70.

[156] Horvath P, Romero DA, Coûté-Monvoisin AC, Richards M, Deveau H, Moineau S, Boyaval P, Fremaux C, Barrangou R. Diversity, activity, and evolution of CRISPR loci in *Streptococcus thermophilus*. J. Bacteriol. 2008;190(4) 1401–1412.

Chapter 2

Lactic Acid Bacteria as Starter-Cultures for Cheese Processing: Past, Present and Future Developments

J. Marcelino Kongo

Additional information is available at the end of the chapter

1. Introduction

The identification of solutions to improve the life and health of consumers, providing safe and nutritious foods, is the major concern in Food Science. Toward that goal, preservation methods such as salting, drying, high/low temperature application, fermentation, and more recently, pulsed electric field, high pressure and radiation - alone or in combination – may be applied. The chosen method will depend on the type of raw materials, availability of the method, cost, effectiveness and degree of change it causes to the flavor and nutritional features of the food product. Fermentation, also called biopreservation, is a cheap, widely accessible method that meets today's increasing consumer's demand for minimally processed/preserved food products. Biopreservation with lactic acid bacteria (LAB) is indeed one of the oldest and highly efficient forms of non-thermal processing method. Cheese production is based on LAB ability to ferment sugars, especially glucose and galactose, so to produce lactic acid and aroma substances that give typical flavors and tastes to fermented products. LAB also release antimicrobial metabolites so called bacteriocins, which are considered safe and natural preservatives, with great potential to be used on their own, or synergistically with other methods in food preservation.

2. Lactic acid bacteria in dairy processing

Milk is a highly perishable food raw material, therefore, its transformation in cheese or other form of fermented dairy product, provides an ideal vehicle to preserve its valuable nutrients (Table 1), making them available throughout the year. It is known that while unprocessed milk can be stored for only a few hours at room temperatures, cheeses may reach a shelf-live up to 5 years (depending on variety).

Animal	Fat	Casein	Lactose	Albumin	Ash	Water
			%			
Cow	3.75	3.0	4.75	0.4	0.75	87.3
Goat	6.0	3.3	4.6	0.7	0.84	84.5
Ewe	9.0	4.6	4.7	1.1	1.0	79.6
Camel	3.0	3.5	5.5	1.7	1.5	84.8
Buffalo	6.0	3.8	4.5	0.7	0.75	85

Note: In cheese, these nutrients will appear at a concentration approximately ten times higher, while the water content decreases.

Table 1. Approximate composition of milk from various species of mammals

Fermentation with lactic acid bacteria (LAB) is a cheap and effective food preservation method that can be applied even in more rural/remote places, and leads to improvement in texture, flavor and nutritional value of many food products. LAB have a long and safe history of application and consumption namely in cheese processing (Aquilanti et al., 2006, Caplice & Fitzgerald, 1999, Giraffa et al., 2010, Ray, 1992; Wood, 1997; Wood & Holzapfel, 1995) thus being generally regarded as safe (GRAS). Increasing knowledge of LAB physiology, together with new developments in processing technology, is leading to their application beyond traditional starter culture application, namely in new food safety roles and direct health applications.

2.1. LAB as starter-cultures in cheese processing

Cheese-making is based on application of LAB in the form of defined or undefined starter cultures that are expected to cause a rapid acidification of milk through the production of lactic acid, with the consequent decrease in pH, thus affecting a number of aspects of the cheese manufacturing process and ultimately cheese composition and quality (Briggiler-Marco et al., 2007).

The earliest productions of cheeses were based on the spontaneous fermentation, resulting from the development of the microflora naturally present in the raw milk and its environment. The quality of the end product was a reflex of the microbial load and spectrum of the raw material. Spontaneous fermentation was later optimized through backslopping, i.e., inoculation of the raw material with a small quantity of whey from a previously performed successful fermentation, and the resulting product characteristics depended on the best-adapted strains dominance (Leroy & De Vuyest, 2004). Today, backslopping is still used to produce many artisanal raw-milk cheeses, namely those bearing the PDO (Protected Designation of Origin) status, which are considered to be an important source of LAB genetic diversity, as well as being crucial from an economic and even ecologic point of view, since production of said cheeses (usually processed on a small-scale) contributes to local employment and maintains people functioning as "guardians of local environment" in regions that otherwise would be deserted.

The starter-culture applied in this, so-called, natural fermentation, is usually a poorly-known microflora mix that although having a predominance of LAB, may also contain non-LAB microorganisms, and its microbial diversity and load is usually variable over time. In fact, studies directed to characterize traditional cheeses show that those made from raw milk harbor a diversity of LAB (Bernardeau et al., 2008) depending on geographical region, where a few may show particular interesting technological features that upon optimization may have industrial applications (Buckenhiiskes, 1993). For example, because wild strains need to withstand the competition of other microorganisms to survive in their hostile natural environment, they often produce antimicrobials substances called bacteriocins (Ayad et al., 2002), which are natural antibacterial proteins that can be incorporated directly into fermented foods as such (food–grade) or indirectly as starter culture (Bernardeau et al., 2008). Although nisin is today the only bacteriocin that reached commercial status, approved worldwide as a natural food preservative, many other bacteriocins may soon reach similar status. Recently, our work (to be published) with LAB isolates from traditional portuguese raw-milk cheeses, revealed several lactobacilli having antibacterial activity against pathogens such as *Listeria monocytogenes*, *Staphyloccus aureus*, *Salmonella newport* and even *E.coli*. Future studies may allow us using these isolates or their metabolites, applied *in situ or ex situ* fashion, in applications where food safety is a concern.

Moreover, traditional cheeses also obtain their flavor intensity also from the non-starter lactic acid bacteria (NSLAB), which are not part of the normal starter flora but develop in the product, particularly during maturation, as a secondary flora (Beresford, et al., 2001). The isolation and optimization of wild-type strains from traditional products, to be used as starter cultures in cheese processing, is indeed a highly active field of research in Food Science today.

2.2. LAB food safety and cheese technology

Cheese is made in almost every country of the world and there are more than 2000 varieties, made from milk of several mammals, processed industrially or by traditional methods (Figure 1).

However, despite the large number of varieties, the basic steps required in any cheese processing are essentially the same, and slight variations in any of these steps may result in products of different general quality (Figure 2).

Milk treatment. In large-scale cheese processing, the milk is heat-treated, e.g. 73 ºC for 15 seconds, to destroy pathogens and reduce microbial numbers, while in most traditional PDO raw-milk cheeses heat treatment is not applied. Also the milk may be standardized, i.e. the fat content may be increased or reduced, or the casein-to-fat ratio may be adjusted.

Starter-culture addition. The type of commercially available starter preparation to be used will be determined by the cheese recipe. As previously stated, large-scale processing relies on using defined, commercially available starters, while for traditional cheeses, a natural fermentation (whey from the previous lot) is often used.

Figure 1. (Top) - Brine salting of cheeses in a large-scale plant processing 20 tons of cheese a day. (Bottom) - Small-scale unit processing 50 Kg per day of a traditional PDO cheese

Milk treatment

↓

Coagulation

↓

Whey draining

↓

Salting/Pressing

↓

Ripening

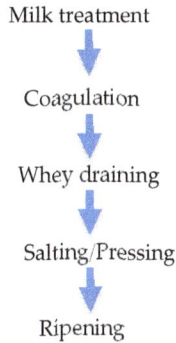

Figure 2. Common steps to most cheese making processes

Coagulation. During coagulation, modifications on the milk protein complex occur under defined conditions of temperature and by action of a coagulant agent, which changes the physical aspect of milk from liquid to a jelly-like mass. Various coagulants are available, e.g. lemon juice, plant rennet or more commonly a proteolytic enzyme such as chymosin (rennin) or – due to high demand from the cheese industry - proteolytic enzymes from the mould *Rhizomucor miehei* obtained via biotechnology. These enzymes have an acidic nature, meaning they have optimum activity in a slightly acidic environment. Therefore, the action of LAB in this phase is crucial as they are required to rapidly release enough lactic acid, to lower the milk pH from 6.7 to near 6.2, (thus creating an appropriate environment for optimum activity of rennin) and later to pH 4.5 as the processing proceeds, creating an inhospitable environment for many unwanted bacteria, thus increasing the end product safety.

Cutting the coagulum. The resulting coagulum may be cut with appropriate knives into curd particles of a defined size, e.g. 1–2 cm, or it may be transferred into containers or cheese moulds. The cutting or ladling of the coagulum is a very important step in the manufacture of some cheese varieties as it determines the rate of acid development and the body (firmness) and texture of the cheese.

Heating or cooking the curds. Heating (37–45 ºC, depending on the type of cheese) the curds and whey affects the rate at which whey is expelled from the curd particles and the growth of the starter microorganisms. During heating, the curds and whey are often stirred to maintain the curd in the form of separate particles.

Whey removal. After heating and stirring, and when the curd particles have firmed and the correct acid development have taken place, the whey is removed allowing the curd particles to mat together.

Milling the curd. In cheeses such as Cheddar, when the curd has reached the desired texture, it is broken up into small pieces to enable it to be salted evenly. Milling the curd can be done either by hand or mechanically. Salting is usually done to enhance the taste of the curd and to increase its safety and shelf life.

Ripening. Finally, for most cheeses, the resulting mass is molded and put to ripening for periods that may vary from 15 days to one, two or more years. Ripening is a slow phase, crucial for the development of aroma and flavor, brought about by the action of the many enzymes released by LAB. During ripening the protein in cheese is broken down from casein to low molecular weight peptides and amino acids. Proteolysis is the major – and certainly the most complex of biochemical events that take place during ripening of most cheese varieties and LAB play an important role in it. This happens while the cheeses are stored in the curing cabinets and in some cases in caves, usually with temperature and humidity controlled (Figure 3).

Figure 3. Cheese ripening in cabinets with controlled temperature and humidity.

During coagulation, the initial step of casein hydrolysis is performed by chymosin (milk coagulant) and proteinases from starter lactic acid bacteria, starter moulds and other microorganisms. The further degradation of high molecular weight peptides produced at the initial step, is subsequently catalised to low molecular weight peptides by endopeptidases from LAB during ripening (see Fig. 4 and 5).

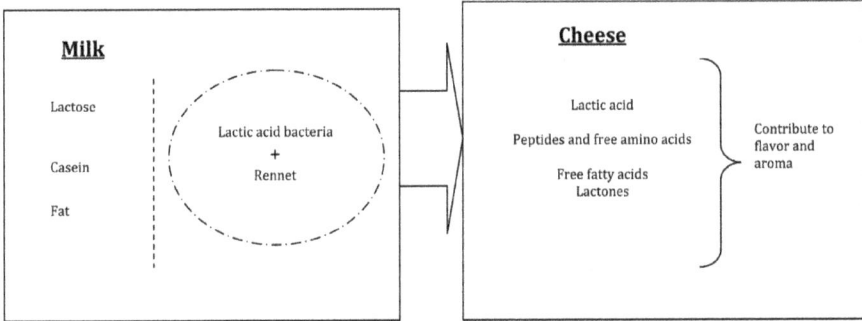

Figure 4. Simplified view of the biochemical changes that lead to texture and flavour changes in cheeses.

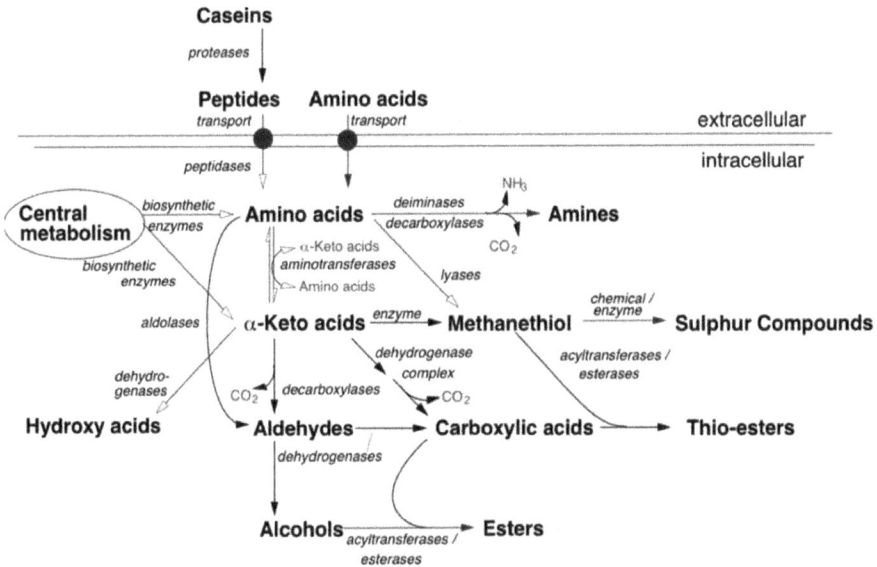

Figure 5. General pathways leading to intracellular meatabolites, and their degradation routes to potential flavour compounds. More specifically, pathways from methionine to flavour compounds (methanethiol, thioesters, sulphur compounds) are shown (Adapted from Kranemburg et al., 2002).

Primary proteolysis in cheese is defined as changes in β-, γ-, αs-caseinpeptides, and other minor proteins that are detected by PAGE (Figure 6). Primary proteolysis leads to the formation of large water-insoluble peptides and smaller water-soluble peptides (Fox, 1993, Mooney et al., 1998). Secondary proteolysis products include those peptides, proteins and amino acids soluble in the aqueous phase of cheese and are extractable as the water-soluble nitrogen (WSN) fraction. The WSN fraction is a complex mixture of large, medium, and small peptides and amino acids. These components result from the action of milk clotting enzymes, milk proteases, starter LAB and contaminating microorganisms (Rank et al., 1985).

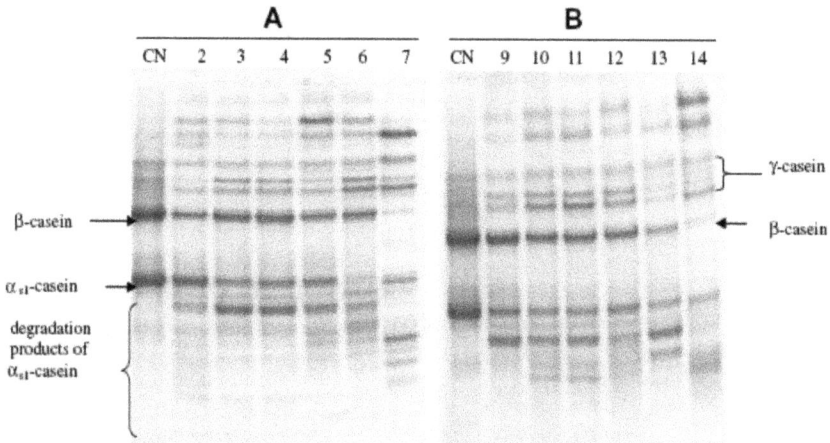

Figure 6. Evolution of proteolysis via urea-polyacrylamide gel electrophoresis in São Jorge cheeses from dairies A and B, by 1, 15, 30, 60, 90 or 130 days of ripening. Lanes 1, 8 and 15, Na- caseinate; lanes 2-6: cheese A; lanes 9-14: cheese B (Kongo et al., 2012).

Typical cheese pH values measured at 3–7 days after manufacture are 4.9–5.5 in most firm and hard ripened varieties, and 4.4–4.8 in fresh lactic and most soft ripened varieties (Table 2 and Figures 7 and 8).

Operations	Swiss type		Gouda		Cheddar		Feta		Cottage	
	Time	pH	Time	pH	Time	pH	Time	pH	Time	pH
Add starter	0	6.60	0	6.60	0	6.60	0	6.60	0	6.60
Add rennet	15	6.60	35	6.55	30	6.55	75	6.50	60	6.50
Cut	45	6.55	70	6.50	75	6.50	115	6.4	300	4.80
Drain or dip into forms	150	6.35	100	6.45	195	6.3	130	NA	360	5.0
Milling	NA	NA	NA	NA	315	5.45	NA	NA	NA	NA
Pressing	165	6.35	130		390	5.40	NA	NA	NA	NA
Demoulding	16 h	5.30	8 h	5.40	10 h	5.20	24 h	4.6	NA	NA
Minimum pH	1 wk	5.20	1 wk	5.20	1 wk	5.10	1 wk	4.4	NA	NA
Retail	6 mo	5.6	6 mo	5.6	4 mo	5.3	6 wk	4.4	2–14 d	5.2

Table 2. Typical pH *vs* time profiles for several cheese varieties (time is in minutes unless otherwise noted).

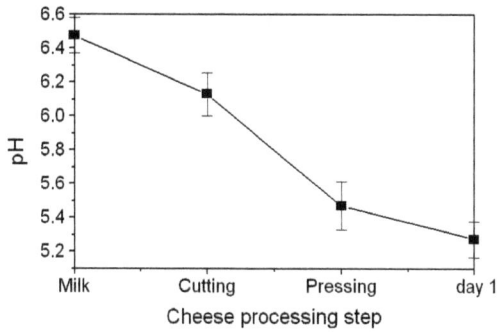

Figure 7. Evolution of pH (average ± standard deviation) in experimental cheese made with a starter culture of authoctonous São Jorge cheese LAB isolates.

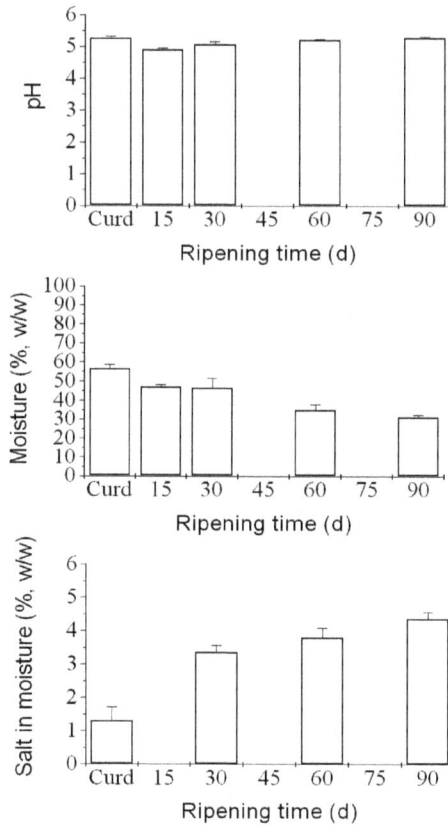

Figure 8. Evolution of physicochemical parameters (average ± standard deviation) throughout ripening of cheeses made with an experimental starter culture.

During processing, the pH history of the cheese is a good indicator of the actual product safety. For example a 'slow vat' allows more time at high pH for undesirable bacteria to grow, while during cheese ripening, unwanted bacteria may grow due to an acidity neutralization resulting from secondary microflora growth such as moulds. For most ripened varieties the combination of a low pH and ripening time, which leads to moisture decrease in the cheese, will in general cause a gradual decline of all groups of bacteria due to increasing inhospitable conditions inside the cheese.

The pH history of a cheese and the hygienic practices applied in its manufacture are thus key factors to guarantee safe products. Thus, the isolation of autochthonous LAB intend to be used for development of specific starter cultures with improved acid production and other antimicrobial activities may be an excellent way towards reaching the goals of simultaneously obtaining safe traditional cheeses, still bearing their unique flavors.

Nowadays, western consumers still enjoy artisan cheeses thanks to their outstanding gastronomic qualities; however, in most industrialized countries the large-scale cheese processing is the most important branch of the food industry. In such cases, there is a strong need to control the fermentation process towards maximum efficiency in terms of yields and standardization of the end product. This, and the need to fulfill the safety assurance of the final product, is usually achieved by, among other improvements, adding a high dosage of pure LAB selected starter cultures, commercially available (today's world starter culture market is more than US$1 billion), as well as by heat treating the raw milk, most commonly by pasteurization.

3. Development of new starter cultures for cheese processing

Traditional raw-milk cheeses are highly valued for their flavors, while large-scale products are often perceived by the consumer as "boring" (Law, 2001) – a consequence of the elimination by pasteurization, of the flora that has a key role in flavor development; and this puts the food industry under pressure to look for alternative LAB cultures capable of improving products flavor (Leroy & De Vuyest, 2004).

Today, the increased understanding of the genomics and metabolomics of food microbes opens up new perspectives for starter-cultures improvements and through genetic engineering it is now possible to express their desirable properties or suppress undesirable features (Del- cour, De Vuyst, & Shortt, 1999; Law, 2001; Mogensen, 1993).

Originally, starter cultures for the cheese industry were maintained by daily propagation, and later, they became available as frozen concentrates and dried or lyophilised preparations, produced on an industrial scale, some of them allowing direct vat inoculation (Sandine, 1996). Because the original starter cultures were mixtures of several undefined microbes, the daily propagation, eventually led to shifts of the ecosystem resulting in the disappearance of certain strains. Because some important metabolic traits in LAB are plasmid-encoded, there was a risk that they would be lost during propagation (Weerkamp et al., 1996). Lactococci are generally used as starter cultures in the production of industrial cheeses and cultured milk products. In traditional cheeses the natural starter cultures may harbor many different species and strains.

On the other hand, cheeses manufactured in a standard (large-scale) processing manner, are considered as safer because of the application of pasteurization and following the standard hygienic practices, including the HACCP. Traditional cheeses have their own specific processing methods, namely the common use of raw milk, however the hygienic procedures and HACCP approaches adapted to their specificities should be applied as well.

Species / subspecies	Main uses / Other comments
Lactococcus	Mesophilic starter used for many cheese types.
Lc. lactis subsp. *lactis*	Used in Gouda, Edam, sour cream and lactic butter.
Lc. lactis subsp. *lactis* biovar diacetylactis	Mesophilic starter used for many cheese types.
Lc. lactis subsp. *cremoris*	
Streptococcus	Thermophilic starter used for yogurt and many cheese types
Sc. thermophilus	particularly hard and semi hard high-cook cheeses.
Lactobacillus	
Lb. acidophilus	Probiotic adjunct culture used in cheese and yogurt.
Lb. delbrueckii subsp. *bulgaricus*	Thermophilic starter for yogurt. and many cheese types, particularly hard and semi hard high-cook cheeses.
Lb. delbrueckii subsp. *lactis*	Used in fermented milks and high-cook cheese.
Lb. helveticus	Thermophilic starter for fermented milks and many cheese types particularly hard and semi hard high-cook cheeses
Lb. casei	Cheese ripening adjunct culture.
Lb. plantarum	Cheese ripening adjunct culture.
Lb. rhamnosus	Cheese ripening adjunct culture.
Leuconostoc	Mesophilic culture used for Edam, Gouda, fresh cheese, lactic
Ln. mesenteroides subsp. *cremoris*	butter and sour cream.
Brevibacterium	Used in smear surface-ripened cheeses, Camembert, Stilton and
Brev. linens	Limburger and as a cheese ripening adjunct culture.
Propionibacterium	Used in Gruyère and Emmental cheeses.
Prop. Acidipropionici	Used in Gruyère and Emmental cheeses.
Prop. freudenreichii subsp. *shermanii*	

Table 3. Main bacteria associated with cheeses or other fermented products (From: Broome et al., 2003).

As previously stated, LAB are only a part of the complete microflora of raw milk (Kongo et al, 2007) and this, associated to other technological methods such as pressing, allows the production of a diversity of traditional cheeses (Parguel, 2011). This raw-milk microflora represents the contamination from the environment (air, utensils, the animal skin), and the load and its diversity will thus vary with local, season and livestock type, influenced by temperature.

These microbial mixes have an interdependent activity when together in their ecosystem and therefore their physiological properties may differ when the biodiversity is disrupted. In fact, it has been shown that certain microbial associations reveal a higher protecting effect against pathogens such as listeria, than when their association diversity is disrupted, (Montel 2010) see Figure 9.

Bacteriocinogenic probiotic bacteria could be beneficial when used as starter cultures in cheese, as they may prolong the shelf-life of the products, while simultaneously providing the consumer with a healthy advantage at a low cost (Gomes et al. 1998). The presence of bacteriocins in foods is, in general, seen as safe for consumers because bacteriocins are inactivated by pancreatic or gastric enzymes (Liu et al., 2011).

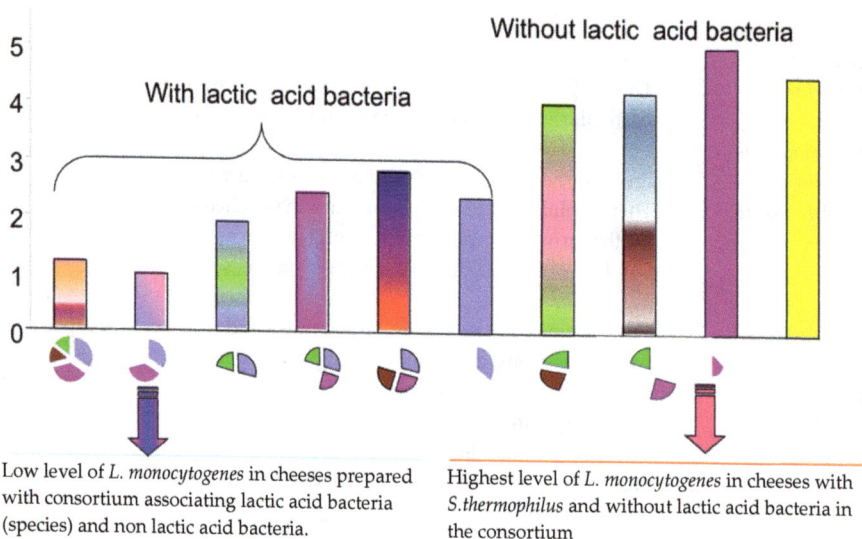

Low level of *L. monocytogenes* in cheeses prepared with consortium associating lactic acid bacteria (species) and non lactic acid bacteria.

Highest level of *L. monocytogenes* in cheeses with *S.thermophilus* and without lactic acid bacteria in the consortium

Figure 9. Level of *L. monocytogenes* in the core of Saint-Nectaire type cheese (28d) (Adapted from Montel & Samelis, (2010).

3.1. EPS-producing cultures and acceleration of cheese ripening

Many LAB produce exopolysaccharides (EPS), which may provide viscosifying, stabilizing, and water-binding effects in cheeses. The growing demand for all-natural, healthy food products, foods with low fat or sugar content and low levels of additives, as well as cost factors has increased the interest of food industry to use LAB polysaccharides. Research has also shown that EPS+ LAB can enhance the functional properties of low fat cheese and that the excellent water- binding properties and moisture retention of EPS can improve the melting properties of low fat Mozzarella cheese. These properties show that EPS have wide technical potentials for development of novel and improved food products with enhanced texture, mouth-feel, taste perception and stability, representing potential sources for economic gains for the dairy industry.

EPS have also the potential to be used as surface carriers of bacteriocins or bacteriocin-producing LAB, and species such as *Leuconostoc mesenteroides*, *Streptococcus mutans* and several lactobacilli (Lactobacillus brevis, *Lactococcus lactis* subsp. *lactis*, *L. lactis* subsp. *cremoris*, *Lactobacillus casei*, *Lb. sake*, *Lb. rhamnosus*,) and thermophilic (*Lb. acidophilus*, *Lb. delbrueckii* subsp. *bulgaricus*, *Lb.helveticus* and *S. thermophilus*) are known to produce EPS. The isolation and characterization of EPS from new wild LAB species, which are ubiquitous in traditional cheeses, is a key strategy towards finding strains with optimized production of EPS.

Finally, cheese ripening is a lengthy and costly process. Therefore, attenuated starter cultures with high autolysis are being sought towards increasing the amount of endogenous peptides, thus accelerating the cheese ageing process as well as enhancing flavour and texture. These cultures may be obtained via application of several techniques such as pulsed electric field, heat treatment, freeze–thawing and lysozyme treatment (Briggs, 2003).

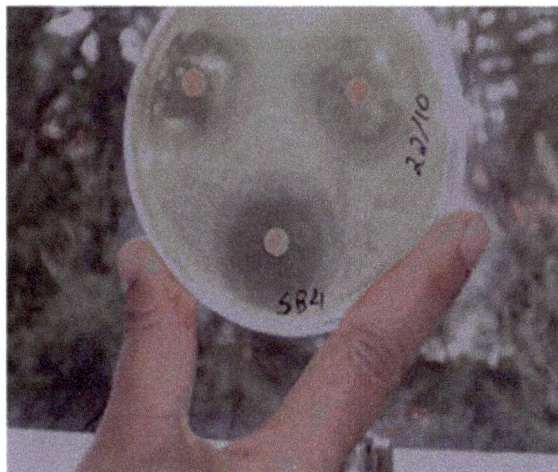

Figure 10. Antilisterial activity of LAB isolates from a traditional cheese.

Thus, the cheese industry in looking for new types of LAB starter-cultures bearing several properties: – cultures that increase microbial safety or offer one or more organoleptic, technological, nutritional (enzymes, or polyunsaturated fatty acids - PUFAs) or health advantages such as probiotic properties, starter cultures with increased resistance to bacteriophage, (recall that high product loss, especially in cheese manufacturing, is often associated with bacteriophages (Parente and Cogan, 2004), cultures that produce EPS and cultures that accelerate cheese ripening.

3.2. Methods used to characterize LAB for starter cultures development

To characterize new LAB isolates, phenotypic methods relying on physiological or bio-chemical criteria have been widely applied (Montel, Talon, Fournaud, & Champomier, 1991, Kongo et al., 2007). These phenotypic profiling methods are very important - especially related to finding the technological features, such as the acidification, proteolytic and lipolytic activity, of a new isolate (see Tables 3 and 4, and Figure 11) and have the advantage of requiring less sophisticated equipment. In most of the cases however, these tests are insufficient for accurate species identification due to the great number of different LAB species with similar phenotypic characteristics (Temmerman et al. 2004).

Characteristics	Lactobacillus	Enterococcus	Lactococcus	Leuconostoc	Pediococcus	Streptococcus
CO2 from glucose	+/-	-	-	+	-	-
Growth at 10 °C	+/-	+	+	+	+/-	-
Growth at 45 °C	+/-	+	-	-	+/	+/-
Growth at 6.5% NaCl	+/-	+	-	+/-	+/	-
Growth at 18% NaCl	-	-	-	-	-	-
Growth at pH 4.4.	+/-	+	+/-	+/-	+	-
Growth at pH 9.6	-	+	-	-	-	-
Type of lactic acid	D, L, DL	L	L	D	L, DL	L

Table 4. Phenotypic characteristics for discrimination of common LAB for dairy processing (modified from Batt, 1995).

	Genus					
	Lactobacillus			Enterococcus		
	Per cent of strains giving indicated reaction*			Per cent of strains giving indicated reaction*		
Enzyme	−	+	++	−	+	++
Alkaline phosphatase	100			100		
Esterase	100			100		
Lipase	100				25	75
Leucine aminopeptidase	12	88				100
Cystine aminopeptidase	100				100	
α-Chymotrypsin	100				100	
Acid-phosphatase	88	24		15	75	
Phosphoamidase		100			100	
β-Galactosidase		100			50	50
β-Glucuronidase		100			25	75
α-Glucosidase	100			100		
β-Glucosaminidase	100			100		

*API-ZYM color: grade 0, negative (−); grades 1 through 3, positive (+); grades 4 and 5, plus positive (++).

Table 5. Enzyme profiling of 14 representative LAB isolates found in Sao Jorge traditional cheese (from Kongo et al., 2007).

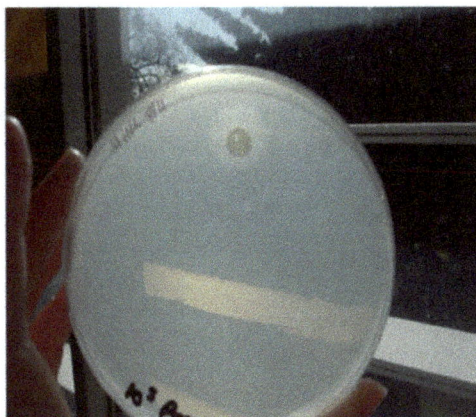

Figure 11. Proteolytic acitivity of a Lactobacillus ssp isolate on milk agar.

Molecular biology (genotypic) methods (Figure 12) on the other hand - largely DNA-based techniques - offer much greater discriminatory power, all the way to differentiation of individual strains (Aymerich et al., 2006, Cocolin et al., 2004, Furet et al., 2004, Prabhakar et al., 2011). Thus, a combination of both phenotypic and genotypic identification techniques (so called polyphasic approach) is preferred (Temmerman et al., 2004, Aquilanti et al., 2006).

Figure 12. Ribotyping as molecular biology technique for identification of LAB to type or strain level. (Kongo et al., 2007)

Finally, it should be mentioned that there are concerns today that commensal bacterial populations from food and the gastrointestinal tract (GIT) of humans and animals, such as LAB, could act as a reservoir for antibiotic resistance genes, and therefore, be transferred to possibly pathogenic bacterial species, complicating the treatment of a disease or infection and leading to the spread of antibiotic-resistant bacteria (Ammor et al., 2007). Thus, before using new isolates as starter cultures or as probiotics, the antibiotic resistance must be addressed.

The European Food Safety Agency (EFSA) proposed a system for a pre-market safety assessment of selected groups of microorganisms, leading to granting a "Qualified Presumption of Safety (QPS)". Therefore, EFSA proposed that a safety assessment of a defined taxonomic group, such as a genus or group of related species could be made based on establishing identity, body of knowledge, possible pathogenicity and end use (European Commission 2007). The 33 *Lactobacillus* species shown in Table 6 are the ones that in 2007 EFSA stated could be considered to have QPS-status. In addition to *Lactobacillus* species, also other LAB species have been granted QPS –status. They include three leuconostocs, (*Ln. citreum, Ln. lactis* and *Ln. mesenteroides*), three pediococci (*P. acidilactici, P. dextrinicus* and *P. pentosaceus*), *Lc. lactis* and *Streptococcus thermophilus*.

Lb acidophilus	Lb farciminis	Lb paracasei
Lb amylolyticus	Lb fermentum	Lb paraplantarum
Lb amylovorus	Lb gallinarum	Lb pentosus
Lb alimentarius	Lb gasseri	Lb plantarum
Lb aviaries	Lb heveticus	Lb pontis
Lb brevis	Lb hilgardii	Lb reuteri
Lb bucheneri	Lb johnsonii	Lb rhamnosus
Lb casei	Lb kefiranofaciens	Lb sakei
Lb crispatus	Lb kefiri	Lb salivarius
Lb curvatus	Lb mucosae	Lb sanfranciscensis
Lb delbrueckii	Lb panis	Lb zeae

Table 6. *Lactobacillus* (Lb) species with QPS- status according to EFSA (from Korhonen, 2010).

Lactobacilli are generally susceptible to antibiotics inhibiting the synthesis of proteins, such as chloramphenicol, erythromycin, clindamycin and tetracycline, and more resistant to aminoglycosides (neomycin, kanamycin, streptomycin and gentamicin. While some species show a high level of resistance to glycopeptides (vancomycin and teicoplanin), susceptibility to bacitracin will vary greatly (Ammor et al, 2007; Coppola et al., 2005).

Antibiotic	Species							
	Lactobacillus obligate homofermentative	Lactobacillus heterofermentative	Lactobacillus plantarum	Enterococcus spp	Pediococcus pp	Leuconostoc	Lactococcus lactis	Streptococcus thermophilus
Ampicillin	4	4	4	8	4	4	4	4
Vancomycin	4	IR	IR	8	IR	IR	4	4
Gentamycin	8	8	64	512	4	4	8	8
Kanamycin	16	16	64	1024	4	8	8	8
Streptomycin	16	16	64	1024	4	8	16	16
Neomycin	16	16	32	1024	8	8	8	8
Erythromycin	4	4	4	4	4	4	4	4
Clindamicin	4	4	4	4	4	4	4	4
Tetracycline	8	8	32	16	4	4	4	4
Chloranphenicol	4	4	8	8	4	4	8	8
Trimethoprim	8	8	8	8	8	8	n.r	n.r

Key: IR, intrinsically resistant

Table 7. Microbiological break points (μg mL^{-1}) categorizing some LAB species as resistant (Adapted from Ammor et al., 2007)

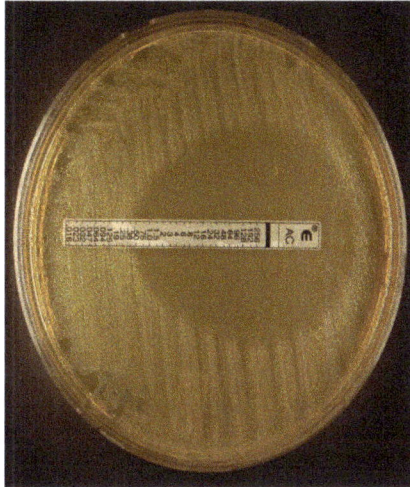

Figure 13. Result of a screening for antibiotic resistance of a *Lactobacillus paracasei* isolate

4. Concluding remarks

LAB are important in cheese processing because (i) they increase food safety through the release of lactic acid and bacteriocins, (ii) produce aromas and flavor and accelerate the maturation process of cheese via their proteolytic and lipolytic activities, bringing economic advantages to the industry, (iii) bring about desirable food textures via release of polysaccharides that increase the viscosity and firmness, and reduce susceptibility to syneresis, (iv) they may be used to deliver polyunsaturated fatty acids (PUFA) and vitamins, leading to dairy products with increased nutritional value, (v) specific probiotic strains contribute to liberation of health-enhancing bioactive peptides improving absorption in the intestinal tract, stimulating the immune system, exerting antihypertensive, antithrombotic effects, or functioning as carriers for minerals.

Novel insights arising from use of Bioinformatics, Systems Biology and Bioengeneering approaches will offer perspectives for the application of a new generation of starter cultures for cheese-making, having enhanced functional features and offering several health, marketing, and technological advantages, contributing to the development of small and medium sized enterprises on the one hand, and product diversification of large companies on the other.

However, there are still many developments to be achieved towards fully realizing the many foreseen potential of LAB or their products. For example extraction and purification of bacteriocins is still difficult as they form micelles or clumps with the nitrogen sources already in the growth medium. On the other hand while genetic engineering may offer many solutions related to optimal use of LAB, they may not be easily allowed by food legislation.

Author details

J. Marcelino Kongo

Instituto de Inovação Tecnológica dos Açores (INOVA)
Currently at Canadian Research Institute of Food Safety

5. References

Ammor, M.S., Florez, A.B. & Mayo, B. (2007) Antibiotic resistance in non- enterococcal lactic acid bacteria and bifidobacteria. *Food Microbiol* 24, 559-570.

Aquilanti, L., Dell'Aquila, L., Zannini, E., Zocchetti, A., & Clementi, F. (2006) Resident lactic acid bacteria in raw milk Canestrato Pugliese cheese. Lett. Appl Microbiol 43 (2006) 161–167.

Ayad, E.H.E., A. Verheul, J.T.M. Wouters & G. Smit, (2002). Antimicrobial-producing wild *Lactococci* isolated from artisanal and non-dairy origins. Int. Dairy J., 12: 145-150.

Aymerich, T., Martín, B., Garriga, M., Vidal-Carou, M.C., Bover-Cid, S., & Hugas, M. (2006) Safety properties and molecular strain typing of lactic acid bacteria from slightly fermented sausages. J Appl Microbiol 100, 40-49.

Batt, C. A., Erlandson, K., & Bsat, N. (1995) Design and implementation of a strategy to reduce bacteriophage infection of dairy starter cultures. Int Dairy J 5, 949–962.

Beresford, T.P., Fitzsimons, N.A., Brennan, N.L., & Cogan, T.M. (2001). Recent advances in cheese microbiology. Int. Dairy J., 11: 259-274.

Bernardeau, M., Vernoux, J. P., Henri-Dubernet, S., Guéguen, M. (2008). Safety assessment of dairy microorganisms: The Lactobacillus genus Int J Food Microbiol 126, 278–285.

Briggiler-Marcó, M., Capra, ML., Quiberoni, A., Vinderola, G., Reinheimer, J.A., & Hynes, E. (2007) Nonstarter Lactobacillus strains as adjunct cultures for cheese making: in vitro characterization and performance in two model cheeses. J Dairy Sci 90, 4532-4542.

Briggs, S. S. (2003). Evaluation of lactic acid bacteria for the acceleration of cheese ripening using pulsed electric fields. MSc Thesis, McGill University, Montreal Quebec, Canada.

Broome, M.C., Powel, I. B & Limsowtin, G. K. Y. (2003). Starter cultures: Specific properties. *In* Enyclopedia of Dairy Sciences. Vol I ed. Regisnki, H. Fuquay, J.W. & Fox, P. F. 269 – 275. London: Academic Press.

Buckenhiiskes, H. J. (1993) Selection criteria for lactic acid bacteria to be used as starter cultures for various food commodities. FEMS Microbiol Rev 12, 253-272

Caplice, E., & Fitzgerald, G. F. (1999). Food fermentations: role of microorganisms in food production and preservation. Int J Food Microbiol 50, 131–149.

Carminati, D., Giraffa, G., Quiberoni, A., Binetti, A.,Suarez, V., & Reinhemer, J. (2010). Advances and Trends in Starter cultures for Dairy Fermentation Chapter 10. *In:* Biotechnology of Lactic Acid Bacteria: Novel Applications. Edited by Mozi, F., Raya, R. R., & Vignolo, G. M. Wiley Blackwell Publisher.

Cocolin, L., Rantsiou, K. Iacumin, L., Urso, R., Cantoni, C., & Comi, G. (2004) Study of the ecology of fresh sausages and characterisation of populations of lactic acid bacteria by molecular methods. Appl Environ Microbiol 4, 1883-1894.

Coppola, R., Succi, M., Tremonte, P., Reale, A., Salzano, G. & Sorrentino, E. (2005) Antibiotic susceptibility of *Lactobacillus rhamnosus* strains isolated from Parmigiano Reggiano cheese. *Lait* 85, 193-204.

Delcour, J., de Vuyst, L., & Shortt, C. (1999) "Recombinant dairy starters, probiotics, and prebiotics: Scientific, technological, and regulatory challenges". Int Dairy J (Special issue) 9, 3–80.

European Commission. (2007) Opinion of the Scientific Committee on a request from EFSA on the introduction of a Qualified Presumption of Safety (QPS) approach for assessment of selected microorganisms referred to EFSA. The EFSA J 587, 1-16.

Fox P.F. (1993) Cheese chemistry physics and microbiology. Vol. 1. Chapman and Hall London, 303-340

Furet, J. P., Quenee, P., & Tailliez, P. (2004) Molecular quantification of lactic acid bacteria in fermented milk products using. J Food Microbiol 103, 131–142.

Giraffa, G., Chanishvili, N., & Widyastuti, Y. (2010) Importance of lactobacilli in food and feed biotechnology. Research Microbiol 161, 480-487.

Gomes, A.M., Malcata, F.X. (1998) Development of probiotic cheese manufactured from goat milk: response surface analysis via technological manipulation. J Dairy Sci 81, 1492-507.

Grappin, R., Beuvier, E. (1997) Possible implications of milk pasteurization on the manufacture and sensory quality of ripened cheese. Int Dairy J 7, 751-761.

Hayaloglu, A.A., Guven, M., & Fox, P.F. (2002) Microbiological, biochemical and technological properties of Turkish White cheese, "Beyaz Peynir". Int Dairy J 12, 635-648.

Klaenhammer, T.R., Barrangou, R., Buck, B.L., Azcarate-Peril, M.A., Alterman, E. (2005) Genomic features of lactic acid bacteria affecting bioprocessing and health, FEMS Microbiol Rev 29, 391-409.

Kongo, J.M., Ho, A.J., Malcata, F.X., & Wiedmann, M. (2007) Characterization of dominant lactic acid bacteria isolated from Sao Jorge cheese, using biochemical and ribotyping methods. J. Appl. Microbiology 103, 1838 –1844.

Kongo, J.M., & Malcata, F.X. (2012) Azorean traditional cheesesmaking: a case study pertaining to a unique food chain. In Food Chains: New Research, edited by Melissa A., Jensen & Danielle W. Mueller. Nova Science Publisher Inc, New York.

Korhonen, J. (2010) Antibiotic Resistance of Lactic Acid Bacteria PhD Disseration, University of Eastern Finland.

Kranenburg, R., Kleerebezem, M., Vlieg, J. H., Ursing, B., Boekhorst, J., Smit, B. A., Ayad, E.H.E., Smit, G., & Siezen, R. J. (2002) Flavour formation from amino acids by lactic acid bacteria: predictions from genome sequence analysis. Int Dairy J 12, 111–121.

Law, B. A. (2001) Controlled and accelerated cheese ripening: the research base for new technology. Int Dairy J 11, 383–398.

Leroy, F., & De Vuyst, L. (2004) Lactic acid bacteria as functional starter cultures for the food fermentation industry. Food Sci Technol 15, 67-78.

Liu, G., Wang, H., Griffiths, M.W. & Li, P. (2011) Heterologous extracellular production of enterocin P in Lactococcus lactis by a food-grade expression system. Europ Food Res Technol 233, 123-129.

McSweeney, P.L.H., & Sousa, M.J. (2000) Biochemical pathways for the production of flavour compounds in cheeses during ripening. A review. Lait 80, 293-324.

Mogensen, G. (1993) Starter cultures. In: Smith, J. (Ed.), Technology of reduced-additive foods. Blackie Academic and Professional, London, pp. 1-25.

Montel M-C., & Samelis, J.(2010) Microbial Safety of Traditional cheeses. NAGREF WP2A Final TRUEFOOD Conference New roots for traditional European foods: Possibilities for success and sustainability. Brussels, 13 April.

Montel, M.C., Talon, R., Fournaud, J., & Champomier, M.C. (1991) A simplified key for identifying homofermentative Lactobacillus and Carnobacterium spp. from meat. J Appl Bacteriol 70, 469–472

Mooney, J.S., Fox, P.F., Healy, A., & Leaver, J. (1998) Identification of the principal water-solubel peptides in Cheddar cheese. Int Dairy J 8, 813-818.

Parente E., & Cogan T.M. (2004). Starter cultures: General aspects. In: General Chemistry, Physics and Microbiology. Vol I, edited by P.F. Fox., P.J. H. McSweeney, T. M. Cogan and T. P. Guninee. Amsterdam Elsevier.

Parguel P (2004). "Milk flores", group Malbuisson (Doubs), pp. 1-7.

Prabhakar V, Kocaoglu-Vurma N, Harper J, Rodriguez-Saona L. (2011). Classification of Swiss cheese starter and adjunct cultures using Fourier transform infrared microspectroscopy. J. Dairy Sci 94, 4374-4382.

Rank, T.C., Grappin, R., & Olson, N.F. (1985) Secondary Proteolysis of Cheese During Ripening: A Review. J Dairy Sci 68, 801- 805.

Ray, B. (1992) The need for food biopreservation. In: B. Ray, & M. Daeschel (Eds.), Food biopreservatives of microbial origin (pp. 1–23). Boca Raton, Florida: CRC Press.

Sandine, W.E., (1996) Commercial Production of Dairy Starter Cultures. In: Dairy Starter Cultures, Cogan, T.M. and J.P. Accolas (Eds.). Wiley-VCH, New York.

Smit, G., Smit, B. A., & Engels, W.J.M. (2005) Flavour formation by lactic acid bacteria and biochemical flavour profiling of cheeses products. FEMS Microbiol Rev 29, 591- 610.

Songisepp, E., Kullisaar, T., Hutt, P., Elias, P., Brilene, T., Zilmer, M., & Mikelsaar, M. (2004) A New Probiotic Cheese with Antioxidative and Antimicrobial Activity. J Dairy Sci 87, 2017–2023.

Temmerman, R., Huys, G. and Swings, J. (2004) Identification of lactic acid bacteria: culture-dependent and culture-independent methods. Trends Food Sci Tech 15, 348-359.

Weerkam, A.H., Klijn, N., Neeter, R., & Smit, G. (1996) Properties of mesophilic lactic acid bacteria from raw milk and naturally fermented raw products. Neth Milk Dairy J 50, 319-322.

Wood, B.J.B. (1997) Microbiology of fermented foods. London: Blackie Academic & Professional.

Wood, B.J.B. & W.H. Holzapfel, (1995.) The Genera of Lactic Acid Bacteria. Vol. 2, Blackie Academic and Professional, Glasgow.

Wouters, J. T. M., Ayad, E. H. E., Hugenholtz, J., & Smit, G. (2002) Microbes from raw milk for fermented dairy products. Int Dairy J 12, 91–109.

Yvon, M., & Rijnen, L. (2001) Cheese flavour formation by aminoacid catabolism. Int Dairy J 11, 185–201.

Chapter 3

Redox Potential: Monitoring and Role in Development of Aroma Compounds, Rheological Properties and Survival of Oxygen Sensitive Strains During the Manufacture of Fermented Dairy Products

F. Martin, B. Ebel, C. Rojas, P. Gervais, N. Cayot and R. Cachon

Additional information is available at the end of the chapter

1. Introduction

Lactic acid bacteria can be found in a diversity of ecosystems, which is consistent with their ability to adapt to highly variable environments. Among the various parameters that characterize these environments (temperature, pH, water activity), redox is relatively recent. It has however already been addressed indirectly in studies relating to the impact of oxidative stress on lactic acid bacteria. Indeed, the concept of oxidation has often been associated with the presence of oxygen; however, oxidoreductive effects on microorganisms must not be limited to oxygen.

A broader vision could be proposed concerning the adaptation of lactic acid bacteria to extracellular redox. The metabolism of lactic acid bacteria, chemosynthetic organisms, involves a series of dehydrogenation (oxidation) and hydrogenation (reduction) reactions. This metabolism follows the principle of conservation of energy and matter, and therefore requires the availability of a terminal electron acceptor. In lactic acid bacteria, a carbon metabolic intermediate is reduced (mainly pyruvate). In homofermentative lactic acid bacteria, redox coenzymes (NAD^+/NADH) enable coupling between oxidation and reduction reactions. During anaerobic glycolysis, glucose is oxidized to 2 moles of pyruvate with the formation of 2 moles of NADH, which then further reduces pyruvate to form lactic acid. Consequently, the typical equation of homolactic fermentation is: 1 glucose $\rightarrow$ 2 lactate.

Such a perfect matching, theoretically consistent, must be qualified according to the environmental conditions, including the redox state of the extracellular medium. The adaptation of lactic acid bacteria to extracellular redox depends on their ability to positively or negatively interfere with oxidants (electron acceptors) or reducing molecules (electron donors). Carbon and electron flow management by the cell will thus be highly dependent on the ability of the microorganisms to interact with the redox environment.

Potentially, all biochemical reactions in the cell, and therefore the enzymatic activity, may be influenced by the redox state of the environment. Dissolved oxygen is an oxidizer and can reach concentrations of 8 mg.L^{-1} of medium (equilibrium with air). Despite the strict anaerobic metabolism of some lactic acid bacteria, the majority are aerotolerant and can react with dissolved oxygen at varying levels. Lactic acid bacteria provided with NADH oxidase can reduce oxygen to water (reduction reaction coupled with the re-oxidation of NADH). This process influences both the intracellular and extracellular redox environment, and will result in a change in the metabolism, cellular physiology and physico-chemical environment surrounding the microorganism.

Changes in the extracellular environment can be monitored by measuring the redox potential (E_h). This parameter plays a key role in the quality of fermented dairy products, but is still rarely taken into consideration or is completely ignored during the manufacturing process. The reasons for this lack of interest can be attributed to difficulties associated with its measurement and control. Over the past ten years, several studies advocate the monitoring and control of E_h in fermented products using lactic acid bacteria selected for their reducing ability, redox molecules, or heat treatment. In terms of food applications, the variation in E_h must involve compounds that do not alter the product characteristics. So, modifying the E_h using gas, which enables the product characteristics to be maintained, may be advantageously exploited in industry.

The aim of this chapter is to present the latest knowledge concerning the adaptation of lactic acid bacteria to their redox environment, and the interest of modifying E_h using gas for lactic acid bacteria applications in the food industry.

2. Redox potential

E_h, like pH, is a parameter of the state of biological media which indicates the capacity to either gain or lose electrons. During oxidation, electrons are transferred from an electron donor to an electron acceptor, which is reduced. Electrochemical measurement of E_h is not new but has attracted little attention as a parameter for controlling fermentation processes due to the sensitivity of its measurement. However, E_h is already indirectly taken into account in industry through oxygen, of which the inhibitory effect on lactic acid bacteria is well-known. Indeed, oxygen modifies the growth capacity of microorganisms and the formation of end products, and so may contribute to the quality of fermented products [1, 2].

2.1. Definition of E_h

Oxidation is a reaction in which a molecule, atom or ion, loses electrons.

Reduction is a reaction in which a molecule, atom or ion, gains electrons.

An **oxidant** (also known as an oxidizing agent, oxidizer or oxidiser) can be defined as a substance that removes electrons from another reactant in a redox reaction.

A **reductant** (also known as a reducing agent or reducer) can be defined as a substance that donates an electron to another species in a redox reaction.

In the same way pH defines acid-base characteristics of a solution, E_h defines the reducing and oxidizing characteristics.

Presented below is the reduction half-reaction of an oxidant (Ox) to its corresponding reduced species (Red):

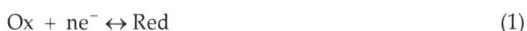

$$Ox + ne^- \leftrightarrow Red \tag{1}$$

The Nernst equation gives the relationship between the redox potential and the activities of the oxidised and reduced species:

$$E_h = E_h^0 + 2.3 \times \left(\frac{RT}{nF}\right) \times \log\left(\frac{[Ox]}{[Red]}\right) \tag{2}$$

where:

E_h = redox potential (mV) (in relation to a normal hydrogen electrode).

E_h^0 = standard redox potential (mV) (in relation to a normal hydrogen electrode) at pH 0

F = Faraday constant (96500 C.mol^{-1})
n = number of electrons exchanged
R = gas constant (8.31 J.mol^{-1}.K^{-1})
T = temperature in K
$2.3 \times \dfrac{RT}{F} = 59$ mV (at 25 °C)

However, chemical reactions in aqueous media involve protons, and the following half-reaction:

$$Ox + mH^+ + ne^- \leftrightarrow Red + H_2O \tag{3}$$

From Equation (2) it can be written:

$$E_h = E_h^0 - 2.3 \times \left(\frac{mRT}{nF}\right) \times pH + 2.3 \times \left(\frac{RT}{nF}\right) \times \log\left(\frac{[Ox]}{[Red]}\right) \tag{4}$$

m = number of protons involved in the reaction

Equation (4) is used to determine $E_h^{0'}$ defined as the standard redox potential at pH 7, which is closer to biochemical and biological processes (Figure 1).

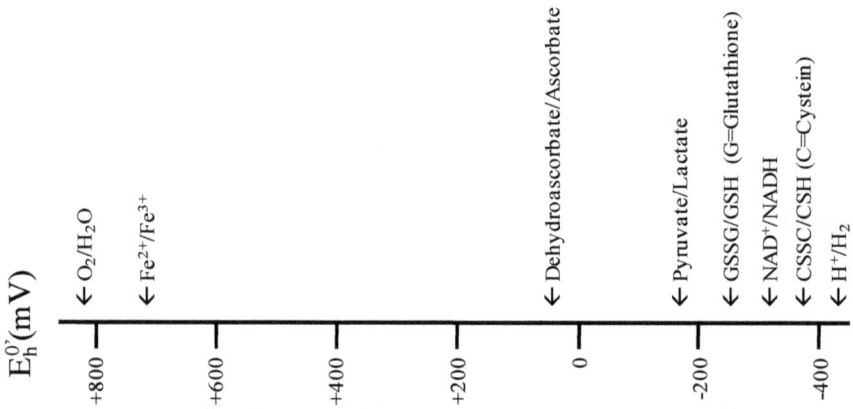

Figure 1. Standard reduction potential $E_h^{0'}$ (mV) of some important half-reactions involved in biological processes at 25 °C and pH 7.

2.2. Measurement of E_h

The first technique for measuring E_h is based on the use of coloured indicators (redox indicators), which are mostly indophenols or indigo derivatives with a reversible structure between oxidized (coloured) and reduced (colourless) state. However, the use of coloured indicators for measuring E_h, including biological media or food, is limited. Indeed, these molecules behave as electron donors and acceptors; they affect and can change the equilibrium. These compounds can also catalyse or inhibit biological reactions and may be toxic to microorganisms. Furthermore, in some cases it is difficult to appreciate a significant colour change and some E_h indicators also change colour with the pH of the medium. For these reasons, redox indicators are rarely used. They are more often used as indicators of redox thresholds, especially in the manufacture of strictly anaerobic culture media (resazurin) where maintaining a minimum level of reduction is essential for the growth of anaerobic microorganisms. Resazurin is also used to evaluate the reducing activity of starter cultures, for sterility testing and for the detection of microorganisms in dairy milk.

The second method commonly used in microbiology is a potentiometric technique which, contrary to redox indicators, is a direct method. The principle consists in measuring a potential difference determined between an inert electrode (usually made of platinum or gold) in contact with a redox couple in solution and a reference electrode. Electron exchange with the reduced and oxidised species takes place at the inert electrode. The inert electrode is made of stainless metals with a high enough standard potential to be electrochemically stable. These metals act as electron conductors between the measuring medium and the reference electrode. The reference system is the standard hydrogen electrode, but in practice two other references are used: the calomel electrode and the silver / silver chloride (Ag/AgCl) electrode. The redox potential is expressed in volts or millivolts. Redox values should always be expressed in relation to the hydrogen electrode. Consequently, potential measurements (E_m) using other references must be adjusted according to the reference potential of the hydrogen electrode (E_r):

$$E_h = E_m + E_r \qquad (5)$$

For example, E_r of the Ag/AgCl electrode is equal to 207 mV at 25 °C [3]. According to data from Galster [3], we propose the following equations linking E_r and temperature for the two reference electrodes:

$$Ag / AgCl \left(KCl\ 3M \right) \qquad E_r = 207 + 0.8 \times (25 - T) \qquad (6)$$

$$Calomel \left(Saturated\ KCl \right) \qquad E_r = 244 + 0.7 \times (25 - T) \qquad (7)$$

Before use, the redox electrodes must be polished with fine aluminium powder to restore the platinum surface, and controlled in tap water. Three measurements in tap water should be compared and need to be within the confidence interval around their mean value (calculated at 20 mV, 95% confidence level) to ensure correct measurement [4].

Equation (4) shows the dependence of E_h on pH. It is possible to overcome pH dependency by applying the Leistner and Mirna equation [5]:

$$E_{h7} = E_{h\beta} - \alpha \times (7 - \beta) \qquad (8)$$

where:

E_{h7} = redox potential (mV) at pH 7
$E_{h\beta}$ = redox potential (mV) at pH β
β = pH of medium
α = Nernst E_h–pH correlation factor (mV/pH unit).

To calculate Eh7 in biological media, the Nernst factor (α) must be determined experimentally by measuring Eh variation as a function of pH using an acid or a base. This value may vary according to the nature of the oxido-reducing molecules in the media. For example, the Nernst factor is 40 mV/pH unit in milk [6].

2.3. Use of gas to modify E_h

Gas applications in the food industry are numerous: modified atmosphere packaging (MAP), beverage distribution, cooling, freezing or carbonation. The advantage of using gases such as hydrogen (H_2), nitrogen (N_2) or carbon dioxide (CO_2) to modify E_h is that they are not directly toxic to microorganisms. There are no safety issues for the product with these gases and they can be used sequentially. Finally, their use is authorized at European level. Of the gases used in the food industry, in this chapter we will focus more particularly on nitrogen and hydrogen.

Nitrogen (N_2) is odourless, colourless, tasteless, non-toxic, and non-flammable. It is used to extend the life of packaged products (authorized additive E941). It is used to expel oxygen from the packaging before it is closed, which prevents oxidative phenomena involving pigmentation, flavours and fatty acids. It is also used for rapid freezing and refrigeration of food during transport.

Hydrogen (H_2) has major potential in food as it is colourless, odourless and has no known toxic effects. It is already used in the food industry for the hydrogenation of liquid oils and their transformation into solid products such as margarine or peanut butter. Hydrogen is a powerful reducer in solution, even at very low concentrations. It has been used to demonstrate the effect of E_h on the heat-resistance of bacteria [7]. Hydrogen is a special reducing agent: it imposes an E_h value on the medium associated with the introduction of the H^+/H_2 couple ($E_h^{0'}$ = -414 mV). This E_h value is highly dependent on the concentration of this couple that mainly influences the stability of the E_h imposed.

With the prospect of food use, hydrogen has the advantage over chemical reducing agents of not changing the product formulation, and therefore not altering the taste. Its industrial use has been rarely seen in this context because of its low flammability limit of 4% in air at 20 °C [8], this is why N_2-H_2 (96%-4%) is preferred to pure hydrogen. Its use in food technology is authorised at the European level (E949).

3. Effect of E_h on a fermented dairy product: Yoghurt

3.1. Reminder regarding the manufacture of yoghurt

We chose to focus on the key steps in the manufacture of yoghurt, which are:

- Delivery of milk: The raw material can be either fresh milk, reconstituted milk (from skim milk powder), or a mixture. In all cases, it is generally accepted that a quality product can be made from an extremely high quality raw material. With this in mind, it is essential that when the milk and other raw materials are received methods are established to detect any potential defects as early as possible. Two parameters must therefore be analysed as soon as the milk is received:
  - Its microbiology: to ensure consumer health, prevent the degradation of milk components that persist in the finished product and eliminate any possible competition between the starter culture and the endogenous flora that may involve bacteriophages.
  - Its chemistry: a rapid analysis of the chemical composition of the milk is necessary in order to identify any problems such as colostrum and late-lactation milk. Furthermore, these data concerning the chemistry of the milk can be useful in the standardization of the mixture.
- Standardization of the mixture: each component in milk plays a role. Fat has an effect on the smoothness and the feeling of softness in the mouth, lactose is the raw material used by

lactic acid bacteria for acidification, proteins act on the texture and minerals help stabilize the gel. These components vary in cow's milk according to race, diet, stage of lactation of the animal and season, which is why, during yoghurt manufacture, it is necessary to standardize the milk fat and protein content to meet the nutritional and organoleptic characteristics of the product and obtain consistent quality throughout the year.

- Homogenization: homogenization has two main effects on milk fat and proteins. The fragmentation of the fat globules prevents the separation of the lipid phase and the rest of the mixture, thus preventing the cream rising to the top during fermentation. Homogenization also stabilizes the proteins.
- Heat treatment: This eliminates most of the microbial flora originally present in the milk, including pathogenic or spoilage flora. It denatures the whey proteins, improves the consistency and viscosity of fermented milks and prevents whey separation. The risk of syneresis is reduced.
  Heat treatment of milk also has a positive effect on enzyme activity by providing a supportive environment. The environment becomes reductive through the elimination of a high proportion of oxygen. This medium is more conducive to fermentation that takes place under anaerobic conditions.
- Cooling: After heat treatment, the mixture must be cooled to temperatures approaching 43 °C for inoculation and incubation of the starter culture.
- Fermentation: "Yoghurt" refers to a product fermented by *Streptococcus thermophilus* (*S. thermophilus*) and *Lactobacillus delbrueckii* ssp. *bulgaricus* (*Lb. bulgaricus*). In general, milk is fermented at 40-45 °C, the optimum growth temperature, with an incubation time of 2 and a half hours. However, a longer incubation period of 16-18 hours can be used at a temperature of 30 °C, or until the desired acidity is attained [9].

3.2. Yoghurt strains: Lactobacillus delbrueckii ssp. bulgaricus and Streptococcus thermophilus

The association of *S. thermophilus* and *Lb. bulgaricus* is called proto-cooperation. Each species produces one or more substances, initially absent from the culture medium, that stimulate the growth of the other species [10]. During the symbiosis observed in yoghurt, the growth phases of these two bacterial species are staggered. Initially, growth of *S. thermophilus* is observed which is then slowed by the inhibitory effect of the lactic acid produced; the growth rate of *Lb. bulgaricus* then increases [11].

S. thermophilus is a strain that often shows little proteolytic activity, due to general low activity or absence of a wall protease. Its growth is limited because the peptides and amino acids initially present in milk are insufficient to cover its needs. In contrast, *Lb. bulgaricus* membrane protease degrades milk caseins releasing small peptides and amino acids which can be used by *S. thermophilus* intracellular peptidases [12].

The cooperation between these two strains also involves the production by *S. thermophilus* of pyruvic acid, formic acid, and carbon dioxide (CO_2 obtained from the decarboxylation of milk urea by urease) which stimulates the growth of *Lb. bulgaricus* [9, 13]. However, formic acid is released late in fermentation and in small quantities. The two bacterial species also consume the formic acid resulting from the heat treatment of milk [14].

Some authors have also demonstrated that the association of *S. thermophilus* and *Lb. bulgaricus* affects the production of volatile compounds involved in flavour development in yoghurt [15]. *S. thermophilus* produces more acetaldehyde, acetoin and diacetyl than *Lb. bulgaricus*, contrary to the rest of the bibliography concerning acetaldehyde [9, 16, 17]. Quantities of these molecules and other carbonyl compounds are not crucial per se for yoghurt flavour, but there are relationships between them that give yoghurt its distinctive flavour.

Finally, Ebel *et al.* [18] showed that during the manufacture and storage of a fermented dairy product, the populations of *Lb. bulgaricus* and *S. thermophilus* are the same whatever the E_h of the milk.

3.3. Texture

3.3.1. A look at yoghurt texture

The transformation of milk into yoghurt is called acid gelation. This gelation is a phenomenon that results in a remarkable change in the physical state of the system which changes from a liquid to a system with the characteristics of a solid. Several phases in the formation of a gel can be distinguished:

- The "solution" phase, where the polymer forms a solution: the macromolecules are not held together;
- The "gel" phase occurs when enough chains have joined together to form a network or gel, with dominant elastic rheological behaviour;
- Sometimes, additional aggregation of associated areas is observed; the gel becomes increasingly rigid and syneresis may occur with time: the gel shrinks and exudes some of the liquid phase.

The slow acidification of milk is due to bacteria that metabolize lactose and produce lactic acid. While casein micelles are stable at normal milk pH and room temperature, this supramolecular structure becomes unstable and leads to the formation of a gel with the slow progressive acidification of milk.

- From pH 6.7 to pH 5.8, the casein micelles seem to retain their integrity, shape and size.
- From pH 5.8 to pH 5.5, the micelles get closer together due to the decrease in the potential ζ and begin to form groups of micelles.
- From pH 5.5 to pH 5.0, significant changes in shape and size take place: micelle aggregates appear and these particles partially fuse. This is the phase transition between the solution and the acid gel.
- When the pH reaches 5, the solubilisation of micellar calcium, which occurs steadily from pH 6.8, is complete. From pH 5 to pH 4.8, rearrangements of the aggregates take place. At pH 4.9, gelation is complete. At pH 4.6, the acid gel is definitively formed. Aggregation of casein micelles at pH 4.6 is irreversible. Hydrophobic interactions are facilitated at this pH due to reduced electrostatic repulsion, leading to micelle aggregation.

Rheology is used to characterize the texture of yoghurt that specifically targets the mechanical properties. The rheological characterization of a product involves the application of a shear stress and measurement of the deformation, or application of a deformation (compression, stretching or shear) and measurement of its ability to withstand this distortion. Yoghurt can be defined as a viscoelastic fluid. It therefore has both the viscous properties of a liquid and the elastic properties of a solid.

3.3.2. Effect of E_h on a model acid skim milk gel

It has been shown that dairy products are affected by E_h [4, 19]. Delbeau et al. [19] showed that the use of gas to change the E_h of milk can modify the sensory properties of a fermented dairy product. However, we do not know if these modifications are due to the impact of E_h on physicochemical phenomena, lactic acid bacteria, or both. For this purpose, Martin et al. [20] wanted to determine to what extent chemical phenomena affect acid milk gelation under different E_h conditions. Glucono-δ-lactone (GDL) was used to acidify milk to avoid variations caused by microorganisms sensitive to E_h.

Martin et al. [20] studied the effects of E_h on model acidified skim milk gels obtained using GDL and prepared under different gaseous conditions. The milk prepared in air is an oxidizing medium; nitrogen, which is a neutral gas, can be used to remove oxygen from milk - even so the milk Eh remains oxidizing in these conditions - and hydrogen leads to a reducing E_h (below 0). Martin et al. [20] focused on the effect of gas bubbling on gel structure through viscoelastic properties and measurement of whey separation (Table 1).

Gaseous conditions applied to milk	pH		E_{h7} (mV)		η (Pa.s)	WS (g/100g of GDL-gel)
	At t=0	At t=3.5 hours	At t=0	At t=3.5 hours		
Air	6.80 ± 0.03	4.6[a] ± 0.0	405 ± 22	414 ± 8	0.039[a] ± 0.000	4.74[a] ± 1.42
Air bubbling	6.70 ± 0.04	4.6[a] ± 0.0	433 ± 6	430 ± 5	0.032[c] ± 0.001	1.26[b] ± 0.26
N₂ bubbling	6.8 ± 0.06	4.6[a] ± 0.0	283 ± 13	288 ± 11	0.035[b] ± 0.001	1.93[b] ± 0.33
N₂ – H₂ bubbling	6.73 ± 0.04	4.6[a] ± 0.0	- 349 ± 6	- 83 ± 18	0.032[c] ± 0.001	0.59[c] ± 0.12

[a-c]: different letters indicate that groups were significantly different at an α risk of 5% (ANOVA test). Values in the same column should be compared.

Table 1. Characteristics of gel structure depending on the different E_h conditions (milk acidified using GDL):
• Apparent viscosity η at 500 1/s of GDL-gel at pH 4.6 and 4 °C. Measurements were carried out 24 hours after addition of GDL.
• Evolution of average whey separation (WS) over 28 days in GDL-gels.
Values are means from triplicate experiments (mean value ± standard deviation).

The apparent viscosity of each gel was characterized at pH 4.6, 4 °C, 24 hours after addition of GDL under the different E_h conditions (Table 1). For GDL-gels, apparent viscosity ranged from 0.032 to 0.039 Pa.s. GDL-gels produced in air had the highest apparent viscosity, whereas values obtained with air and $N_2 - H_2$ bubbling were similar and significantly lower than those obtained with N_2 bubbling. So, for GDL-gels, the viscosity was affected by bubbling. Martin et al. [20] showed that the type of gas used for bubbling has a significant influence but no clear trend can be deduced from these results in terms of the influence of an oxidizing or reducing environment.

The gel structure was then observed during storage for up to 28 days. The mean whey separation values of GDL-gels produced under different E_h conditions are presented in Table 1. For each gaseous condition, the authors observed that whey separation occurred from the very first day of storage and the volume of whey separation was relatively constant during the 28 days of storage [20]. Whey separation ranged from 0.59 to 4.74 g / 100 g of GDL-gels. The highest whey separation was obtained with air but this value was lower than values reported in the literature: 18.48% of GDL-gels in the work by Lucey et al. [21] and 10% in a study by Fiszman et al. [22]. One explanation for is that in the study by Lucey et al. [21] the method used to measure whey separation was to remove the gels from their flasks and thus whey separation could have been over-estimated. Whey separation obtained with gas bubbling was lower (1.26 g / 100 g with air bubbling, 1.93 g / 100 g with N2 bubbling and 0.59 g / 100 g with N_2 - H_2 bubbling). The lowest whey separation was observed with GDL-gels made under $N_2 - H_2$. Adjusting the E_h of milk to reducing conditions (under $N_2 - H_2$) could be a possible way of significantly decreasing the phenomenon of whey separation.

3.3.3. Effect of E_h on a non-fat yoghurt

In a second step, the authors proposed studying the extent to which lactic acid bacteria affect acid milk gelation under different E_h conditions [23]. Indeed, oxygen modifies the growth capacity of bacteria and the formation of end products. So, E_h may contribute to the quality of fermented products [2, 24, 25]. Martin et al. [23] wanted to determine the effects of E_h on yoghurts made under various gaseous conditions. In this study they focused on exopolysaccharide production and gel structure (Table 2). The same gaseous conditions as in the study on the effect of E_h on model acid skim milk gels were chosen.

Lb. bulgaricus and S. thermophilus produce exopolysaccharides (EPS) which can contribute to improving the texture and viscosity of fermented dairy products [26]. In standard yoghurts (produced in air) the concentration of EPS was 63.60 mg.L^{-1}, in accordance with the literature (50 to 350 mg.L^{-1}) [27, 28]. The concentration was lower in yoghurts produced with air bubbling (15.22 mg.L^{-1}) than in yoghurts produced with N_2 bubbling, which was lower than those made with $N_2 - H_2$ bubbling. The EPS concentration of yoghurts made in Air and with $N_2 - H_2$ bubbling were similar. In reducing E_h conditions, lactic acid bacteria produced the same amount of EPS as in ambient air. This result has already been observed in the literature. Indeed, Lactobacillus sake 0-1 was reported to have optimal EPS production in anaerobic conditions [29], while higher EPS yields were correlated with a lower oxygen tension [30].

Gaseous conditions applied to milk	pH		E_{h7} (mV)		C_{EPS} (mg/L)	η (Pa.s)	WS (g/100g of yoghurt)
	At t=0	At t=3.5 hours	At t=0	At t=3.5 hours			
Air	$6.80^a \pm 0.0$	$4.6^a \pm 0.0$	$425^a \pm 20$	$171^a \pm 2$	$63.60^b \pm 3.72$	$0.046^a \pm 0.00$	$1.98^a \pm 0.54$
Air bubbling	$6.80^a \pm 0.0$	$4.6^a \pm 0.0$	$435^a \pm 3$	$241^a \pm 8$	$15.22^a \pm 0.74$	$0.046^a \pm 0.00$	$1.76^a \pm 0.31$
N_2 bubbling	$6.81^a \pm 0.0$	$4.6^a \pm 0.0$	$285^b \pm 11$	$139^b \pm 5$	$25.29^c \pm 0.40$	$0.035^b \pm 0.00$	$1.03^{ab} \pm 0.27$
$N_2 - H_2$ bubbling	$6.81^a \pm 0.0$	$4.6^a \pm 0.0$	$-345^c \pm 4$	$-309^c \pm 10$	$62.70^b \pm 0.75$	$0.021^c \pm 0.01$	$0.59^b \pm 0.12$

a, b, c: different letters indicate that groups were significantly different at an α risk of 5% (ANOVA test). Values in the same column should be compared.

Reprinted from Journal of Food Res. Int., Vol 431, Martin F, Cayot N, Vergoignan C, Journaux L, Gervais P, Cachon R, Impact of oxidoreduction potential and gas bubbling on rheological properties of non-fat yoghurt, Pages No. 218-223, Copyright (2010), with permission from Elsevier.

Table 2. Characteristics of gel structure depending on the different E_h conditions (milk acidified using lactic starters):

• Concentrations of exopolysaccharides (C_{EPS}) in yoghurts after one day of storage.
• Apparent viscosity η at 500 1/s of yoghurt at pH 4.6 and 4 °C. Measurements were made 24 hours after addition of starter culture.
• Evolution of average whey separation over 28 days (WS) in yoghurts.
Values are means from triplicate experiments.

The apparent viscosity of each yoghurt was characterized at pH 4.6 and 4 °C, 24 hours after addition of bacteria under the different Eh conditions (Table 2). The apparent viscosity ranged from 0.021 to 0.046 Pa.s. Yoghurts produced in air and with air bubbling had the highest apparent viscosity. The apparent viscosity of yoghurts made with N_2 bubbling was lower (0.035 Pa.s) than other oxidizing conditions (0.046 Pa.s), and values obtained with N_2 – H_2 bubbling were the lowest (0.021 Pa.s). Apparent viscosity is clearly affected by the gas type. A reducing environment reduces the apparent viscosity of yoghurt.

Apparent viscosity depends on the solid fraction in the gel as well as the relationships between the different solid elements. In yoghurt, solid particles include milk proteins, lactic acid bacteria and their EPS. Indeed, the gel of yoghurts produced under N_2 – H_2 conditions is weaker despite greater EPS production [23]. It is a common assumption that EPS produced by bacteria contribute to the rheological properties of yoghurt [31-33] but, as reported by Hassan et al. [34], van Marle [35] and Martin et al. [23], no correlation between the viscosity of yoghurt and EPS concentrations was found.

Whey separation of yoghurts produced under different E_h conditions over 28 days of storage was then studied [23] (Table 2). Concerning GDL-gels, whey separation of yoghurts occurred from the very first day of storage and the volume of whey separation was relatively constant over the 28 days of storage. Whey separation ranged from 0.59 to

1.98 g/100 g of yoghurt. The highest whey separation was obtained with air and air bubbling and these values are in accordance with the literature [22]. Whey separation obtained with N_2 bubbling (1.03 g / 100 g) and N_2 - H_2 bubbling (0.59 g / 100 g) was lower. So, the more reducing the environment, the lower the whey separation. Adjusting the E_h of milk to reducing conditions (under N_2 – H_2) could be a possible way of significantly decreasing the phenomenon of whey separation.

3.4. Aroma compounds

3.4.1. A look at yoghurt aroma compounds

The typical flavours of fermented milk are mainly due to a blend of the following compounds: lactic acid, carbon compounds such as acetaldehyde, acetone, acetate and diacetyl, non-volatile acids such as pyruvic, oxalic and succinic acids, volatile acids such as acetic, propionic and formic acids and products from the thermal degradation of proteins, lipids or lactose.

Ott et al. [36] identified 91 aroma compounds (GC-olfactometry) in yoghurt among which 21 were detected more frequently and would thus have a major impact on flavour. Acetaldehyde is found in significant quantities and is responsible for the characteristic smell of yoghurt. Diacetyl, pentane-2,3-dione, and dimethyl sulphide also have a major impact on yoghurt flavour [36, 37].

Acetaldehyde was firstly reported by Pette et al. [38] as the main aromatic compound in yoghurt. During manufacture, production of this compound is only highlighted when a certain level of acidification is reached (pH 5.0). Concentrations found in the final product are 0.7 to 15.9 mg.kg^{-1}. The maximum amount is obtained at pH 4.2 and stabilizes at pH 4.0. The production of acetaldehyde and other flavour compounds by S. thermophilus and Lb. bulgaricus occurs during yoghurt fermentation and the final amount is dependent on specific enzymes which are able to catalyse the formation of carbon compounds from the various milk constituents.

Three metabolic pathways producing acetaldehyde were identified and some pathways may take place simultaneously [39]:

- From glucose in the glycolytic pathway,
- From the degradation of DNA,
- From L-threonine with threonine aldolase.

However, 90% of acetaldehyde produced by Lb. bulgaricus comes from glucose and 100% in the case of S. thermophilus [39].

Diacetyl and pentane-2,3-dione also have a significant impact on the final aroma of yoghurt: 1 mg of diacetyl and 0.1 mg of pentane-2, 3-dione per kg of yoghurt are produced by lactic acid bacteria during fermentation. These diketones are produced by decarboxylation of their precursors, 2-acetolactate and 2-aceto-hydroxybutyrate [39]. These compounds are thermally unstable and in the presence of oxygen are converted into their corresponding diketones [40, 41]. Moreover, during storage at 4 °C, the concentration of the two diketones increases slightly [41] due to the basal metabolic activity of the bacteria.

Agitating a mixed culture of *Lactococcus* and *Leuconostoc* promotes diacetyl production by allowing oxidative decarboxylation of 2-acetolactate [42, 43]. In unstirred cultures, the redox potential of the medium decreases rapidly at the start of fermentation. Only acetoin and 2-acetolactate are produced. The authors also showed that controlled oxygenation of the *Lactococcus lactis* ssp. *lactis* culture medium favoured diacetyl production by increasing the activity of diacetyl synthase [44].

Neijssel *et al.* [40] showed that the distribution of carbon flux from pyruvate depended on the NADH / NAD$^+$ ratio, intracellular redox potential or the concentration of metabolites and particularly that of pyruvate. Finally, the authors suggested adding air or oxygen to milk in order to increase the amount of diacetyl in cheese [45].

References [36] and [37] are the only articles that mention dimethyl sulphide as a compound having a significant impact on the flavour of yoghurt. The metabolic pathways involved in the synthesis of sulphur compounds are not well-known in yoghurt. However, the literature mentions these synthetic pathways in the development of cheese flavour.

In general, the majority of sulphur aromatic compounds come from methionine [46]. Methanethiol is easily oxidized to dimethyl disulphide and dimethyl trisulphide [47]. The appearance of these compounds is the direct result of the methanethiol content and is modulated by the low redox potential in Cheddar. Dimethyl sulphide is produced by a metabolic pathway that does not involve methanethiol, but that is different to that of dimethyl sulphide and trimethyl disulphide from methionine [48].

Studies have also shown that when the redox potential decreases, methanethiol and hydrogen sulphide concentrations increase [45]. Moreover, the cheeses to which reducing compounds (dithiothreitol or glutathione) were added contained higher amounts of sulphur compounds and had better qualitative and quantitative flavour performances [45]. It therefore seems that a reducing environment is essential for the production of aroma compounds by bacteria. If a cheese is exposed to air, the redox increases and this leads to the oxidation of sulphur compounds, resulting in lower quality aromatics.

3.4.2. Impact on aroma biosynthesis by lactic acid bacteria

Studies on aroma biosynthesis by LAB usually take into account environmental factors such as pH and temperature. However, the E_h of the medium has not yet been considered, although it is supposed to affect bacterial metabolism [49, 50]. Martin *et al.* [51] determined to what extent E_h can affect the metabolic pathways involved in the production of aroma compounds in *Lb. bulgaricus* and *S. thermophilus*. Four aroma compounds (acetaldehyde, dimethyl sulphide, diacetyl and pentane-2,3-dione) were chosen as metabolic tracers of lactic acid bacteria metabolism. The same gaseous conditions as in the study of the effect of E_h on model acid skim milk gels and non-fat yoghurt were chosen. The amounts of each of the four aroma compounds extracted using a headspace solid-phase micro-extraction technique (HS-SPME) and analysed using gas chromatography coupled with mass spectrometry (GC-MS) during 28 days of storage are reported in Table 3.

Firstly, the authors focused on the impact of these different E_h conditions on the biosynthesis of these four aromas by bacteria after one day of storage [51]. In the standard yoghurt (made in ambient air), diacetyl was observed in the highest concentrations, and acetaldehyde the lowest. This result is contrary to the literature where the lowest concentrations were reported for dimethyl sulphide (0.013-0.070 mg.kg^{-1}; measured using dynamic and trapped headspace GC [37, 41]). In the same way, published concentrations were generally higher for acetaldehyde (0.7-15.9 mg.kg^{-1}) than in our standard yoghurt (0.18 mg.kg^{-1}). In the literature, the concentrations of diacetyl (0.31-17.3 mg.kg^{-1}) and 2,3-pentanedione (0.02-4.5 mg.kg^{-1}) were lower than in our standard yoghurt (162 mg.kg^{-1} and 115 mg.kg^{-1} respectively). An explanation for these differences can be put forward: the quantification technique used by Ott *et al.* [41] and Imhof *et al.* [37] was dynamic and trapped headspace GC. This technique requires Tenax® traps which may be saturated, as we showed in a preliminary experiment. Furthermore, in our study, to enable a more complete extraction of the aroma compounds, a saturated solution of NaCl was added to the yoghurt. Finally, we did not use the same species of LAB as Ott and Imhof, which may have resulted in different quantities of the various aroma compounds.

Yoghurts made with air bubbling had significantly higher concentrations of acetaldehyde and diacetyl compared to standard yoghurts. The concentration of dimethyl sulphide was significantly lower and that of pentane-2,3-dione was the same.

With N_2 bubbling, the concentration of acetaldehyde was similar to that in yoghurts made with air bubbling, whereas the concentration of dimethyl sulphide was lower. The concentration of diacetyl was the same as in standard yoghurts and the concentration of pentane-2,3-dione was not significantly different from that in yoghurts made in air (bubbling or not).

The authors also demonstrated that oxidative E_h conditions clearly increased the production of aroma compounds [51]. These results are consistent with the bibliography. Oxidative conditions stimulated the production of volatile sulphur compounds such as dimethyl sulphide, and aldehydes such as acetaldehyde [49]. In the presence of oxygen, the oxidative decarboxylation of 2-acetolactate and 2-aceto-hydroxybutyrate to diacetyl and pentane-2,3-dione respectively was also favoured [40, 42, 44, 52]. For diacetyl, our result can be explained by the fact that in anaerobic conditions lactic acid bacteria dehydrogenate the NADH produced during glycolysis via lactate dehydrogenase (LDH) activity. Boumerdassi *et al.* [44] confirmed that oxygen increases NADH oxidase activity [53], which causes NADH re-oxidation at the expense of LDH, butanediol dehydrogenase and acetoin dehydrogenase activity [54]. Then, excess pyruvate is partially eliminated through acetolactate production, which increases diacetyl production [44].

Finally, bubbling with $N_2 - H_2$ (reducing conditions), the concentration of acetaldehyde and pentane-2,3-dione was the same as in standard yoghurts. The concentration of dimethyl sulphide was the same as in yoghurts made without oxygen and the concentration of diacetyl was significantly lower than under the other three E_h conditions.

Then, [51] kept the yoghurts in Hungate tubes at 4 °C for 28 days in order to prevent exposure of the contents to oxygen, and the gaseous conditions applied to the milk are thus assumed to be constant during storage.

Gaseous conditions and storage period (days)	Aroma compound (mg.kg)			
	ACH	DMS	DY	PTD
Ambient air				
1	0.18[a] ± 0.02	10.16[a] ± 0.59	162.08[a] ± 13.49	115.25[a] ± 33.70
7	0.13[ab] ± 0.02	10.16[a] ± 0.99	115.55[cb] ± 5.13	84.33[ab] ± 1.86
14	0.10[b] ± 0.00	9.52[a] ± 0.42	91.63[c] ± 8.02	71.56[ab] ± 4.49
21	0.18[a] ± 0.00	13.06[b] ± 1.21	141.29[ab] ± 11.77	65.19[b] ± 5.49
28	0.16[ab] ± 0.01	11.82[ab] ± 0.48	112.66[c] ± 7.56	51.79[b] ± 2.95
Air bubbling				
1	0.28[a] ± 0.00	5.27[a] ± 0.53	299.90[a] ± 18.37	123.47[a] ± 3.23
7	0.14[b] ± 0.01	6.34[ab] ± 0.30	127.58[bc] ± 3.93	83.98[b] ± 2.20
14	0.11[b] ± 0.01	6.85[b] ± 0.32	104.23[b] ± 4.66	78.10[b] ± 3.15
21	0.24[a] ± 0.02	9.33[c] ± 0.55	143.84[c] ± 6.69	67.89[c] ± 0.80
28	0.13[b] ± 0.02	6.78[b] ± 0.30	101.91[b] ± 3.03	54.36[d] ± 4.19
N2 bubbling				
1	0.35[a] ± 0.04	3.72[a] ± 0.22	147.79[a] ± 9.91	110.87[a] ±4 .49
7	0.22[bc] ± 0.02	5.68[ab] ± 0.34	103.13[bc] ± 6.50	75.67[b] ± 1.98
14	0.18[c] ± 0.01	6.16[b] ± 0.17	78.80[c] ± 4.60	58.16[c] ± 0.48
21	0.28[ab] ± 0.04	10.34[c] ± 1.37	122.27[ab] ± 8.94	51.95[c] ± 2.29
28	0.18[c] ± 0.00	7.16[b] ± 0.77	84.76[c] ± 9.92	38.72[d] ± 3.03
N2 – H2 bubbling				
1	0.17[ab] ± 0.00	2.71[a] ± 0.57	102.73[a] ± 9.10	76.99[a] ± 5.90
7	0.13[bc] ± 0.02	5.49[b] ± 0.53	89.61[ab] ± 7.70	66.10[ac] ± 2.29
14	0.11[c] ± 0.01	6.07[bc] ± 0.43	74.35[b] ± 5.42	57.60[bc] ± 0.86
21	0.19[a] ± 0.01	7.48[c] ± 0.53	112.09[a] ± 7.82	52.10[bd] ± 0.59
28	0.12[c] ± 0.01	7.22[c] ± 0.63	78.36[b] ± 9.30	40.60[d] ± 7.93

[a, b, c, d]: different letters indicate that groups were significantly different at an α risk of 5% (ANOVA test).

Reprinted from Journal of J. Dairy Sci., Vol 942, Martin F, Cachon R, Pernin K, De Coninck J, Gervais P, Guichard E, Cayot N, Effect of oxidoreduction potential on aroma biosynthesis by lactic acid bacteria in nonfat yogurt, Pages No. 614-622, Copyright (2011), with permission from Elsevier.

Table 3. Evolution of average amounts of aroma compounds (mg.kg^{-1}) quantified in headspace of yoghurts made under different E_h conditions (ambient air, bubbling with air, bubbling with N2 and bubbling with N2 – H2) during 28 days of storage. ACH: Acetaldehyde (A); DMS: Dimethyl sulphide (B); DY: Diacetyl (C); PTD: pentane-2,3-dione (D). Values are means of experiments carried out in triplicate. Values in the same column should be compared.

During the 28 days of storage, for the standard yoghurt, the quantities of acetaldehyde and dimethyl sulphide produced were relatively stable, while diketone concentrations significantly decreased.

For yoghurts made with air bubbling, the aroma profiles remained almost constant. During storage, the concentration of acetaldehyde decreased slightly whereas that of dimethyl sulphide increased slightly. The diketone concentration significantly decreased.

For yoghurts made without oxygen (bubbling with N_2), the quantities of acetaldehyde, diacetyl and pentane-2,3-dione decreased during storage while that of dimethyl sulphide increased.

For yoghurts produced under reducing conditions (bubbling with $N_2 - H_2$), the aroma profiles during storage were the same as those made without oxygen. The concentration of acetaldehyde, diacetyl and pentane-2,3-dione decreased while that of dimethyl sulphide increased.

Furthermore, during storage, different profiles were observed for the four aromas depending on the E_h conditions [51]. Under oxidizing conditions (+170 to +245 mV), the concentration of acetaldehyde was relatively stable during storage, which is in accordance with the literature [9, 55, 56] and the concentration of dimethyl sulphide was also stable. On the contrary, under reducing conditions (-300 to -349 mV), the concentration of acetaldehyde decreased and that of dimethyl sulphide increased. The metabolic pathways involved in the biosynthesis of sulphur compounds are still unclear. Under reducing conditions, it seems that another pathway promotes the production of dimethyl sulphide and that acetaldehyde may be reduced to ethanol. For diketones, whatever the E_h conditions, the concentration decreased during storage. Diacetyl and pentane-2,3-dione can be reduced respectively to acetoin and pentane-2,3-diol [57].

4. Impact of E_h on other dairy products

4.1. Probiotic dairy products

The use of gas to modify E_h seems to be an interesting way of varying the organoleptic properties of dairy products as well as improving the survival of oxygen sensitive strains during storage in fermented dairy products containing probiotics. Indeed, these microorganisms are mainly anaerobes. Oxygen, which is a powerful oxidant, has a drastic effect on E_h values and the viability of probiotic bacteria during manufacturing and storage [58-60]. So, many studies modify the redox potential to protect probiotics from oxygen toxicity in dairy products [1, 61-65]. However, these techniques sometimes have deleterious effects on the organoleptic properties of fermented milk. An alternative to these methods could be the use of gases. Indeed, Ebel et al. [18] showed that fermented dairy products made from milk gassed with N_2 and more particularly those made from milk gassed with $N_2 - H_2$, were characterized by a significant increase in Bifidobacterium bifidum survival during storage (Figure 2).

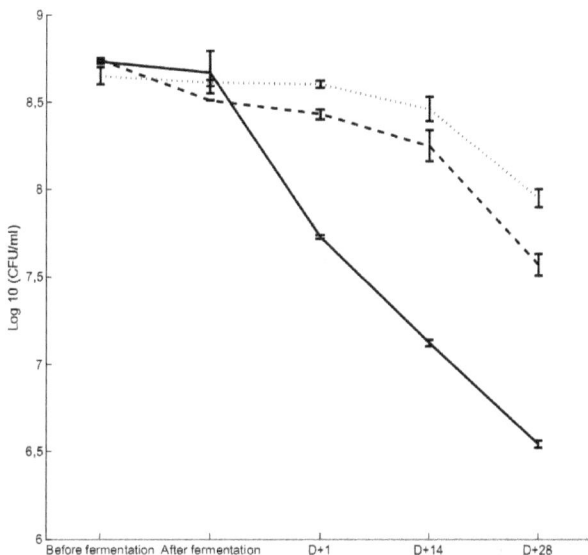

Reprinted from Journal of J. Dairy Sci., Vol 945, Ebel B, Martin F, Le LDT, Gervais P, Cachon R, Use of gases to improve survival of *Bifidobacterium bifidum* by modifying redox potential in fermented milk, Pages No. 2185-2191, Copyright (2011), with permission from Elsevier.

Figure 2. Evolution of a population of *Bifidobacterium bifidum* during fermentation and storage. Different gaseous conditions were applied to the milk: control (solid line), gassed with N_2 (dashed line), or gassed with $N_2 - H_2$ (dotted line).

After 28 days of storage, a difference in bacterial counts of 1.2 log and 1.5 log was observed between the control milk and after bubbling with N_2 or $N_2 - H_2$ respectively. No differences were highlighted during the fermentation process. It is interesting to note that this technique was set up without affecting the fermentation kinetics and survival of *S. thermophilus* and *Lb. bulgaricus*. The use of gas is a possible way of improving probiotic survival during storage without affecting acidification properties of yoghurt strains and consequently organoleptic properties.

4.2. Cheese

Controlling E_h in cheese seems essential in governing aroma characteristics. Indeed, a reducing E_h is necessary for the development of the characteristic flavour of certain fermented dairy products such as cheeses, notably through the production of thiol compounds [45, 66]. It has also been reported that Cheddar has a reducing E_h and is an indicator of the establishment of the conditions required for the formation of aroma compounds [67]. As shown previously, E_h can modify the metabolic pathways of aroma production by lactic bacteria [51]. Kieronczyk *et al.* [49] demonstrated that reducing E_h conditions can stimulate carboxylic acid production in cheese, while oxidative E_h conditions improve the production of volatile sulphur compounds and aldehydes. By ripening cheese under reducing E_h conditions, the production of volatile fatty acids increased [68]. Adjusting the E_h of the milk before cheese ripening could be a possible way of modifying the metabolism of lactic bacteria.

5. Conclusion

Pasteur defined fermentation as "life without air". In lactic acid bacteria, some exogenous electron acceptors may interfere significantly with the fermentative metabolism by acting on different cellular activities. A better understanding of the adaptive mechanisms to extracellular redox is still lacking, but the results in the literature show that lactic acid bacteria may use passive or active mechanisms. A remarkable feature in lactic acid bacteria is their ability to reduce the redox environment to low E_h values.

With the prospect of food applications, changing E_h using pure or a mixture of gases has the advantage of maintaining product safety as opposed to the use of oxidizing or reducing molecules. This chapter demonstrates the importance of E_h both on the physico-chemistry of milk gels and bacterial metabolism and viability. The use of gas to modify E_h seems to be an interesting way of varying the organoleptic properties of dairy products.

Author details

F. Martin, B. Ebel, C. Rojas, P. Gervais, N. Cayot and R. Cachon
Unité Procédés Alimentaires et Microbiologiques, UMR A 02.102, AgroSup Dijon/Université de Bourgogne, 1 esplanade Erasme, Dijon, France

6. References

[1]   Dave RI, Shah NP (1997) Effectiveness of ascorbic acid as an oxygen scavenger in improving viability of probiotic bacteria in yoghurts made with commercial starter cultures. Int. dairy j. 7(6-7): 435-443.

[2]   Rödel W, Scheuer R (2000) Redox potential of meat and meat produits. IV. Recording criteria of quality in meat and meat products by measuring the redox potential. Fleischwirtschaft. 2: 46-48.

[3]   Galster H (1991) pH Measurement: fundamentals, methods, applications, instrumentation. New York: VCH Publishers.

[4]   Abraham S, Cachon R, Colas B, Feron G, De Coninck J (2007) E_h and pH gradients in camembert cheese during ripening: Measurements using microelectrodes and correlations with texture. Int. dairy j. 178: 954-960.

[5]   Leistner L, Mirna A (1959) Das redoxpotential von pökelladen. Fleischwirtschaft. 8: 659-666.

[6]   Cachon R, Jeanson S, Aldarf M, Divies C (2002) Characterisation of lactic starters based on acidification and reduction activities. Lait. 823: 281-288.

[7] George SM, Richardson LCC, Pol IE, Peck MW (1998) Effect of oxygen concentration and redox potential on recovery of sublethally heat-damaged cells of *Escherichia coli* O157:H7, *Salmonella enteritidis* and *Listeria monocytogenes*. J. appl. microbiol. 84: 903-909.

[8] Medard (2002) Encyclopédie des gaz de l'air liquide 3rd ed. Amsterdam, The Netherlands: Lavoisier.

[9] Tamine AY, Robinson RK (1999) Yoghurt: science and technology. Washington DC: CRC Press.

[10] Fredrickon AG (1977) Behavior of mixed cultured of microorganisms. Annu. rev. microbiol. 31: 63-87.

[11] Rasic JL, Kurmann JA. Fermented fresh milk products (1978) In: Rasic JL, Kurmann JA, editors. Yoghurt scientific grounds, technology, manufacture and preparations. Copenhangen, Denmark: Technical Dairy Publishing House.

[12] Moreira M, Abraham A, De Antoni G (1999) Technological properties of milks fermented with thermophilic acid bacteria at suboptimal temperature. J. dairy sci. 83: 395-400.

[13] Zourari A, Accolas JP, Desmazeaud M (1992) Metabolism and biochemical characteristics of yoghurt bacteria: a review. Lait. 72: 1-34.

[14] Loones A. Laits fermentés par les bactéries lactiques (1994) In: De Roissart H, Luquet FM, editors. Bactéries lactiques. Paris, France: Lorica, Uriage. p. 133-154.

[15] Courtin P, Rul F (2003) Interactions between microorganisms in a simple ecosystem: yogurt bacteria as a study model. Lait. 84: 125-134.

[16] El-Soda MA, Abou-Donia SA, El-Shafy HK, Mashaly R, Ismail AA (1986) Metabolic activities and symbiosis in zabady isolated cultures. Egypt. j. dairy sci. 14: 1-10.

[17] Xanthopoulos V, Petridis D, Tzanetakis N (2001) Characterisation and classification of *Streptococcus thermophilus* and *Lactobacillus delbrueckii* subsp. *bulgaricus* strains isolated from traditional Greek yogurts. J. food sci. 66: 747-752.

[18] Ebel B, Martin F, Le LDT, Gervais P, Cachon R (2011) Use of gases to improve survival of *Bifidobacterium bifidum* by modifying redox potential in fermented milk. J. dairy sci. 945: 2185-2191.

[19] Cachon R, Feron G, Delbeau C, Ibarra D, Ledon H (2006) Process for modifying the sensory characteristics of a fermented milk product and its maturation during conservation of said product. EP1649755. L'Air Liquide, France.

[20] Martin F, Cayot N, Marin A, Journaux L, Cayot P, Gervais P, Cachon R (2009) Effect of oxidoreduction potential and of gas bubbling on rheological properties and microstructure of acid skim milk gels acidified with glucono-δ-lactone. J.dairy sci. 92: 5898-5906.

[21] Lucey JA, Tamehana M, Singh H, Munro PA (1998) A comparison of the formation, rheological properties and microstructure of acid skim milk gels made with a bacterial culture or glucono-δ-lactone. Food res. int. 312: 147-155.

[22] Fiszman SM, Lluch MA, Salvador A (1999) Effect of addition of gelatin on microstructure of acidic milk gels and yoghurt and on their rheological properties. Int. dairy j. 912: 895-901.

[23] Martin F, Cayot N, Vergoignan C, Journaux L, Gervais P, Cachon R (2010) Impact of oxidoreduction potential and of gas bubbling on rheological properties of non-fat yoghurt. Food res. int. 431: 218-223.

[24] Dave RI, Shah NP (1997) Effectiveness of ascorbic acid as an oxygen scavenger in improving viability of probiotic bacteria in yoghurts made with commercial starter cultures. Int. dairy j. 76-7: 435-443.

[25] van Dijk C, Ebbenhorst-Selles T, Ruisch H, Stolle-Smits T, Schijvens E, van Deelen W, Boeriu C (2000) Product and Redox Potential Analysis of Sauerkraut Fermentation. J. agr. food chem. 482: 132-139.

[26] Ruas-Madiedo P, Hugenholtz J, Zoon P (2002) An overview of the functionality of exopolysaccharides produced by lactic acid bacteria. Int. dairy j. 12(2-3): 163-171.

[27] Cerning J, Bouillanne C, Desmazeaud MJ, Landon M (1986) Isolation and characterization of exocellular polysaccharide produced Lactobacillus bulgaricus. Biotechnol. letters. 89: 625-628.

[28] Cerning J, Bouillanne C, Desmazeaud MJ, Landon M (1988) Exocellular polysaccharide production by Streptococcus thermophilus. Biotechnol. letters. 104: 255-260.

[29] van den Berg D, Robijn GW, Janssen AC, Giuseppin M, Vreeker R, Kamerling JP, Vliegenthart J, Ledeboer AM, Verrips CT (1995) Production of a novel extracellular polysaccharide by Lactobacillus sake 0-1 and characterization of the polysaccharide. Appl. environ. microb. 618: 2840-4.

[30] De Vuyst L, Vanderveken F, van de Ven S, Degeest B (1998) Production by and isolation of exopolysaccharides from Streptococcus thermophilus grown in a milk medium and evidence for their growth-associated biosynthesis. J. appl. microbiol. 846: 1059-1068.

[31] Hess SJ, Roberts RF, Ziegler GR (1997) Rheological properties of nonfat yogurt stabilized using Lactobacillus delbrueckii ssp. bulgaricus producing exopolysaccharide or using commercial stabilizer systems. J. dairy sci. 802: 252-263.

[32] van Marle ME, van den Ende D, de Kruif CG, Mellema J (1999) Steady shear viscosity of stirred yoghurts with varying ropiness. J. rheol. 43(6): 1643-1663.

[33] Rawson HL, Marshall VM (1997) Effect of 'ropy' strains of Lactobacillus delbrueckii ssp. bulgaricus and Streptococcus thermophilus on rheology of stirred yogurt. Int. j. food sci. technol. 323: 213-220.

[34] Hassan AN, Frank JF, Farmer MA, Schmidt KA, Shalabi SI (1995) Formation of yogurt microstructure and three-dimensional visualization as determined by confocal scanning laser microscopy. J. dairy sci. 7812: 2629-2636.

[35] van Marle ME. (1998) Structure and rheological properties of yoghurt gels and stirred yoghurts. Enschede, the Netherlands: Twente.

[36] Ott A, Fay LB, Chaintreau A (1997) Determination and origin of the aroma impact compounds of yogurt flavor. J. agric. food chem. 453: 850-858.

[37] Imhof R, Glättli H, Bosset JO (1994) Volatile organic aroma compounds produced by thermophilic and mesophilic mixed strain dairy starter cultures. LWT - Food sci. technol. 275: 442-449.

[38] Pette J, Lolkema H (1950) Yoghurt, III: Acid production and aroma formation in yoghurt. Neth. milk dairy j. 4: 261-273.

[39] Biavati B, Vescovo M, Torriani S, Bottazzi V (2000) Bifidobacteria: history, ecology, physiology and applications. Ann. microbiol. 502: 117-131.

[40] Neijssel OM, Snoep JL, Teixeira de Mattos MJ (1997) Regulation of energy source metabolism in streptococci. J. appl. microbiol. 83S1: 12S-19S.

[41] Ott A, Germond JE, Baumgartner M, Chaintreau A (1999) Aroma comparisons of traditional and mild yogurts: headspace gas chromatography quantification of volatiles and origin of α-diketones. J. agric. food chem. 476: 2379-2385.

[42] Monnet C, Schmitt P, Divies C (1994) Method for assaying volatile compounds by headspace gas chromatography and application to growing starter cultures. J. dairy sci. 777: 1809-1815.

[43] Starrenburg M, Hugenholtz J (1991) Citrate fermentation by *Lactococcus* and *Leuconostoc* spp. Appl. environ. microb. 57: 3535-3540.

[44] Boumerdassi H, Desmazeaud M, Monnet C, Boquien CY, Corrieu G (1996) Improvement of diacetyl production by *Lactococcus lactis* ssp. *lactis* CNRZ 483 through oxygen control. J. dairy sci. 795: 775-781.

[45] Green ML, Manning DJ (1982) Development of texture and flavour in cheese and other fermented products. J. dairy res. 4904: 737-748.

[46] Tachon S, Michelon D, Chambellon E, Cantonnet M, Mezange C, Henno L, Cachon R, Yvon M (2009) Experimental conditions affect the site of tetrazolium violet reduction in the electron transport chain of *Lactococcus lactis*. Microbiology+. 1559: 2941-2948.

[47] Singh H, Drake MA, Cadwallader KR (2003) Flavor of cheddar cheese: a chemical and sensory perspective. Compr. rev. food sci. f. 2: 139-162.

[48] Bonnarme P, Arfy K, Dury C, Helinck S, Yvon M, Spinnler HE (2001) Sulfur compound production by *Geotrichum candidum* from L-methionine: importance of the transamination step. FEMS microbiol. lett. 205: 247-252.

[49] Kieronczyk A, Cachon R, Feron G, Yvon M (2006) Addition of oxidizing or reducing agents to the reaction medium influences amino acid conversion to aroma compounds by *Lactococcus lactis*. J. appl. microbiol. 1015: 1114-1122.

[50] Riondet C, Cachon R, Waché Y, Alcaraz G, Diviès C (1999) Changes in the proton-motive force in *Escherichia coli* in response to external oxidoreduction potential. Eur. j. biochem. 2622: 595-599.

[51] Martin F, Cachon R, Pernin K, De Coninck J, Gervais P, Guichard E, Cayot N (2011) Effect of oxidoreduction potential on aroma biosynthesis by lactic acid bacteria in nonfat yogurt. J. dairy sci. 942: 614-622.

[52] Hugenholtz J, Kleerebezem M (1999) Metabolic engineering of lactic acid bacteria: overview of the approaches and results of pathway rerouting involved in food fermentations. Curr. opin. biotech. 105: 492-497.

[53] Condon S (1987) Responses of lactic acid bacteria to oxygen. FEMS microbiol. lett. 463: 269-280.

[54] Bassit N, Boquien CY, Picque D, Corrieu G (1993) Effect of Initial Oxygen Concentration on Diacetyl and Acetoin Production by *Lactococcus lactis* subsp. *lactis* biovar diacetylactis. Appl. environ. microb. 596: 1893-1897.

[55] Gardini F, Lanciotti R, Elisabetta Guerzoni M, Torriani S (1999) Evaluation of aroma production and survival of *Streptococcus thermophilus, Lactobacillus delbrueckii* subsp. *bulgaricus* and *Lactobacillus acidophilus* in fermented milks. Int. dairy j. 92: 125-134.

[56] Imhof R, Bosset JO (1994) Quantitative GC-MS analysis of volatile flavour compounds in pasteurized milk and fermented milk products applying a standard addition method. LWT - Food sci technol. 273: 265-269.

[57] Ott A, Germond JE, Chaintreau A (2000) Vicinal diketone formation in yogurt: 13C precursors and effect of branched-chain amino acids. J. agr food chem. 483: 724-731.

[58] Storz G, Imlay JA (1999) Oxidative stress. Curr. opin. microbiol. 22: 188-94.

[59] Talwalkar A, Kailasapathy K (2004) A review of oxygen toxicity in probiotic yogurts: Influence on the survival of probiotic bacteria and protective techniques. Compr. rev. food sci. f. 33: 117-124.

[60] Talwalkar A, Kailasapathy K (2004) The role of oxygen in the viability of probiotic bacteria with reference to *L. acidophilus* and *Bifidobacterium* spp. Curr. issues intest. microbiol. 51: 1-8.

[61] Bolduc MP, Bazinet L, Lessard J, Chapuzet JM, Vuillemard JC (2006) Electrochemical modification of the redox potential of pasteurized milk and its evolution during storage. J agric. food chem. 5413: 4651-4657.

[62] Bolduc MP, Raymond Y, Fustier P, Champagne CP, Vuillemard JC (2006) Sensitivity of bifidobacteria to oxygen and redox potential in non-fermented pasteurized milk. Int. dairy j. 169: 1038-1048.

[63] Dave RI, Shah NP (1997) Effect of cysteine on the viability of yoghurt and probiotic bacteria in yoghurts made with commercial starter cultures. Int. dairy j. 78-9: 537-545.

[64] Lourens-Hattingh A, Viljoen BC (2001) Yogurt as probiotic carrier food. Int. dairy j. 111-2: 1-17.

[65] Mortazavian AM, Ehsani MR, Mousavi SM, Reinheimer JA, Emamdjomeh Z, Sohrabvandi S, Rezaei K (2006) Preliminary investigation of the combined effect of heat treatment and incubation temperature on the viability of the probiotic micro-organisms in freshly made yogurt. Int. j. dairy technol. 59(1): 8-11.

[66] Kristoffersen T (1985) Development of flavor in cheese. Milchwissenschaft. 40(4): 197-199.

[67] Urbach G (1995) Contribution of lactic acid bacteria to flavour compound formation in dairy products. Int. dairy j. 58: 877-903.

[68] Ledon H, Ibarra D (2006) Method for modifying hygienic, physico-chemical and sensory properties of cheese by controlling the redox potential. US 2009/0214705 A1. L'Air Liquide, France.

Meat Products

The Role of Lactic Acid Bacteria in Safety and Flavour Development of Meat and Meat Products

Lothar Kröckel

Additional information is available at the end of the chapter

1. Introduction

Lactic acid bacteria (LAB) are widespread in nature and commonly occur on all kind of plant materials, on mucous membranes, in saliva and, in feces. Consequently and unavoidably they are part of the contamination flora of fresh meats after slaughter. Under certain conditions, e.g. in packaged refrigerated meats or raw sausage meats, they are able to compete efficiently with accompanying microorganisms for nutrients and may reach substantial viable counts. Their metabolic activities may ultimately result in either a desired preservative effect due to the repression of pathogenic and spoilage microorganisms, a desired tasty meat product, such as raw fermented sausage, or in meat spoilage through undesired transformations of raw and cooked meats. Heterofermentative LAB of the *Carnobacterium*, *Leuconostoc* and *Weissella* genera are usually more involved in meat spoilage than the homofermentative *Lactobacillus* and *Pediococcus* genera. Therefore, commercially available meat starter cultures for dry-fermented sausage production exclusively belong to the latter two. Homofermentative LAB produce almost exclusively lactic acid from fermentable carbohydrates present in meats, which is relatively mild and palatable, while heterofermentative species produce significant amounts of less desirable fermentation end products, such as CO_2 gas, ethanol, acetic acid, butanoic acid and acetoin. However, under certain conditions *Lactobacillus* spp. may also produce significant amounts of acetic acid, ropy slime and, discolouration (greening) of meats [1,2].

In food industry starter and protective cultures are currently used in a number of products to safeguard the microbial and sensory quality. Lactic acid bacteria (LAB) are the main players in the natural transformation of agricultural primary products into safe, delicious and shelf stable foods for human consumption. In meat products there are three basic fields of application for the targeted use of such cultures: raw fermented sausages, raw cured hams, and pasteurised, sliced prepackaged meats (cold cuts) [3-8]. The use of protective cultures in prepackaged, refrigerated sliced Bologna-type sausage and cooked ham against pathogenic listeria is a much discussed, sustainable technology for improving the microbial safety and quality of these products. It helps to avoid chemical preservatives, such as sodium lactate/potassium acetate additives, or repasteurisation in package after slicing and packaging, which both have a negative impact on sensory product quality leaving a numb mouthfeel or warmed-over flavour, resp. [9,10].

2. Meat and meat products

2.1. Raw fermented sausages

The importance of starter and protective cultures for the manufacturing of safe and high-quality fermented sausages has been known for a long time and, lactobacilli play an important role in their production [11,5]. *Lb. sakei* and *Lb. curvatus* are quite often the predominant LAB in dry-fermented sausage while other lactobacilli, such as *Lb. versmoldensis*, *Lb. plantarum*, *Lb. brevis*, *Lb. farciminis*, *Lb. alimentarius*, *Weissella* species, pediococci, and leuconostocs, usually occur in significantly lower numbers [12]. This has recently been also shown for different traditional salamis from North Italy [13-15]. However, other recipes and ripening conditions may promote other LAB as well. LAB isolated from dry spontaneously fermented sausages from 15 different producers in Spain included mainly *Lb. sakei* (66%), *Lb. curvatus* (26%), and *Lb. plantarum* (8%) [16]. For dry fermented Spanish 'chorizo' sausage *Lb. sakei* (69%), *Lb. curvatus* (16%) and *Pediococcus* (9%) have been reported [17]. From naturally fermented Greek dry salami about 50% of the isolates belonged to *Lb. sakei/curvatus*, 30% to the *Weissella* genus, 10% to *Lb. plantarum* and 3% each to *Lb. farciminis* and *Enterococcus (Ec.) faecium* [18]. In "Alheira", a fermented sausage produced in Portugal, *Lb. plantarum* and *Ec. faecalis* prevailed while other LAB, such as *Lb. paraplantarum*, *Lb. brevis*, *Lb. rhamnosus*, *Lb. sakei*, *Lb. zeae*, *Lb. paracasei*, *Leuconostoc (Leuc.) mesenteroides*, *Pediococcus (Pc.)* *pentosaceus*, *Pc. acidilactici*, *Weissella (Ws.) cibaria*, *Ws. viridescens* and *Ec. faecium*, occurred in lower numbers [19].

The main role of LAB is to convert fermentable sugars in the sausage batter to lactic acid, thereby contributing to product safety by creating unfavourable conditions for pathogens and spoilage organisms. The production of lactic acid has also a direct impact on sensory product quality by providing a mild acidic taste, and by supporting the drying process which requires a sufficient decline in pH. Furthermore, LAB influence the sensory characteristics of the fermented sausages by the production of small amounts of acetic acid, ethanol, acetoin, pyruvic acid, carbon dioxide, and their ability to initiate the production of aromatic substances from proteinaceous precursors [20-22]. The selection criteria for lactic acid bacteria to be used in the production of fermented sausage include (i) fast production of lactic acid (ii) good growth at different temperatures, (iii) homofermentative metabolism, (iv) persistence over the whole fermentation and ripening process, (v) nitrate reduction, (vi) ability to express catalase, (vii) no fermentation of lactose, (viii) formation of flavour, (ix) no formation of peroxide, (x) no formation of biogenic amines, (xi) no formation of ropy slime, (xii) tolerance or even synergy to other microbial components of the starter, (xiii) antagonism against pathogens, (xiv) antagonism against technologically undesirable microorganisms, (xv) improvement of the nutritional value of the sausage and, (xvi) economic factors [23]. Many homofermentative LAB associated with cured meat products are quite resistant to nitrite up to 200 ppm [24]. A new starter culture for raw sausages, 'BITEC Advance LD-20' from Frutarom Savory Solutions, containing *Lb. sakei* and *S. carnosus* is marketed as consistently providing a 'pleasant mild taste' while rapidly deminishing the pH value of the sausage batter. Rapid acidification is important for product safety while a high competitiveness against the spontaneous lactic flora is important for product quality. The culture can be used for firm and fresh raw sausages as well as sausage spreads.

The use of homofermentative lactic acid bacteria is desirable because acetic acid has an unpleasant taste as compared with lactic acid [25].

It must be kept in mind, however, that, although lactic acid production and pH reduction by LAB provide quite unfavorable conditions for pathogenic bacteria thereby preventing them from growing and contributing to their reduction, several pathogenic microorganisms are able to survive in fermented sausages under certain conditions for extended periods, especially during refrigerated storage of sparsely dried sausages. Pathogenic strains of *Escherichia (E.) coli, Listeria (Li.) monocytogenes* and *Yersinia (Y.) enterocolitica* are inactivated better after the initial fermentation and ripening stage if stored at ambient rather than at refrigeration temperature. Inclusion of a maturation period above refrigeration temperatures before distribution may increase the safety of these products [26-29].

2.2. Dry-cured hams

Currently there are only a few publications which clearly substantiate the advantages of starter and protective cultures during raw cured ham production. On the other hand, starter cultures have been more and more implemented by meat industry into the production of dry-cured hams since the early 1980s [6,30]. These cultures are expected to be active under the harsh manufacturing conditions (low temperatures, high salt, lack of oxygen, presence of nitrite). LAB contribute to a moderate pH decrease which promotes the microbial stability as well as product texture, reduce stickiness and pH variations of the raw material. As an example, FSC-111 Bactoferm[R] from Chr.-Hansen A/S contains, besides a staphylococcal strain, also a strain of *Lb. sakei*.

The LAB induced acidification is usually more pronounced with injected or compound meats than with dry-salted ones. Modern turkey hams are produced by squeezing turkey breast over the screw of an extruder in the presence of (g/kg) nitrite curing salt (35), diphosphate (2,5), dextrose (2), water (100), starter culture and a spice compound, and subsequent tumbling until protein release. This mixture is then stuffed into fiber casings and left for 5 days at 2°C. This is followed by a fermentation step of around 16 hours at 22°C and 92-94% relative humidity until a pH below 5.4 is reached. Finally, the product is heated in a cabinet at 47 °C to a core temperature of 40°C. The desired result is a fresh looking product with a slightly hyaline appearance with an optimum safety against undesired and pathogenic microorganisms [30].

2.3. Fresh meats

In chilled vacuum-packaged beef, even close to the freezing point, psychrotrophic LAB are able to attain high population densities. At -1.5 °C LAB grew to 8-9 $\log_{10}$ cfu ml^{-1} drip in 16 weeks with maximum doubling times of around 2-4 days [31]. In this study, *Cb. divergens, Leuc. mesenteroides* and *Lb. delbruckii* dominated the LAB flora after 4, 8-12 and 16 weeks, respectively. At 2 °C other workers have reported *Lb. sakei, Lb. curvatus, Carnobacterium (Cb.) divergens, Cb. maltaromaticum, Leuconostoc spp.* and *Lactococcus raffinolactis* as relevant LAB with t$_d$ of around 19 hours and less [32]. After 25 days maximum LAB numbers of around 7-8 $\log_{10}$ cfu cm^{-2} were reached and after 8 weeks the the meat odour immediately after opening the bags was regarded "definitely off" ("slightly off" between 4-6 weeks).

LAB may be useful as protective cultures during the ripening of vacuum-packaged raw beef and, bioprotective cultures may also help to reduce *E. coli* O157:H7 in frozen ground-beef patties [33,34]. Peptides generated by LAB have been suggested as sensorial and hygienic biomarkers in meat conditioning and fermentation [35].

Today, meat industry is forced to produce meats with a shelf life long enough to fulfill logistic, retail sale and consumer demands. Besides general hygienic considerations, including appropriate temperature control modified-atmosphere packaging (MAP) with 30-40% CO_2 is used to prevent early spoilage. While Gram-negative spoilage bacteria are suppressed, psychrotrophic LAB are not [36-38].

2.4. Cooked meats

Cooked, sliced and prepackaged meat products are popular convenience foods. They are retailed under refrigeration with varying shelf lifes, e.g. at 5 to 7 °C for 14 to 28 days. During slicing and packaging the slices may be contaminated with microorganisms from the production environment. Especially certain psychrotrophic LAB may then attain high cell counts during cold storage and impair the sensory quality of the products [39-42]. More than 2/3 of the refrigerated sliced cooked meats from the German retail market contained LAB counts above 7 $\log_{10}$ cfu g^{-1} one week past the indicated shelf life (Figure 1) [43]. The LAB flora on Bologna-type sausage is mostly dominated by the *Lb. sakei/curvatus* cluster while *Leuc. carnosum* frequently dominates on cooked ham. Occasionally, also *Ws. viridescens, Cb. maltaromaticum* and *Leuc. mesenteroides* ssp. *mesenteroides* may occur in higher numbers. Independent from dominant occurrence, eight LAB species have been identified in German retail samples. The number of samples (n) out of 50 in which these species occurred were *Lb. sakei* (40), *Leuc. carnosum* (22), *Lb. curvatus* (18), *Ws. viridescens* (11), *Leuc. mesenteroides* ssp. *mesenteroides* (8), *Cb. maltaromaticum* (4), *Lactobacillus* sp. (4), *Lactococcus* sp. (4), *Cb. divergens* (2), *Leuc. gelidum* (1), *Leuconostoc* sp. (1) (Figure 2) [43].

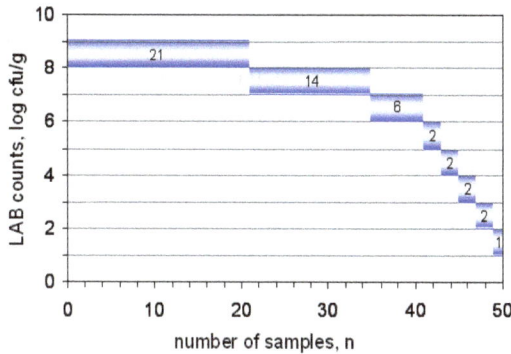

Figure 1. Distribution of samples of refrigerated sliced cooked meats from the German retail market with respect to different LAB counts one week past the indicated shelf life [43].

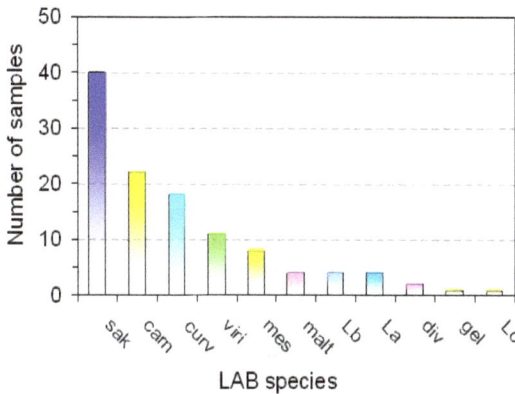

Figure 2. Abundancy of different LAB species in refrigerated sliced cooked meats from German retail (n=50). sak, *Lb. sakei*; carn, *Leuc. carnosum*; curv, *Lb. curvatus*; viri, *Ws. viridescens*; mes, *Leuc. mesenteroides*; malt, *Cb. maltaromaticum*; Lb, *Lactobacillus* sp.; La, *Lactococcus* sp.; div, *Cb. divergens*; gel, *Leuc. gelidum*; Lc, *Leuconostoc* sp. [43].

3. Biopreservation

Biopreservation of meats refers to the control of pathogenic and spoilage microorganisms by a competitive microflora of desired indigenous microorganisms or so-called starter and protective cultures. The development of starter cultures for meats is tightly coupled with the industrialisation of the traditional artisanal processes. The production of safe and tasty fermented sausages by traditional technologies requires expert knowledge and continous attention to guide the fermentation into the desired direction, i.e. to promote the development of the desired microorganisms and to suppress the development of undesired microorganisms. Mistakes are heavily paid for by dangerous and/or low quality outcomes.

Starter cultures, added at the beginning of fermentation, allow a standardization of the product quality and considerably reduce the risk of product defects. However, it should be kept in mind that starter cultures can not replace good manufacturing practice which besides the selection of the appropriate raw materials with acceptable hygienic parameters also includes the implementation and control of appropriate processing conditions. This is especially true with respect to the health risks associated with enterohaemorrhagic *E. coli*. Because of its increased acid tolerance and low infective dose for human infection, additional hurdles besides starter cultures have become very important for the production of safe raw fermented sausages. The hurdles principle for controlling undesired microorganisms in raw sausage fermentation has been illustrated by LEISTNER [27,44,45] and, in the meantime the implementation of HACCP (hazard analyis critical control point) concepts have become mandatory in food production [46].

Protective cultures may be distinquished from starter cultures by their lack of, or their reduced product transformation capabilities. Protective cultures my be used for a number of applications with the main focus on pathogen control, especially of *Li. monocytogenes*, but also of spoilage organisms such as LAB involved in the spoilage of deli meats, or of *Brochothrix thermosphacta* and *Clostridium estertheticum* in vacuum-packaged raw meats [47-50].

Of special interest are strains which excrete powerful anti-listerial bacteriocins *in situ* and, which at the same time have no or only a very weak spoilage potential [21, 51].

A strain of *Lactococcus (Lc.) lactis*, marketed as Bactoferm® Rubis by Chr.-Hansen A/S, is offered as a protective culture to be used instead of chemicals to preserve/stabilize the normal colour of vacuum packed or controlled atmosphere packaged, sliced, cured meat products [52].

The big retail chains and the official food control authorities look at high microbial counts in deli meats, regardless of the responsible microflora, usually with suspicion. The German Society for Hygiene and Microbiology (DGHM), e.g., recommends a maximum of 5×10^6 cfu g^{-1} [53]. In reality, however, many of the prepackaged sliced cold cuts display 10-100 times higher counts at the the end of their indicated shelf lives without being recognized as spoiled by sensory panels. On the other hand, unpleasant tastes and smells (not fresh, sour) are often associated with high LAB counts [54]. But a high count *per se* does not tell how long the product has been exposed to this high count already. Protective LAB cultures are looked at with suspicion because they have to be added in high numbers and, if metabolically too active, may reduce shelf life. Some authors generally view psychrotrophic LAB as spoilage organisms, regardless of their generally moderate role in spoilage [55]. There is no doubt that cold-cuts with protective cultures will differ from products without protective culture. But, as long as this difference is only manifested in a minor sour taste this kind of sensory deviation may be a reasonable price to pay for an increased food safety, especially with respect to *Li. monocytogenes*, without chemical preservatives and the control of more striking spoilage organisms, e.g. such as *Brochothrix thermosphacta*. Food preferences are changing, and presently many consumers tend to prefer products which are as much as possible free of chemical preservatives [8], processing aids and allergenic additives, and which are not overly treated by physical processes, such as heat, high pressure and irradiation. Nevertheless, many consumers also simply do not care, as long as the product is safe and affordable. Thus, protective cultures may be interesting for health and wellness-oriented consumers in countries with higher living standards. But less developed countries could also benefit, especially where cold-chain management is difficult and high-tech processing aids are not readily available. The challenge simply is to find the right LAB cultures for the particular product.

4. Sensory acceptance of bioprotective cultures on prepackaged cold cuts

As already mentioned, the application of bioprotective microbial cultures to prepackaged cold cuts is a much discussed innovative and sustainable technology for improving the microbiological safety and overall quality of these products. It could be an alternative to chemical preservatives or to a second pasteurisation step after packaging which both have a negative sensory impact. Although quite a number of lactic acid bacteria (LAB) have been suggested as protective cultures for sliced cooked meats, there is basically no information on consumer perception of products with added LAB. At the International Green Week Berlin 2010 the concept was introduced for the first time to a broader public and visitors were asked to participate in a sensory preference test [7].

Bologna-type sausages in 70 mm fiber casings were produced and stored at 2 °C until slicing. On the day of packaging the casings were removed and the sausages were briefely submersed in an aqueous suspension of a protective culture consisting of $Lb.$ $sakei$ strain Lb674 (sakacin P positive) and containing 8.5 $\log_{10}$ LAB ml^{-1}. Subsequently, the sausages were sliced, vacuum-packaged in polyethylene bags and kept refrigerated at 5°C until presentation to interested visitors. The consumers reacted predominantly positive on the possibility of safeguarding cold cuts with bioprotectants. Up to day 15 after packaging the inoculated samples reached a relative preference score (achieved points versus achievable points) of more than 45% (max. 60%) as compared to 60-70% for the freshly sliced samples without added LAB. Thereafter, the overall liking of the inoculated prepackaged sausage gradually decreased (Figure 3). The results indicate a potential market for more natural, microbiologically safe and sound cold cuts as a specialized segment of the convenience sector. As stated above, a mild acidic note may not be completely avoided when using protective cultures. But, this 'disadvantage' should be balanced against the risk of an uncontrolled growth of listeria on the one hand and the demand of many consumers for less chemical preservatives or thermal treatments on the other hand.

5. Probiotics

The steeply increasing business in the industrialised countries with health and wellness oriented foods in the 1990s, starting with probiotics in dairy products, has also raised interest in the development of probiotic meat products [56]. The concept of probiotics requires the intake of relevant amounts by the consumer of living probiotic microorgansims,

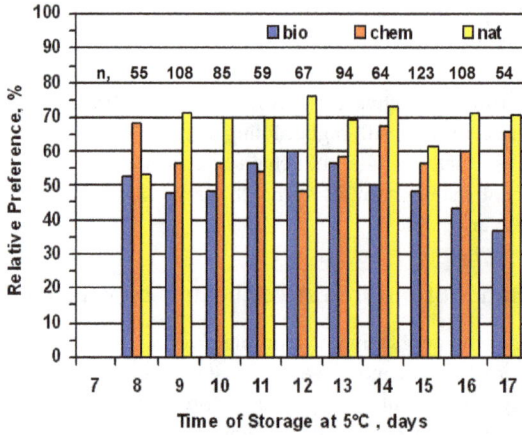

Figure 3. Consumer preference of vacuum-packaged Bologna-type sausage with *Lb. sakei* protective culture (bio) in comparison to non-packaged, sliced on-the-spot sausages without (nat) and with chemical (chem) preservatives presented at the International Green Week Berlin 2010. n, number of responses [7].

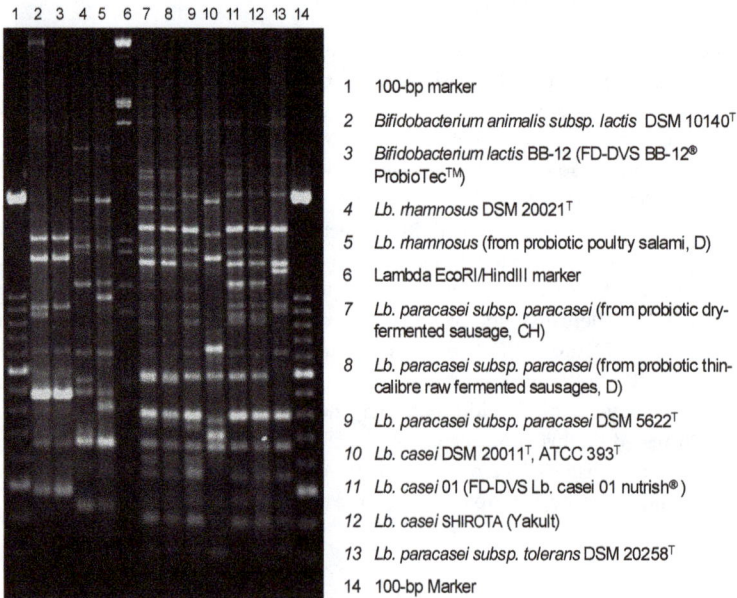

1   100-bp marker

2   *Bifidobacterium animalis subsp. lactis* DSM 10140^T

3   *Bifidobacterium lactis* BB-12 (FD-DVS BB-12® ProbioTec™)

4   *Lb. rhamnosus* DSM 20021^T

5   *Lb. rhamnosus* (from probiotic poultry salami, D)

6   Lambda EcoRI/HindIII marker

7   *Lb. paracasei subsp. paracasei* (from probiotic dry-fermented sausage, CH)

8   *Lb. paracasei subsp. paracasei* (from probiotic thin-calibre raw fermented sausages, D)

9   *Lb. paracasei subsp. paracasei* DSM 5622^T

10   *Lb. casei* DSM 20011^T, ATCC 393^T

11   *Lb. casei* 01 (FD-DVS Lb. casei 01 nutrish®)

12   *Lb. casei* SHIROTA (Yakult)

13   *Lb. paracasei subsp. tolerans* DSM 20258^T

14   100-bp Marker

Figure 4. Genetic fingerprints of probiotic LAB and related reference strains using BOX-PCR [57].

and raw fermented sausages were considered as an appropriate vehicle for these probiotics. However, these environments are quite different from the human gastrointestinal (GI) tract, and the strains under consideration have to cope with and survive in the presence of nitrite, sodium chloride, reduced pH and water activity, various processing steps and, eventually, long-term storage. Due to the manufacturing process raw fermented sausages contain high numbers of lactic acid bacteria which, however, are not regarded as probiotics. On the other hand, most of the known probiotic bacteria are unable to establish themselves in the raw sausage environment. Exceptions thereof are microbial cultures belonging to the *Lactobacillus plantarum group* and to the *Lactobacillus casei* group [57-59]. The use of protective and probiotic cultures may be a useful and effective strategy to prevent or reduce pathogens in the food chain, improve food safety and consumer health.

Within a project investigating the possibilities for manufacturing high quality and microbiologically sound products from meat of mother sheep, salami-type raw fermented sausages were produced with added conventional (*Lb. sakei, Lb. plantarum*) and probiotic lactic starter cultures (*Lb. paracasei*). The products were subjected to microbiological and sensory evaluation for up to nine months. All sausage batches with added cultures resulted in microbiological safe and sensory appealing products. The *Lb. sakei* culture survived during the whole storage period on a high level (> 10^8 cfu/g) while the two other cultures (*Lb. plantarum, Lb. paracasei*) partly reached the threshold of 10^6 cfu g^{-1} already after 3 months and were replaced by indigenous lactic acid bacteria of the *Lb. sakei / curvatus* group. For some batches, however, an acceptable number of probiotic bacteria could still be detected after nine months. Overall, *Lb. paracasei* showed a better survival in the ripened sausage than *Lb. plantarum* [7].

One problem for official authorities involved in consumer protection is to verify the presence of the indicated probiotics at sufficiently high levels. In the absence of simple and relyable identification procedures this may be a challenging task. In such cases genetic fingerprinting of isolates recovered on suitable agar media at relevant dilutions is the method of choice (Figure 4) [57, 60]. In the past, *Lb. rhamnosus* and *Lb. paracasei* ssp. *paracasei* have been used in fermented sausages, and labelling was quite confusing (Table 1). As can be seen, *Lb. paracasei* survived in relatively high numbers even in very dry salami. More recently, also other LAB species have been suggested as probiotics, and microencapsulation of strains has been used to overcome survival problems in the sausage environment. Still, human verification studies for probiotic administration are quite rare [61].

6. Functional starter cultures

In fermented sausage production classical starter cultures are usually also protective cultures, especially with respect to the acid-sensitive microflora. Modern cultures may provide additional protective action, e.g. by producing bacteriocins inhibitory to listeria and/or undesired LAB, or they may possess an additional probiotic functionality. Strains combining these traits have been termed also 'functional starter cultures' [62].

7. Bacteriocin production

Strains from many LAB species excrete anti-listerial bacteriocins, of which nisin produced by *Lc. lactis* and pediocin produced by *Pc. acidilactici* are the most wellknown. Besides,

Sausage no.	1	2	3	4
Type	soft, quickly ripened, smoked salami	very hard, air-dried salami	quickly ripened, thin-calibre, smoked sausage	smoked, dry-fermented salami with 30% weightloss
Characteristics	pH 4.7; a_w 0.954	pH 5.6	pH 4.9	n. d.
Origin	D	CH	D	D
Claims	'probiotic poultry salami with three probiotic cultures (*Bifidus, Lb. casei, Lb. acidophilus*)'	,probiotic', with beef and pork	'probiotic culture in high numbers (5×10^8 cfu/g)', with beef and pork	'Probiotic !!!, naturally ripened', with beef and vegetable fat (no lard)
advertised culture detectable	no	not labeled	yes	not labeled
detected potentially probiotic culture	*Lb. rhamnosus*	*Lb. paracasei* subsp. *paracasei*	*Lb. paracasei* subsp. *paracasei*	*Lb. paracasei* subsp. *paracasei*
Viable counts (cfu/g) of probiotic culture	$1\text{-}6 \times 10^7$	$4\text{-}9 \times 10^7$	3×10^7	1×10^7

n. d., not determined; D, Germany; CH, Switzerland.

Table 1. Detection of probiotic cultures in probiotic raw fermented sausages from retail [57].

bacteriocin-producing LAB with anti-listerial activity naturally occur on a wide range of ready-to-eat foods, including meats [63-65]. From a meat point of view the sakacins of *Lb. sakei* are the most interesting because of the high competitivity of this species in the meat environment [22,49]. The pediocin producer *Pc. acidilactici* is commonly used by the Spanish meat industry as a starter culture [66].

8. Hydrogen peroxide production

The demonstration of hydrogen peroxide formation by meat-borne lactic acid bacteria is of considerable importance for the characterization of individual strains, the selection of suitable starter and protective cultures for various applications for meat and meat products as well as for the search of potential microbiological causes for undesired sensory deviations (discolourations/'greening', rancidity). Many LAB are able to form hydrogen peroxide as a by-product of O_2-dependent metabolic pathways. Dependent on the environment, this trait may be desired or undesired [1,23,67,69,104].

In foods and feed it may contribute to the inhibition of an undesired accompanying microbiota [67]. The H_2O_2 formed by LAB acts bacteriostatic on GRAM-positive bacteria and bactericidal on Gram-negatives [12,68].

In a recent study a novel agar medium ('Prussian Blue' (PB) agar) was applied for the first time to lactic acid bacteria relevant to meat and meat products [69]. The PB agar detects H_2O_2 through the formation of Prussian Blue (Figure 5). It principally delivers similar results as the traditional manganese dioxide agar. However, it is more sensitive and, it is also more easily prepared and delivers results more quickly. A representative number of strains was used in the evaluation of the new medium (Table 2).

As to the production of H_2O_2, the study revealed large differences within the Lb. sakei/curvatus group. The bacteriocin producers frequently seemed to be relatively weak peroxide producers, while many commercial starter cultures were recognized as more or less strong peroxide producers. More recent field isolates of Lb. sakei/curvatus from prepackaged sliced Bologna-type sausage gave an essentially similar picture. In this case, however, only one of ten isolates of Lb. curvatus gave rise to a positive reaction.

Species	strains[a]	PB	MnO$_2$
Lb. sakei (Lb. bavaricus)	DSM 20494	-	nd
Lb. sakei ssp. carnosus (Lb. curvatus ssp. melibiosus)	DSM 15740	+	++
Lb. sakei ssp. carnosus	Lb1047	-	+
Lb. sakei ssp. carnosus	DSM 15831^T	++	++
Lb. sakei ssp. carnosus (ssp. sakei)	23K	+++	++
Lb. sakei ssp. carnosus	Lb790	+++	++
Lb. sakei ssp. sakei	DSM 20017^T	+++	+
Lb. brevis	DSM 20054^T	++	-
Lb. farciminis	DSM 20180^T	-	nd
Lb. hilgardii	DSM 20176^T	-	-
Weissella paramesenteroides	DSM 20288^T	+++	nd
Weissella minor	DSM 20014^T	+++	nd
Leuconostoc carnosum	Lb1259	+	+
Leuconostoc carnosum	Lb1054	++	+
Leuconostoc carnosum	Lb1045	++	+

[a] strains have been obtained from the German Collection of Microorganisms (DSM) and the strain collections of MRI Location Kulmbach (Lb) and INRA at Jouy-en-Josas (23K)

Table 2. Reaction of different LAB species on PB agar with BHI or MRS base, and on MnO$_2$ agar. -, no production of H_2O_2; +/++/+++, moderate to strong production of H_2O_2; nd, not determined [69].

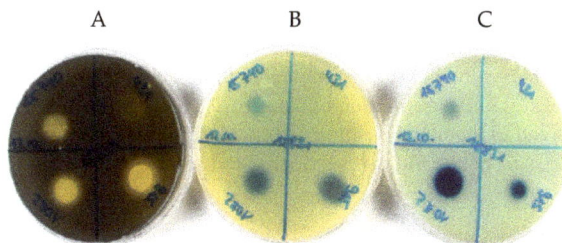

A   B   C

Figure 5. Reaction of different LAB species on MnO$_2$ agar (A) and on PB agar with MRS (B) or BHI (C) base. Production of H_2O_2 is indicated by bright and blue halos, resp. [69].

9. Formation of biogenic amines

Several LAB may produce biogenic amines by decarboxylation of amino acids, e.g. *Lb. buchneri, Lb. brevis, Lb. curvatus, Lb. hilgardii, Cb. maltaromaticum, Cb. divergens* [70]. Examples are such as tyramine and histamine during sausage fermentation. Strains of *Lb. plantarum, Lb. brevis* and *Lb. casei/paracasei,* and *Ec. faecium* and *Ec. faecalis* were identified as tyramine/histamine producers in the sausages [71]. Suitable starter cultures may contribute to reduction of biogenic amines in fermented sausages [72].

10. Identification of LAB

Identification of meat associated LAB is still wideley performed with phenotypic methods only, e.g API 50 CH [73]. These are, however, not always satisfying and may lead to misidentifications [74]. Nowadays, the application of PCR-DGGE and 16S rRNA gene sequencing allow the identification of a large number of strains in a quick and fast way [21,75]. Also various genomic fingerprinting methods are available. Nevertheless, conventional approaches remain important, especially when dealing with previously unknown species. Modern identification procedures rely on polyphasic approaches, integrating several lines of evidence to obtain a comprehensive description of a new species or of a microbiota [76].

11. Important LAB in meats

11.1. The Lb. sakei/curvatus cluster

In his 1983 review on lactic acid bacteria of meat and meat products EGAN mentions that according to recent findings of KANDLER and co-workers *Lb. sakei* (then *Lb. sake*) and *Lb. curvatus* were very common on German meat products [1]. Presently, two subspecies of *Lb. sakei* are known of which ssp. *carnosus* is the one characteristic for meats. It is common in fermented meat products, and is regularly found in vacuum-packaged meat and fermented plant material (sauerkraut). The subspecies *sakei* has been isolated from the Japanese sake starter and is regularly found in fermented meat products, vacuum-packaged meat, fermented plant material (sauerkraut), and human feces. The two subspecies can not be separated based on their physiological and biochemical characteristics [12]. The genomes of *Lb. sakei* 23K from a French dry-fermented sausage and *Lb. curvatus* CRL705 from an Argentinean artisanal fermented sausage have been sequenced [77,78]. Both genomes are highly similar. *Lb. curvatus* CRL705 lacks several genes present in *Lb. sakei* such as those related to fatty acid biosynthesis FASII, sucrose utilization, the arginine deiminase pathway, and citrate metabolism. The ones unique in *Lb. curvatus* CRL705 include genes for proteins and enzymes involved in the metabolism of carbohydrates, DNA, and fatty acids, as well as in the oxidative stress response and in bacteriocin production.

11.2. Lb. plantarum

The LAB species *Lb. plantarum* displays a high flexibility and versatility, and is able to colonize several ecological niches such as vegetables, meats, fish, milk substrates, and the human GI tract.

This is the basis of many applications in the food and health areas. As a starter culture for salamis *Lb. plantarum* is used since decades. More recently also probiotic strains have been described. With a size of 3.3 Mb its genome is one of the largest of LAB. A recent study on the phenetic and genetic diversity of the species revealed a high phenetic diversity which generally correlated with the origin of the isolates, e.g. from meat fermentations, kimchi, sourdough, egg plants and cheese. Four main clusters were determined: (i) meat, (ii) vegetable, (iii) sourdough, (iv) mixed sources with high meat content. On the genome level there were seven main clusters. The core genome contains more than 2000 genes, 121 genes being specific for *L. plantarum*. None of the strains could grow in milk, or at 4°C, or in the presence of 10% NaCl. A limited number grew at 17°C, or at 6% NaCl [79]. One of the earliest and most successful starter cultures for raw fermented sausages on the German market, "DuploFerment 66", contains a strain of *Lb. plantarum*. This is also the case for the "Saga II" starter from the US. In contrast to the first one, the latter strain does not grow at 10°C. Both strains are homofermentative for lactate and grow at 42°C but not at 8°C [25]. They provide rapid acidification of the raw sausage batter. On the other hand, *Lb. plantarum* is not very well adapted to meat and fails to maintain sufficiently high cell numbers to outcompete indigenous LAB. Sometimes it even does not grow in the meat batter [80,81]. In Italian natural fermented sausage the initial dominant populations of *Lb. plantarum* were accompanied by *Lb. sakei* and *Lb. curvatus* from the 10th day of fermentation and were finally competed out by the latter [21]. But, in certain traditional Greek fermented sausages *Lb. plantarum* and *Lb. plantarum/pentosus* may predominate [82,83].

11.3. Lb. brevis

In combination with *Pc. pentosaceus*, *Lb. brevis* has been used as an indigenous starter culture for a Vietnamese fermented meat product [84]. While *Lb. brevis* strongly acidifies the product, *Pc. pentosaceus* acts as a mild acidifier. The combination of both species resulted in a product with an intermediate taste (not too mild and not too sour) preferred by the sensory panel. Meat isolates of *Lb. brevis* may produce bacteriocins with antagonistic activity against *Li. monocytogenes* [85].

11.4. Lb. versmoldensis

This species was first reported in 2003 as the dominant LAB in some German raw fermented poultry salamis. The species was present in high numbers and frequently dominated the lactic acid bacteria (LAB) populations of the products [86]. Later, the species has been isolated also from Scandinavian fermented meats, Egyptian Domiati cheese and Japanese traditional fermented fish products [87-89]. There are no studies to date on the general behaviour of this species in meat ecosystems. The genome of strain KCTC 3814, an isolate from poultry salami, has been recently sequenced by the Korea Research Institute of Bioscience & Biotechnology [90].

11.5. Carnobacteria

Carnobacteria are non-aciduric and, therefore, are preferentially isolated from meats with elevated pH. *Cb. divergens* and *Cb. maltaromaticum* frequently constitute a major component of the microflora of packaged raw meats as well as of refrigerated, prepackaged, sliced cooked deli meats. Meat spoilage by *Cb. maltaromaticum* has been associated with "dairy", "spoiled-meat", and "mozarella cheese" perception [31,91,92]. The major volatiles on meat, acetoin, 1-octen-3-ol and butanoic acid, are volatile organic compounds with low sensory impacts. Butanoic acid in stored beef was also associated with *Cb. divergens*. It has a rancid cheese-like odor and can derive from leucine metabolism, microbial consumption of free amino acids via the Stickland reaction or from tributyrin hydrolysis.

The metabolites from leucine degradation are involved in dry fermented sausage aroma. The catabolism of leucine by a strain of *Cb. maltaromaticum* was studied directly in the growth medium with H-3-labelled leucine to investigate the effect of five parameters: phase of growth, pH, oxygen, glucose and alpha-ketoisocaproic acid. Leucine catabolism was most important during the exponential phase of growth. The addition of alpha-ketoisocaproic acid at 1%, glucose at levels of 0.5% to 2% and shaking of the growth medium increased leucine catabolism. At pH 5.4 and 7.2, the main metabolites detected were 3-methyl butanal, 3-methyl butanol and alpha-ketoisocaproic acid. At pH 6.5, the leucine catabolism was maximum and was characterised by a high production of 3-methyl butanoic acid [93].

Positive and negative effects of carnobacteria in the environment and in foods have recently been reviewed [94]. Because *Cb. divergens* and *Cb. maltaromaticum* show good growth in refrigerated meats and some of the strains produce potent anti-listerial bacteriocins, they may have some role as bioprotectants in meat environments. However, carnobacteria are associated with unpleasant spoilage metabolites in meats, such as acetic and butanoic acid as well as gas production in vacuum packed beef. An undesirable trait is also their ability to produce the biogenic amine tyramine from tyrosine. Carnobacteria are not regarded as human pathogens, but *Cb. maltaromaticum* is a well known fish pathogen and catagorised as a safety-level-2 microorganism. The genome of *Cb. maltaromaticum* ATCC 35586 carries putative virulence genes which probably play a role in fish pathogenesis [95]. Since carnobacteria are inhibited by acetate they do not grow well on routine LAB media such as MRS. A selective enumeration medium using a combination of three antibiotics (gentamicin, nalidixic acid, vancomycin) and an alkaline pH value (8.8) has recently been proposed for *Cb. maltaromaticum* from cheese [96].

11.6. Leuconostoc

Leuc. gelidum is a major spoilage organism in Finnish fresh meats [97]. Certain strains of *Leuc. gelidum* may produce yellow discolourations on prepackaged refrigerated German 'Weisswurst' and cold cuts (Figure 6, 7) [98]. Recently, the genome of a plant isolate of *Leuc. gelidum* has been sequenced [99].

The responsible pigment for the intensive 'neon-like' yellow discolouration is a bacterial carotenoid, the non-polar C30-carotenoid 4,4'-di-apo-7,8,11,12-tetra-hydro-lycopene. On fat-containing substrates this compound does not only stain the bacterial cells but also the substrate and, in the case of 'Weisswurst' does stain the natural casing (porc intestine) of the sausage as well as the sausage surface beneath. This triterpenoid is an intermediate in the microbial synthesis of 4,4'-diaponeurosporene which represents the main carotenoid in pigmented enterococci, Leuc. citreum and Lb. plantarum. Identification of the pigment was achieved by using UV-VIS spectroscopy in combination with available data from literature [100].

A report from Canada also described the yellow discolouration phenomenon on cooked sliced meats which had been stored for an extended time period under refrigeration [101]. These authors, employees of a big Canadian food company (then Canada Packers Inc.), tentatively identified an *Enterococcus* sp. as the causative agent.

Figure 6. Yellow discolourations on prepackaged refrigerated German 'Weisswurst' after targeted inoculation with *Leuc. gelidum* and incubation at 5°C for 14 days [98].

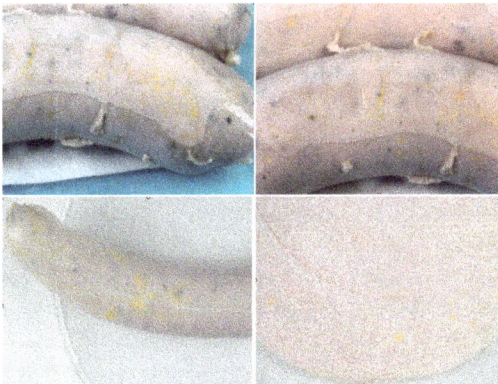

Figure 7. Yellow discolourations on pre-packaged meat products produced by *Leuc. gelidum*. A and B, 'Weisswurst' from organic production; C, grill sausage from conventional production; D, sliced cooked turkey breast from conventional production [98].

Leuc. gasicomitatum has been recognized as a specific spoilage organism in cold-stored Finnish MAP meats. It emerged as a spoilage problem of tomato-marinated, raw broiler meat strips. Due to CO_2 production the packages already showed clear bulging more than a week before the expected shelf life [102]. It is a psychrotrophic species and, because of its dominance in marinated meats and fish as well as in vegetable sausages, probably of plant origin. But, it was also detected in minced meat and high-oxygen modified-atmosphere packaged raw, beef steaks injected with sugar-salt solutions, so-called moisture-enhanced or value-added meats [97,103]. Recently, the genome of the type strain *Leuc. gasicomitatum* LMG 18811[T] has been sequenced [55].

11.7. Weissella

Weissella spp. are heterofermenters producing CO_2, ethanol and/or acetate from glucose. The species *Ws. viridescens*, *Ws. halotolerans* and *Ws. hellenica* have been associated with meat and meat products. *Ws. viridescens* is considered as heat resistant and may cause green discolouration in cured meats [104]. This species is frequently isolated from refrigerated sliced cooked meats [43] and was reported to produce cavities in the muscles of hams after cooking [105].

11.8. Pediococcus

The homofermentative pediococci are mostly applied for rapid and strong acidification at elevated temperature, especially in US summer sausage fermentation. Usually *Pc. acidilactici* and *Pc. pentosaceus* are the species involved. *Pediococcus* sp. are among the most common starter cultures in the US [11,21]. A pediocin producing *Pc. acidilactici* is also commonly used by the Spanish meat industry as a starter culture [66].

11.9. Enterococcus

In mediterranian traditional dry-fermented sausages enterococci are found in relevant numbers and are believed to contribute to the characterisic product flavor. *Ec. faecalis*, e.g., is common in Portuguese 'alheira' [19].

On the other hand, the presence of enterococci in foods is debatable, since some strains carry antibiotic resistances and virulence determinants relevant in human medicine [22,106]. Also, *Ec. faecium* and *Ec. faecalis* were identified as tyramine/histamine producers in the sausages [71]. The use of *Ec. faecium* strains has been suggested to control the growth of undesirable microorganisms such as listeria on material and environmental surfaces in meat plants [107].

12. Outlook

Meat and meat products provide a concentrated source of protein of high biological value and can make a valuable contribution to human diets. However, they are also highly perishable commodities which rapidly spoil and may even allow the growth of food-borne pathogenic microorganisms if no suitable preservative actions are taken. Meat fermentation involving beneficial LAB has become an important and sustainable preservation technology, and today a number of suitable species and strains are successfully applied as starter and protective cultures in various fermented meats all over the world. These cultures not only prevent the growth of common food pathogens but also of undesirable food spoilage bacteria, including heterofermentative LAB. The answer to the question which strains we should use for which products largely depends on consumer expectations and technological needs. Much has been learned over the years, however, we are still far from understanding the complex metabolic interactions of LAB in meats.

Systems biology has become an important approach in LAB microbiology and will become even stronger in the future [108]. It links quantitative microbial physiology with population dynamic modelling and ecological theories. In comparative systems biology of LAB, the so-called "omics"-techniques ("genomics", "proteomics", "transcriptomics", "metabolomics") and mathematical and statistical methods are of crucial importance [109, 110]. Comparative analyses between various species is expected to deliver understandable models of the metabolism of these species. Whole genome sequencing has made a quantum leap in the past few years and it is likely that very soon all genomes of meat associated LAB species and even of different strains will be available for comparative studies. Diversity and differences within each of the species at the strain level will have to be considered. The ripening, packaging and storage of meats could benefit from improved systems knowledge of the diverse meat microcosms with respect to microbial survival and growth, as well as desired and unwanted microbial transformations of meat components to ensure high-quality, healthy, safe and tasty products. The beneficial aspects of LAB in meat preservation could be explored using systems techniques and will decrease our dependence on chemical preservatives. Likewise, the impact of microbes on meat spoilage could be better managed with a systems understanding of the interplay of microbes, raw materials, additives and processing technologies.

In a global perspective, the role of starter and protective cultures for the safety and quality of meats is expected to increase. Although the chemical preservatives currently applied to prevent the growth of pathogens and spoilage bacteria in deli meats perfectly serve this purpose, there is an increasing consumer demand for more natural products. This is in part reflected by the so-called clean label strategies of the big manufacturers. Many chemical additives not only contribute to the sodium burden of the meats, but also leave an undesirable numb mouthfeel which negatively effects the sensory perception of the meat aroma. Innovations in fermented meat production will benefit from an improved knowledge of systems microbiology of LAB in the various meat environments on the one hand, and the gastrointestinal environment on the other. A future challenge will be to link intraspecies diversity to a specific sensory profile [21]. The application of probiotic starter microorganisms in dry-fermented sausages remains appealing for the wellness-oriented consumers even if immediate health claims should be difficult to establish. In this sense beneficial LAB will vitally contribute to a sustainable and diversified food production.

Author details

Lothar Kröckel
Max Rubner-Institute Location Kulmbach, Department of Safety and Quality of Meat, Kulmbach, Germany

13. References

[1] Egan AF (1983) Lactic Acid Bacteria of Meat and Meat Products. Antonie van Leeuwenhoek 49: 327-336.

[2] Björkroth J, Ridell J, Korkeala H (1996) Characterization of *Lactobacillus sake* Strains Associated with Production of Ropy Slime by Randomly Amplified Polymorphic DNA (RAPD) and Pulsed-Field Gel Electrophoresis (PFGE) patterns. Int. j. food microbiol. 31: 59–68.

[3] Marshall VM (1987) Lactic Acid Bacteria: Starters for Flavour. FEMS microbiol. lett. 46: 327-336.

[4] Caplice E, Fitzgerald GF (1999) Food Fermentations: Role of Microorganisms in Food Production and Preservation. Int. j. food microbiol. 50: 131-149.

[5] Lücke FK (2000) Utilization of Microbes to Process and Preserve Meat. Meat sci. 56: 105-115.

[6] Schlafmann K, Meusburger AP, Hammes WP, Braun C, Fischer A, Hertel C (2002) Starter Cultures to Improve the Quality of Raw Ham. Fleischwirtschaft 82 (11): 108-114.

[7] Kröckel L, Dederer I, Troeger K (2011) Starter and Protective Cultures for Meat Products. Fleischwirtschaft 91 (3): 93-98.

[8] McIntyre L, Hudson JA, Billington C, Withers H (2012) Biocontrol of Foodborne Bacteria. In: McElhatton A, Sobral PJA, editors. Novel Technologies in Food Science – Integrating Food Science and Engineering Knowledge into the Food Chain 7. Springer Science+Business Media. pp. 183-204.

[9] Katla T, Moretro T, Sveen I, Aasen M, Axelsson L, Rorvik LM, Naterstad K (2002) Inhibition of *Listeria monocytogenes* in Chicken Cold Cuts by Addition of Sakacin P and Sakacin P-Producing *Lactobacillus sakei*. J. appl. microbiol. 93: 191-196.

[10] Vermeiren L, Devlieghere F, Vandekinderen I, Rajtak U, Debevere J (2006) The Sensory Acceptability of Cooked Meat Products Treated with a Protective Culture Depends on Glucose Content and Buffering Capacity: A Case Study with *Lactobacillus sakei* 10A. Meat sci. 74: 532-545.

[11] Hammes WP, Knauf HJ (1994) Starters in the Processing of Meat Products. Meat sci. 36: 155-168.

[12] Hammes WP, Hertel C (2009) Genus I. *Lactobacillus* Beijerinck 1901, 212AL. In: De Vos P, Garrity GM, Jones D, Krieg NR, Ludwig W, Rainey FA, Schleifer KH, Whitman WB, editors. The Firmicutes. Bergey's Manual of Systematic Bacteriology, 2nd Ed., Vol. 3. Springer, Dordrecht, Heidelberg London, New York; pp. 465- 511.

[13] Urso R, Comi G, Cocolin L (2006) Ecology of Lactic Acid Bacteria in Italian Fermented Sausages: Isolation, Identification and Molecular Characterization. Syst. appl. microbiol. 29: 671-680.

[14] Cocolin L, Dolci P, Rantsiou K, Urso R, Cantoni C, Comi G (2009) Lactic Acid Bacteria Ecology of Three Traditional Fermented Sausages Produced in the North of Italy as Determined by Molecular Methods. Meat sci. 82: 125-132.

[15] Cocolin L, Dolci P, Rantsiou K (2011) Biodiversity and Dynamics of Meat Fermentations: The Contribution of Molecular Methods for a Better Comprehension of a Complex Ecosystem. Meat sci. 89: 296-302.

[16] Hugas M, Garriga M, Aymerich T, Monfort JM (1993) Biochemical Characterization of Lactobacilli from Dry Fermented Sausages. Int. j. food microbiol. 18: 107-113.

[17] Santos EM, González-Fernández C, Jaime I, Rovira J (1998) Comparative Study of Lactic Acid Bacteria House Flora Isolated in Different Varieties of `Chorizo'. Int. j. food microbiol. 39: 123-128.

[18] Samelis J, Maurogenakis F, Metaxopoulos J (1994) Characterisation of Lactic Acid Bacteria Isolated from Naturally Fermented Greek Dry Salami. Int. j. food microbiol. 23: 179-196.

[19] Albano H, van Reenen CA, Todorov SD, Cruz D, Fraga L, Hogg T, Dicks LMT, Teixeira P (2009) Phenotypic and Genetic Heterogeneity of Lactic Acid Bacteria Isolated from "Alheira", a Traditional Fermented Sausage Produced in Portugal. Meat sci. 82: 389-398.

[20] Adams MR, Nicolaides L (1997) Review of the Sensitivity of Different Foodborne Pathogens to Fermentation. Food control 8: 227-239.

[21] Cocolin L, Rantsiou K (2012) Meat Fermentation. In: Hui YH, editor. Handbook of Meat and Meat Processing, Second Edition. CRC Press, pp. 557-572.

[22] Fontana C, Fadda S, Cocconcelli PS, Vignolo G (2012) Lactic Acid Bacteria in Meat Fermentations. In: Lahtinen S, Ouwehand AC, Salminen S, von Wright A, editors. Lactic Acid Bacteria – Microbiological and Functional Aspects, 4th Ed. CRC Press, Taylor & Francis, Boca Raton, London, New York. pp. 247-264.

[23] Buckenhüskes HJ (1993) Selection Criteria for Lactic Acid Bacteria to be Used as Starter Cultures for Various Food Commodities. FEMS microbiol. rev. 12: 253-271.

[24] Dodds KL, Collins-Thompson DL (1984) Nitrite Tolerance and Nitrite Reduction in Lactic Acid Bacteria Associated with Cured Meat Products. Int. j. food microbiol. 1: 163-170.

[25] Nordal J, Slinde E (1980) Characteristics of Some Lactic Acid Bacteria Used as Starter Cultures in Dry Sausage Production. Appl. environ. microbiol. 40: 472-475.

[26] Nissen H, Holck A (1998) Survival of *Escherichia coli* O157:H7, *Listeria monocytogenes* and *Salmonella kentucky* in Norwegian Fermented, Dry Sausage. Food microbiol. 15: 273–279.

[27] Leistner L (2000) Basic Aspects of Food Preservation by Hurdle Technology. Int. j. food microbiol. 55:181–186.

[28] Lindqvist R, Lindblad M (2009) Inactivation of *Escherichia coli*, *Listeria monocytogenes* and *Yersinia enterocolitica* in Fermented Sausages During Maturation/Storage. Int. j. food microbiol. 129: 59-67.

[29] Heir E, Holck AL, Omer MK, Alvseike O, Hoy M, Mage I, Axelsson L (2010) Reduction of Verotoxigenic *Escherichia coli* by Process and Recipe Optimisation in Dry-Fermented Sausages. Int. j. food microbiol. 141: 195–202.

[30] Erkes M (2011) Einsatz von Kulturen bei Rohpökelwaren. Fleischwirtschaft 91 (11): 39-43.

[31] Jones RJ (2004) Observations on the Succession dynamics of Lactic Acid Bacteria Populations in Chill-Stored Vacuum-Packaged Beef. Int. J. food microbiol. 90: 273– 282.

[32] Schillinger U, Lücke FK (1987) Lactic Acid Bacteria on Vacuum-Packaged Meat and their Influence on Shelf Life. Fleischwirtschaft 67: 1244-1248.

[33] Castellano P. Gonzalez C, Carduza F, Vignolo G (2010) Protective Action of *Lactobacillus curvatus* CRL705 on Vacuum-Packaged Raw Beef. Effect on Sensory and Structural Characteristics. Meat sci. 85: 394-401.

[34] Castellano P, Belfiore C, Vignolo G (2011) Combination of Bioprotective Cultures with EDTA to Reduce *Escherichia coli* O157:H7 in Frozen Ground-Beef Patties. Food control 22: 1461-1465.

[35] Fadda S, Lopez C, Vignolo G (2010) Role of Lactic Acid Bacteria During Meat Conditioning and Fermentation: Peptides Generated as Sensorial and Hygienic Biomarkers. Meat sci. 86: 66-79.

[36] Borch E, Kant-Muermans ML, Blixt Y (1996) Bacterial Spoilage of Meat and Cured Meat Products. Int. j. food microbiol. 33: 103–120.

[37] Korkeala HJ, Bjørkroth KJ (1997) Microbiological Spoilage and Contamination of Vacuum-Packaged Cooked Sausages. J. food prot. 60: 724–731.

[38] Nattress FM, Jeremiah LE (2000) Bacterial Mediated Off-Flavours in Retail-Ready Beef after Storage in Controlled Atmospheres. Food res. int. 33: 743-748

[39] Björkroth KJ, Vandamme P, Korkeala HJ (1998) Identification and Characterization of *Leuconostoc carnosum*, Associated with Production and Spoilage of Vacuum-Packaged, Sliced, Cooked Ham. Appl. environ. microbiol. 64: 3313-3319.

[40] Kröckel L (1998) Lactic Acid Bacteria as Protective Cultures in the Preservation of Meat. In: Adria-Normandie, editor. Les Bactérie Lactic – Quelles Souches? Pour quels Produits? Lactic Acid Bacteria – Which Strains for which Products? – Actes du colloque LACTIC 97, Caen, 10-12 Sept 1997 – Adria Normandie, Villers-Bocage, pp. 229-242.

[41] Laursen BG, Bay L, Cleenwerck I, Vancanneyt M, Swings J, Dalgaard P, Leisner JJ (2005) *Carnobacterium divergens* and *Carnobacterium maltaromaticum* as Spoilers or Protective Cultures in Meat and Seafood: Phenotypic and Genotypic Characterization. Syst. appl. microbiol. 28: 151-64.

[42] Lücke FK, Raabe C, Hampshire J (2007) Changes in Sensory Profile and Microbiological Quality During Chill Storage of Cured and Uncured Cooked Sliced Emulsion-Type Sausages. Arch. Lebensmittelhyg. 58: 57-63.

[43] Kröckel L (2008) Mikrobiologische Qualität Vorverpackter Aufschnittware - Aktuelle Untersuchungen von Erhitztem, Vorverpacktem Brühwurst- und Kochschinkenaufschnitt auf *Listeria monocytogenes*. Fleischwirtschaft 88 (11): 112-116.

[44] Leistner L (1995) Principles and Applications of Hurdle Technology. In: Gould GW, editor. New Methods of Food Preservation. Springer; pp. 1–21.

[45] Leistner L, Gorris LG (1995) Food Preservation by Hurdle Technology. Trends food sci technol 6: 41-46.

[46] Hui YH (2012) Hazard Analysis and Critical Control Point System. In: Hui YH, editor. Handbook of Meat and Meat Processing, 2nd Ed. CRC Press, pp. 741-767.

[47] Jacobsen T, Budde BB, Koch, AG (2003) Application of *Leuconostoc carnosum* for Biopreservation of Cooked Meat Products. J. appl. microbiol. 95: 242–249.

[48] Vermeiren L, Devlieghere F, Debevere J (2004) Evaluation of Meat Born Lactic Acid Bacteria as Protective Cultures for the Biopreservation of Cooked Meat Products. Int. j. food microbiol. 96: 149–164.

[49] Jones RJ, Wiklund E, Zagorec M, Tagg JR (2010) Evaluation of Stored Lamb Biopreserved Using a Three-Strain Cocktail of *Lactobacillus sakei*. Meat sci. 86: 955-959

[50] Jones RJ, Zagorec M, Brightwell G, Tagg JR (2009) Inhibition by *Lactobacillus sakei* of Other Species in the Flora of Vacuum Packaged Raw Meats During Prolonged Storage. Food microbiol. 26: 876-881.

[51] Ananou S, Maqueda M, Martínez-Bueno M, Valdivia E (2007) Biopreservation, an Ecological Approach to Improve the Safety and Shelf-Life of Foods. In: Méndez-Vilas A, editor. Communicating Current Research and Educational Topics and Trends in Applied Microbiology, pp. 475-486. Available: http://www.formatex.org/ microbio/pdf/Pages475-486.pdf. Accessed 2012 Apr 3.

[52] www (2012) Ingredients with Potential: Chr. Hansen A/S - Bactoferm Rubis. Available: http://www.safoodcentre.com.au/__data/assets/pdf_file/0015/131451/ innova_ingredients_with_potential.pdf. Accessed: 2012 Apr 4

[53] Anonymous (2011) Veröffentlichte Mikrobiologische Richt- und Warnwerte zur Beurteilung von Lebensmitteln (Stand: November 2011) Eine Empfehlung der Fachgruppe Lebensmittelmikrobiologie und -hygiene der Deutschen Gesellschaft für Hygiene und Mikrobiologie (DGHM); http://www.dghm.org/m_275. Accessed 2012 Apr 3.

[54] Kröckel L (2010) Mikrobiologisch-Genetische Ressourcen bei Fleisch - Biodiversität und Nachhaltige Nutzung bei der Herstellung von Fleisch-Erzeugnissen. Forschungsreport Ernährung, Landwirtschaft, Verbraucherschutz 1/2010: 24-26.

[55] Johansson P, Paulin L, Säde E, Salovuori N, Alatalo ER, Bjørkroth KJ, Auvinen P (2011) Genome Sequence of a Food Spoilage Lactic Acid Bacterium, *Leuconostoc gasicomitatum* LMG 18811(T), in Association with Specific Spoilage Reactions. Appl. environ. microbiol. 77: 4344-4351.

[56] Hugas M, Monfort JM (1997) Bacterial Starter Cultures for Meat Fermentation. Food chemistry 59: 547-554.

[57] Kröckel L (2006a) Use of Probiotic Bacteria in Meat Products. Fleischwirtschaft 86:109–13.

[58] Ammor MS, Mayo B (2007) Selection Criteria for Lactic Acid Bacteria to be used as Functional Starter Cultures in Dry Sausage Production: an Update. Meat sci. 76: 138-146.

[59] De Vuyst L, Falony G, Leroy F (2008) Probiotics in Fermented Sausages. Meat sci. 80: 75-78.

[60] Huang CH, Lee FL (2009) Development of Novel Species-Specific Primers for Species Identification of the *Lactobacillus casei* Group Based on RAPD Fingerprints. J. sci. food agric. 89: 1831-1837.

[61] Khan MI, Arshad MS, Anjum FM, Sameen A, Aneeq-ur-Rehman, Gill WT (2011) Meat as a Functional Food with Special Reference to Probiotic Sausages. Food res. int. 44: 3125-3133.

[62] Leroy F, Verluyten J, De Vuyst L (2006) Functional Meat Starter Cultures for Improved Sausage Fermentation. Int. j. food microbiol. 106: 270-285.

[63] Abee T, Kröckel L, Hill C (1995) Bacteriocins: Modes of Action and Potentials in Food Preservation and Control of Food Poisoning. Int. j. food microbiol. 28:169-185.

[64] Kelly WJ, Asmundson RV, Huang CM (1996) Isolation and Characterization of Bacteriocin-Producing Lactic Acid Bacteria from Ready-to-Eat Food Products. Int. j. food microbiol. 33: 209-218.

[65] Hugas M (1998) Bacteriocinogenic Lactic Acid Bacteria for the Biopreservation of Meat and Meat Products. Meat sci. 49: S139-S150.

[66] Nieto-Lozano JC, Reguera-Useros JI, Peláez-Martínez MC, Hardisson de la Torre A (2002) Bacteriocinogenic Activity from Starter Cultures Used in Spanish meat Industry. Meat sci. 62: 237-243.

[67] Lücke FK, Popp J, Kreutzer R (1986) Bildung von Wasserstoffperoxid durch Laktobazillen aus Rohwurst und Brühwurstaufschnitt. Chem. mikrobiol. technol. lebensm. 10: 78-81.

[68] Condon S. (1987) Responses of lactic acid bacteria to oxygen. FEMS Microbiol. rev. 46, 269-280.

[69] Kröckel L (2011) Evaluation of a Novel Agar Medium for the Detection of Hydrogen Peroxide Producing Lactic Acid Bacteria. Fleischwirtschaft 91 (10): 97-101.

[70] Silla-Santos MH (1996) Biogenic Amines: their Importance in Foods. Int. j. food microbiol. 29: 213–231.

[71] Komprda T, Sládková P, Petirová E, Dohnal V, Burdychová R (2010) Tyrosine- and Histidine-Decarboxylase Positive Lactic Acid Bacteria and Enterococci in Dry Fermented Sausages. Meat sci. 86: 870-877.

[72] Lu S, Xu X, Zhou G, Zhu Z, Meng Y, Sun Y (2010) Effect of Starter Cultures on Microbial Ecosystem and Biogenic Amines in Fermented Sausage. Food control 21: 444-449.

[73] Samappito W, Leenanon B, Levin RE (2011) Microbiological Characteristics of "Mhom", a Thai Traditional Meat Sausage. The Open Food Science Journal 5: 31-36.

[74] Sohier D, Coulon J, Lonvaud-Funel A (1999) Molecular Identification of *Lactobacillus hilgardii* and Genetic Relatedness with *Lactobacillus brevis*. Int. j. syst. bacteriol. 49: 1075-1081.

[75] Rantsiou K, Cocolin L (2006) New Developments in the Study of the Microbiota of Naturally Fermented Sausages as Determined by Molecular Methods: A Review. Int. j. food microbiol. 108: 255-267.

[76] Vandamme P, Pot B, Gillis M, de Vos P, Kersters K, Swings J (1996) Polyphasic taxonomy, a consensus approach to bacterial systematics. Microbiol rev. 60: 407-38.

[77] Chaillou S, Champomier-Verges MC, Cornet M, Crutz-Le Coq AM, Dudez AM, Martin V, Beaufils S, Darbon-Rongere E, Bossy R, Loux V, Zagorec M (2005) The Complete Genome Sequence of the Meat-Borne Lactic Acid Bacterium *Lactobacillus sakei* 23K. Nat. biotechnol. 23: 1527-1533.

[78] Hebert EM, Saavedra L, Taranto MP, Mozzi F, Magni C, Nader MEF, Font de Valdez G, Sesma F, Vignolo G, Rayaa RR (2012) Genome Sequence of the Bacteriocin-Producing *Lactobacillus curvatus* Strain CRL705. J. bacteriol. 194: 538-539.

[79] Siezen RJ, Tzeneva VA, Castioni A, Wels M, Phan HTK, Rademaker JLW, Starrenburg MJC, Kleerebezem M, Molenaar D, van Hylckama Vlieg JET (2010) Phenotypic and Genomic Diversity of *Lactobacillus plantarum* Strains Isolated from Various Environmental Niches. Environ. microbiol. 12: 758-773.

[80] Marchesini B, Bruttin A, Romailler N, Moreton RS, Stucchi C, Sozzi T (1992) Microbiological Events During Commercial Meat Fermentations. J. appl. bacteriol. 73: 203-209.

[81] Kröckel L (unpublished observations)

[82] Drosinos EH, Mataragas M, Xiraphi N, Moschonas G, Gaitis F, Metaxopoulos J (2005) Characterization of the Microbial Flora from a Traditional Greek Fermented Sausage. Meat sci. 69: 307-317.

[83] Drosinos EH, Paramithiotis S, Kolovos G, Tsikouras I, Metaxopoulos I (2007) Phenotypic and Technological Diversity of Lactic Acid Bacteria and Staphylococci Isolated from Traditionally Fermented Sausages in Southern Greece. Food microbiol. 24: 260-270.

[84] Ho TNT, Nguyen NT, Deschamps A, Hadj Sassi A, Urdaci M, Caubet R (2009) The Impact of *Lactobacillus brevis* and *Pediococcus pentosaceus* on the Sensorial Quality of "Nem Chua" – a Vietnamese Fermented Meat Product. Int. food res. j. 16: 71-81.

[85] Coventry MJ, Wan J, Gordon JB, Mawson RF, Hickey MW (1996) Production of Brevicin 286 by *Lactobacillus brevis* VB286 and Partial Characterization. J. appl. bacteriol. 80: 91-8.

[86] Kröckel L, Schillinger U, Franz CMAP, Bantleon A, Ludwig W (2003) *Lactobacillus versmoldensis* sp. nov., isolated from raw fermented sausage. Int. j. syst. env. microbiol. 53: 513-517.

[87] Klingberg TD, Axelsson L, Naterstad K, Elsser D, Budde BB (2005) Identification of Potential Probiotic Starter Cultures for Scandinavian-Type Fermented Sausages. Int. j. food microbiol. 105: 419-431.

[88] El-Baradei G, Delacroix-Buchet A, Ogier JC (2007) Biodiversity of Bacterial Ecosystems in Traditional Egyptian Domiati Cheese. Appl. environ. microbiol. 73: 1248-1255.

[89] An C, Takahashi H, Kimura B, Kuda T (2010) Comparison of PCR-DGGE and PCR-SSCP Analysis for Bacterial Flora of Japanese Traditional Fermented Fish Products, Aji-Narezushi and Iwashi-Nukazuke. J. sci. food agric. 90: 1796-1801.

[90] Kim DS, Choi SH, Kim DW, Kim RN, Nam SH, Kang A, Kim A, Park HS (2011a) Genome Sequence of *Lactobacillus versmoldensis* KCTC 3814. J. bacteriol. 193: 5589-5590.

[91] Casaburi A, Nasi A, Ferrocino I, Di Monaco R, Mauriello G, Villani F, Ercolini D (2011) Spoilage-Related Activity of *Carnobacterium maltaromaticum* Strains in Air-Stored and Vacuum-Packed Meat. Appl. environ. microbiol. 77: 7382-7393.

[92] Ercolini D, Ferrocino I, Nasi A, Ndagijimana M, Vernocchi P, La Storia A, Laghi L, Mauriello G, Guerzoni ME, Villani F (2011) Monitoring of Microbial Metabolites and Bacterial Diversity in Beef Stored Under Different Packaging Conditions. Appl. environ. microbiol. 77: 7372-7381.

[93] Larrouture-Thiveyrat C, Montel MC (2003) Effects of environmental factors on leucine catabolism by *Carnobacterium piscicola*. Int. j. food microbial. 81: 177–184.

[94] Leisner JJ, Groth Laursen B, Prevost H, Drider D, Dalgaard P (2007) *Carnobacterium*: Positive and Negative Effects in the Environment and in Foods. FEMS microbiol. rev. 31: 592–613.

[95] Leisner JJ, Hansen MA, Larsen MH, Hansen L, Ingmera H, Sørensen SJ (2012) The Genome Sequence of the Lactic Acid Bacterium *Carnobacterium maltaromaticum* ATCC 35586 Encodes Potential Virulence Factors. Int J. food microbiol. 152: 107–115.

[96] Edima HC, Cailliez-Grimal C, Revol-Junelles AM, Tonti L, Linder M, Millière JB (2007) A Selective Enumeration Medium for *Carnobacterium maltaromaticum*. J. microbiol. meth. 68: 516-21.

[97] Vihavainen E, Björkroth J (2007) Spoilage of Value-Added, High-Oxygen Modified-Atmosphere Packaged Raw, Beef Steaks by *Leuconostoc gasicomitatum* and *Leuconostoc gelidum*. Int. j. food microbiol. 119: 340–345.

[98] Kröckel L (2006b) Yellow Discolourations of Prepackaged Refrigerated German Weisswurst are Due to *Leuconostoc gelidum*. Fleischwirtschaft 86 (9): 129-133.

[99] Kim DS, Choi SH, Kim DW, Kim RN, Nam SH, Kang A, Kim A, Park HS (2011b) Genome Sequence of *Leuconostoc gelidum* KCTC 3527, Isolated from Kimchi. J. bacteriol. 193: 799–800.

[100] Kröckel L (2007) Ein Bakterielles Carotinoid Färbt Vorverpackte, Kühl Gelagerte Weißwurst Gelb. Mitteilungsblatt der Fleischforschung Kulmbach 46, Nr. 178: 223-230.

[101] Whiteley AM, D'Souza MD 1989. A Yellow Discoloration of Cooked Cured Meat Products - isolation and characterization of the causative organism. J. food protect. 52: 392– 395.

[102] Björkroth KJ, Geisen R, Schillinger U, Weiss N, De Vos P, Holzapfel WH, Korkeala HJ, Vandamme P (2000). Characterization of *Leuconostoc gasicomitatum* sp. nov., Associated with Spoiled Raw Tomato-Marinated Broiler Meat Strips Packaged Under Modified-Atmosphere Conditions. Appl. environ. microbiol. 66: 3764–3772.

[103] Nieminen TT, Vihavainen E, Paloranta A, Lehto J, Paulin L, Auvinen P, Solismaa M, Björkroth KJ (2011) Characterization of Psychrotrophic Bacterial Communities in Modified Atmosphere-Packed Meat with Terminal Restriction Fragment Length Polymorphism. Int. J. food microbiol. 144: 360–366.

[104] Björkroth J, Dicks LMT, Holzapfel WH (2009) Genus III. *Weissella*. In: De Vos P, Garrity GM, Jones D, Krieg NR, Ludwig W, Rainey FA, Schleifer KH, Whitman WB, editors. The Firmicutes. Bergey's Manual of Systematic Bacteriology, 2nd Ed., Vol. 3. Springer, Dordrecht, Heidelberg London, New York; pp. 643- 654.

[105] Comi G, Iacumin L (2012) Identification and Process Origin of Bacteria Responsible for Cavities and Volatile Off-Flavour Compounds in Artisan Cooked Ham. Int. j. food sci. technol. 47, 114-121.

[106] Mathur S, Singh R (2005) Antibiotic Resistance in Food Lactic Acid Bacteria - A Review. Int. j. food microbiol. 105: 281-295.

[107] Ammor S, Tauveron G, Dufour E, Chevallier I (2006) Antibacterial Activity of Lactic Acid Bacteria Against Spoilage and Pathogenic Bacteria Isolated from the Same Meat Small-Scale Facility: 2 - Behaviour of Pathogenic and Spoilage Bacteria in Dual Species Biofilms Including a Bacteriocin-Like-Producing Lactic Acid Bacteria. Food control 17: 462-468.

[108] Teusink B, Bachmann H, Molenaar D (2011) Systems Biology of Lactic Acid Bacteria: a Critical Review. Microb. Cell. Fact. 10 (suppl 1): S11.

[109] McLeod A, Zagorec M, Champomier-Vergès MC, Naterstad K, Axelsson L (2010) Primary Metabolism in *Lactobacillus sakei* Food isolates by Proteomic Analysis. BMC microbiol. 10: 120.

[110] Nyquist OL, McLeod A, Brede DA, Snipen L, Aakra Å., Nes IF (2011) Comparative Genomics of *Lactobacillus sakei* with Emphasis on Strains from Meat. Molec. genetics and genomics 285: 297-311.

Chapter 5

Potential of Fermented Sausage-Associated Lactic Acid Bacteria to Degrade Biogenic Amines During Storage

Jirasak Kongkiattikajorn

Additional information is available at the end of the chapter

1. Introduction

Biogenic amines (BAs) are organic bases with aliphatic, aromatic or heterocyclic structures that can be found in several foods, in which they are mainly produced by microbial decarboxylation of amino acids, with the exception of physiological polyamines. BAs may be of endogenous origin at low concentrations in non-fermented food such as fruits, vegetables, meat, milk and fish. High concentrations have been found in fermented foods as a result of a contaminating microflora exhibiting amino acid decarboxylase activity (Silla-Santos, 1996). However, BAs can also trigger human health problems leading to palpitations, hypertension, vomiting, headaches and flushing if food containing high concentrations are ingested. In fermented foods, some lactic acid bacteria (LAB) are able to convert available amino acid precursors into BAs via decarboxylase or deiminase activities during or following ripening processes. For this reason, amino acid catabolism by LAB can affect both the quality and safety of fermented foods (Verges et al., 1999). The amount and type of BAs formed depends on the nature of food and particularly on the kind of microorganisms present. Enterobacteriaceae and certain LAB are particularly active in the production of BA (Beutling, 1996). These amine-producing microorganisms either may form part of the food associated population or may be introduced by contamination before, during or after processing of the food product. Therefore, microorganisms naturally present in raw materials, introduced throughout the processing or added as starter culture can critically influence BA production during the manufacture of fermented products (Bover-Cid et al., 2000).

Nham is a Thai-style fermented pork sausage. Nham ripening generally takes 3-5 days and relies mainly on adventitious microorganisms, which are normally found in raw materials.

LAB produce organic acids from carbohydrates and cause the pH drop, which contribute to Nham formation. *Micrococcus* and *Staphylococcus* are capable of reducing nitrate to nitrite, which is important in producing the characteristic pigmentation. Also, as a source of lipolytic and proteolytic enzymes, they may contribute to flavor production. Therefore, the acidification and the proteolytic process occurring during Nham ripening make the environment particularly favorable for BAs production.

During meat ripening, microbial growth, acidification and proteolysis provide favourable conditions for BA production. The species of lactobacilli most commonly found in meat and meat products are *Lactobacillus sake* and *Lactobacillus curvatus*, which together with *Lactobacillus bavaricus* and *Lactobacillus plantarum* constitute the main microbial flora isolated from fermented sausages. Other bacteria that can be found in relatively high numbers include enterococci (*E. faecalis* and *E. faecium*), which also contribute to the ripening process. However, the presence of enterococci might also reflect a given level of contamination or a poor curing process. Salt-tolerant, nitrate-reducing coagulase-negative staphylococci are also detected in relatively high numbers in ripened meat products. *Staphylococcus xylosus* is the main species found in Spanish fermented sausages, although *S. carnosus* can also be used as a starter culture. BAs can be degraded through oxidative deamination catalyzed by amines oxidase (AO) with the production of aldehyde, ammonia and hydrogen peroxide. Monoamine oxidases (MAOs) and diamine oxidases (DAOs) had been described from some genus of the family Enterobacteriaceae (Yamashita et al., 1993). The potential role of microorganisms with AO activity had become a particular interest in the last few years to prevent or reduce BA accumulation in food products, especially fermented foods. Mah and Hwang (2009) investigated the effect of *Staphylococcus xylosus* to inhibit BA formation in a salted and fermented anchovy. Reduction of tyramine during ripening of fermented sausages was achieved when *Micrococcus varians* was applied as starter culture (Leuschner and Hammes, 1998). Inoculation of *L. plantarum* in sauerkraut effectively suppressed the production of tyramine, putrescine and cadaverine (Kalac et al., 2000).

BAs are physiologically inactivated by AO, which are enzymes found in bacteria, fungi, plant and animal cells able to catalyse the oxidative deamination of amines with production of aldehydes, hydrogen peroxide and ammonia (Cooper, 1997). The sequential action (in the presence of an electron acceptor, such as O_2) of an AO and an aldehyde dehydrogenase leads to the production of an acid and ammonia, which can be used to support microbial growth (Parrot et al., 1987). MAO and DAO activity has been described in higher organisms as well as in bacteria (Murooka et al., 1976, 1979; Ishizuka et al., 1993). There are relevant differences between microbial AO in terms of substrate specificity and location, as stated by Cooper (1997). DAOs can oxidase several BA, such as putrescine and histamine, and their activity can be affected by substrate inhibition; aminoguanidine, antihistaminic drugs and foodborne inhibitors, such as ethanol, carnosine, thiamine, cadaverine and tyramine, reduce their activity (Lehane and Olley, 2000). The potential role of microorganisms involved in food ripenings with AO activity has been investigated with the aim to prevent or reduce the accumulation of BA in foods. Leuschner et al. (1998) tested in vitro the potential amine degradation by many bacteria isolated from foods and, in particular, in strains belonging to the genera *Lactobacillus,*

Pediococcus, Micrococcus, as well as to the species *S. carnosus* and *Brevibacterium linens*. They found that this enzymatic activity can be present at very different quantitative levels. Tyramine oxidase activity of several microbial strains was strictly dependent on pH (with an optimum at 7.0), temperature and NaCl, as well as glucose and hydralazine concentration. Moreover, this enzyme was characterised by a higher potential activity under aerobic conditions. Temperature has also an important effect on histamine degradation (Dapkevicius et al., 2000). The highest degradation rate of this amine was observed at 37 °C, but at 22°C and 15 °C, degradation was still considerable. The AO responsible for this degradation has its optimum temperature at 37°C and retains about 50% of its maximum activity at 20 °C (Schomburg and Stephan, 1993). Many *S. xylosus* strains isolated from artisanal fermented sausages in southern Italy showed the ability to degrade BA in vitro (Martuscelli et al., 2000). Among the strains tested, *S. xylosus* S81 completely oxidised histamine, but it degraded, under the adopted conditions, also a part of tyramine. Even if the AO activity in vitro of microorganisms is not quantitatively reproducible in vivo (due to the more severe conditions and, in particular, to the low O_2 tension, pH and salt concentration), reduction of histamine in dry sausages has been observed in the presence of AO-positive staphylococcal starter cultures (Leuschner and Hammes, 1998). In addition, important reduction of the concentration of tyramine and putrescine in the presence of AO positive *S. xylosus* starter cultures have been observed by Gardini et al. (2002). In other words, BA presence in foods is the consequence of a complex equilibrium between the composition of the food and the enzymatic activities of the microbial population. Together with the decarboxylating aptitude of the starter cultures, the presence and relative activity of AO should be considered as an important characteristic in the selection of starter cultures used in the production of fermented foods.

Since Nham is normally consumed without cooking, proper acid production is important to determine the quality and safety of Nham for consumption. Depending on the initial number of contamination, the occurrence of pathogens such as *Salmonella* spp., *Staphylococcus aureus*, and *Listeria monocytogenes* was found specially in Nham with pH higher than 4.6. Due to inconsistency of product quality and ambiguous product safety, improved process of Nham ripening has been developed by using a starter culture technology. Starter cultures are applied to improve and stabilize the quality of the final product and to shorten the ripening period of Nham production. Meanwhile, only little information is available on the effect of starter culture on BA reduction in Nham. Therefore, the objective of this study was to investigate the effectiveness of AOs activity of LAB in inhibiting BA accumulation during Nham ripening. In addition, the change of chemical and microbial properties of Nham during ripening and subsequently during 28 days stored at different temperature was investigated.

2. Materials and methods

2.1. Microbiological analysis

Nham sausages (25 g) were aseptically transferred into a stomacher bag, with 225 mL of peptone (0.85% of sodium chloride added) and then homogenized for two minutes. Further decimal dilutions were made and then 100 µL of each dilution was spread onto agar plates.

Aerobic plate count agar was used to determine total aerobic. BA producing bacteria were counted using differential media supplemented with amino acids as precursor of BAs (Joosten and Northolt, 1989). The media contained of tryptone (0.5%), yeast extract (0.5%), sodium chloride (0.5%), glucose (0.1%), Tween 80 (0.05%), $MgSO_4 \bullet 7\ H_2O$ (0.02%), $CaCO_3$ (0.01%), $MnSO_4 \bullet 4H_2O$ (0.005%), $FeSO_4 \bullet 7H_2O$ (0.004%), bromocresol purple (0.006%), amino acid (2%) and agar (2%). The medium contained the precursor amino acids (0.5% tyrosine di-sodium salt and 0.25% L-histidine monohydrochoride, L-ornithine monohydrochoride, L-lysine monohydrochoride, L- phenylalanine, and L-tryptophan), pyridoxal-5-phosphate as a codecarboxylase factor, growing factors and buffer compounds. All plates were then incubated for 48 h at 37 °C. Bacterial colonies which developed on each agar were then enumerated and expressed as log colony forming unit (CFU)/mL. Only bacterial colonies with purple halo in the differential media were counted as BAs producing bacteria.

2.2. Bacterial strains and growth conditions

Bacterial strains isolated from different fermented sausages were tested. LAB were grown in MRS broth.

2.3. Determination of amine degradation

An overnight culture was harvested, washed with 0.05 M phosphate buffer (pH 7) and the cell pellet resuspended in 0.05 M phosphate buffer supplemented with tyramine, histamine, tryptamine, phenylethylamine, putrescine, and cadaverine. The cell concentration was adjusted to 10^6, 10^7 and 10^8 CFU/mL. The cell suspensions (20 mL) were incubated in a 100 ml flask and shaken at 200 rpm. Samples were taken and added to an equal amount of 1 M HCl. The mixture was boiled for 10 min and centrifuged at 9000 g. The supernatant was frozen at -15°C until HPLC analysis.

2.4. Preparation of starter culture

Starter cultures used in this study were *L. plantarum* + *L. sake*, which were isolated from sausage. A loop from a slant tryptic soy agar culture of each culture was inoculated in 10 mL of tryptic soy broth and incubated at 37 °C for 24 h. Five milliliters of the culture was then transferred to 100 mL of tryptic soy broth and incubated at 37 °C for another 24 h. The culture was centrifuged at 10,000 g for 10 min at 4°C and then washed with broth. Broth was prepared by homogenizing 1 part with 9 part of distilled water, filtered, adjusted to pH 7.0 and then autoclaved at 121°C for 15 min. After centrifugation, the cell pellet was resuspended in sterile fish broth, adjusted to approximately 10^7 cell/g and used as starter culture in sauce ripening.

2.5. Nham preparation

Minced pork (56%), pieced cooked pork skin (37%), garlic (3.2%), cooked rice (2%), sodium polyphosphate (0.15%), sodium chloride (1.5%) and sodium erythrobate (0.15%) chili (1%)

were mixed thoroughly, packed into a plastic casing and sealed before incubation. Two separated batches of fermented sausage were prepared without starter culture and with different starter cultures (*L. plantarum* + *L. sake*) of approximately 10^7 cell/g. After incubation the fermented sausages were homogenized for analysis.

2.6. Physical and chemical analyses

The pH was measured directly from samples using a microcomputerized pH meter, inserting the electrode into the middle of the sausage. Moisture was determined by drying the sample at 100–105°C until a constant weight was achieved. The color of Nham was determined by Minolta Model DP-301 colorimeter. Color values (L, a, and b) were measured. A white standard tile was used to calibrate the colorimeter (L= 100.01, a= -0.01, b= -0.02) before measurements. Therefore L measures lightness (luminosity) and varies from white to black. The chromatically (a and b values) gives designations of color as follows; a-value measures redness when positive, gray when zero, and greenness when negative, b-value measures yellowness when positive, gray when zero, and blueness when negative. The titratable acidity (TA) determined as total acid was estimated according to AOAC (2000) and expressed as g/100 g dry matter. TCA (trichloroacetic acid)-soluble peptide of the fermented sausages was measured by the method of Greene and Babbitt (1990). The oligopeptide content in the supernatant was determined according to by the method of Lowry et al (1951). Results were expressed as μmol/g (dry matter). Free α-amino acid was measured using TNBS according to Benjakul and Morrissey (1997) Results were expressed as μmol/g (dry matter).

2.7. Extraction of amino acids and BAs

10 ML of 10% (w/v) trichloroacetic acid (TCA) were added to 3 g-samples, and homogenization of the mixture was effected via shaking for 1 h. The extract was then filtered through Whatman No. 1 filter paper. To remove any fat, the samples were kept at -20 °C for 1 d, and then centrifuged at 7000 g for 15 min. The supernatants were collected and filtered through a 0.25 μm membrane filter.

2.8. Determination of BAs

Amines were determined by the high-performance liquid chromatography (HPLC) method described by Hernández-Jover et al. (1996). The method is based on the formation of ion pairs between amines extracted with 0.6 M perchloric acid from 5 to 10 g of sample, and octanesulphonic acid present in the mobile phase. Separation is preformed using a reversed phase column, then a postcolumn derivatization with o-phthalaldehyde (OPA) is followed by spectrofluorimetric detection. The method allows one to quantify, by an external standard procedure, 6 BAs, i.e., tyramine, histamine, tryptamine, phenylethylamine, putrescine, cadaverine. Samples for BA determination were stored at -15°C until required.

2.9. Determination of amino acids

Free amino acids (FAAs) in samples were determined using HPLC according to the method proposed by Rozan et al. (2000). A 20 µL aliquot of amino acid standard and digested sauce samples were transferred into vials and dried under vacuum. Then 20 µL of drying reagent containing methanol, water and triethylamine (ratio 2:2:1 v/v) was added. Then 20 µL of derivatizing reagent containing methanol, triethylamine, water and phenylisothiocyanate (PITC) (ratio 7:1:1:1 v/v) was added. The derivatized samples were then dissolved in 100 mL of buffer A that was used as mobile phase for HPLC. A Purospher® STAR RP-18e, 5 µm column was used with buffer A (0.1 M ammonium acetate, pH 6.5) and buffer B (0.1 M ammonium acetate containing acetonitrile and methanol, 44:46:10 v/v, pH 6.5) as mobile phase set for linier gradient at the flow rate of 1 mL/min. The injected sample volume was 20 µL and monitored at 254 nm of wavelength.

2.10. Statistical analysis

Data was analysed by one-way ANOVA and differences among treatment means were determined by Duncan's new multiple-range test.

3. Results and discussion

The effect of starter cultures of LAB on BAs and FAAs content was examined during the ripening process of Nham sausages. Microbial counts, pH and proteolysis-related parameters were also studied. The occurrence of amino acid-decarboxylase activity in 7 strains of LAB isolated from Nham sausages was investigated.

Starter culture	Percent degradation (%)					
	Trypta-mine	Phenylet-hylamine	Putre-scine	Cada-verine	Hista-mine	Tyra-mine
Lactobacillus curvatus 1271	0	0	0	0	0	0
Lactobacillus farciminis 1452	0	0	0	0	0	5.7
Lactobacillus kandleri 2439	0	0	0	3.6	0	0
Lactobacillus kefir 2045	0	0	0	0	0	0
Lactobacillus plantarum 9825	0	0	12.6	9.3	0	19.4
Leuconostoc mali 7412	0	0	0	0	0	0
Lactobacillus pentosus 7054	0	0	5.2	4.7	0	0
Lactobacillus reuteri 7498	0	0	4.8	0	0	0
Lactobacillus sake 4127	0	0	17.3	8.2	0	14.5

Table 1. Strains exhibiting the potential to degrade BAs in a buffer system within 24 h at 30°C

The presence of BAs in a decarboxylase synthetic broth was determined by high performance liquid chromatography with OPA derivatization. Among the 9 LAB strains tested, 5 lactobacilli (in particular, L. curvatus) were amine producers and L. plantarum and L.

sake, were non-amine forming strains. The ability of AO exhibiting strains of LAB to degrade amine in vivo during sausage ripening was investigated.

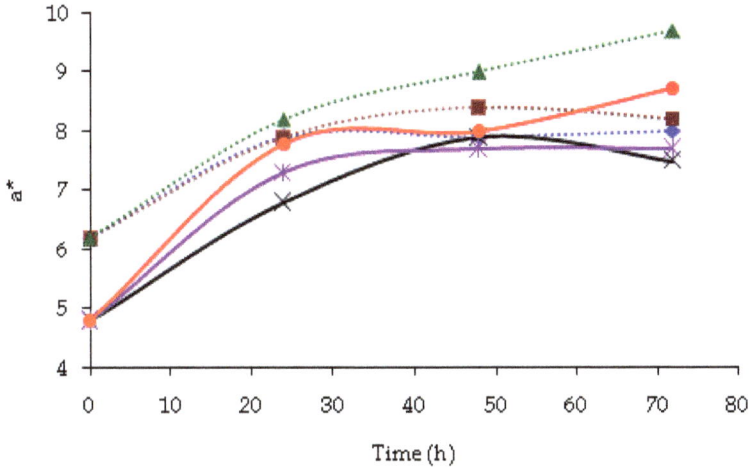

Figure 1. a* Value during ripening of Nham control at 25°C (♦), 30°C (■), 37°C (▲) and Nham with starters (*L. plantarum* + *L. sake*) at 25°C (×), 30°C (✳), 37°C (●).

Fig. 1 showed a* values represent red color of Nham during ripening time and temperature at 25°C, 30°C and 37°C, respectively. The results showed a value increased according to ripening and the a value of Nham control at 72 hours 37°C was higher than the other sample.

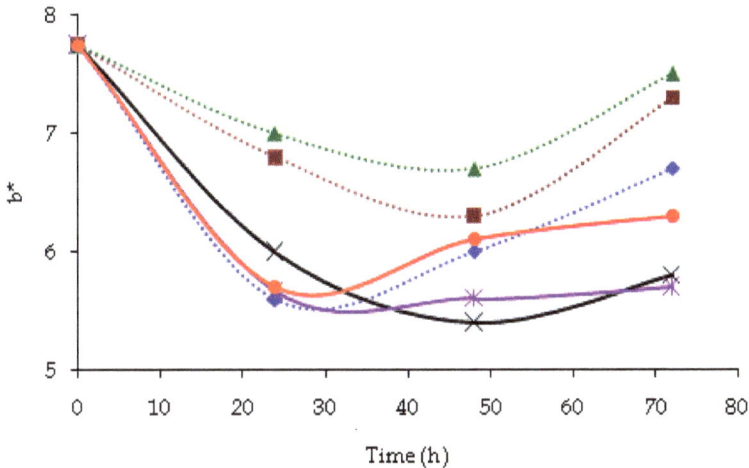

Figure 2. b* Value during ripening of Nham control at 25°C (♦), 30°C (■), 37°C (▲) and Nham with starters (*L. plantarum* + *L. sake*) at 25°C (×), 30°C (✳), 37°C (●).

Fig. 2 shows b* values represent yellow color of Nham during ripening time and temperature at 25°C, 30°C and 37°C, respectively. The results showed b value decreased according to storage and the b value of Nham with starters was lower than that of Nham control.

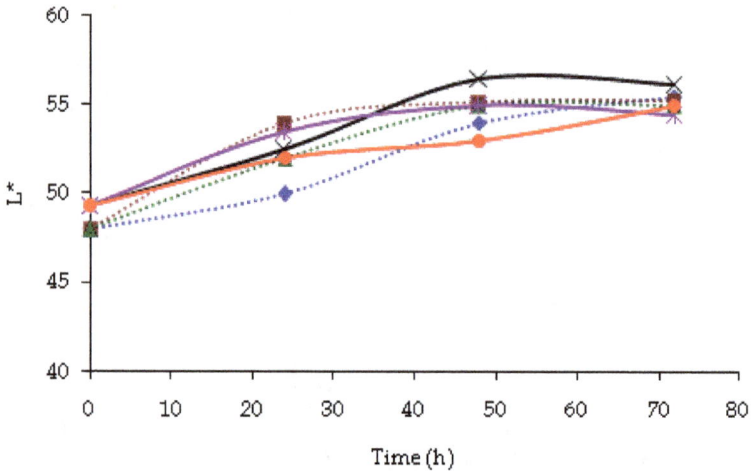

Figure 3. L* Value during ripening of Nham control at 25°C (♦), 30°C (■), 37°C (▲) and Nham with starters   (*L. plantarum* + *L. sake*) at 25°C (×), 30°C (✳), 37°C (●).

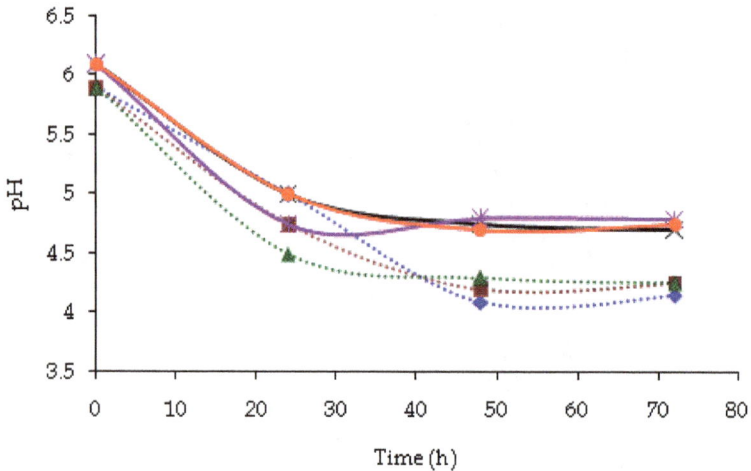

Figure 4. pH during ripening of Nham control at 25°C (♦), 30°C (■), 37°C (▲) and Nham with starters (*L. plantarum* + *L. sake*) at 25°C (×), 30°C (✳), 37°C (●).

Fig. 3 is represent L*values represent white color of Nham during ripening time and temperature at 25°C, 30°C and 37°C, respectively. The results showed L* value increased according to storage during 72 hour of ripening.

Fig. 4 shows that the initial pH of Nham samples ranged from 5.9 to 6.1. It then gradually decreased throughout the ripening process and there was significant difference at each time of sampling (P <0.05). The pH value reached 4.1 to 4.8 at the end of ripening (hour 72). However, there was significant difference (P<0.05) between the pH of Nham control and samples inoculated with starter cultures after 48 hour of ripening.

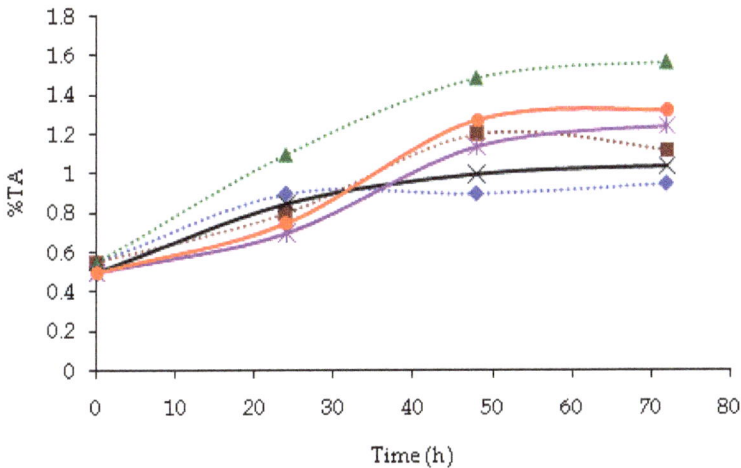

Figure 5. Total acid content during ripening of Nham control at 25°C (♦), 30°C (■), 37°C (▲) and Nham with starters (*L. plantarum* + *L. sake*) at 25°C (×), 30°C (✳), 37°C (●).

Fig. 5 shows that the initial total acid content of Nham samples ranged from 0.5 to 0.55. It then gradually increased throughout the ripening process and there was significant difference at each time of sampling (P <0.05). The total acid content of Nham control and Nham with starters reached 0.95% to 1.57% and 1.04 %to 1.32% at the end of ripening (hour 72). However, there was significant difference (P<0.05) between the total acid content of Nham control and samples inoculated with starter culture after 48 hour of ripening. The results was shown that Nham control fermented at 37°C contained total acid content higher than the other Nham samples.

Fig. 6 shows that TCA-soluble peptide content of Nham samples, the initial content was 9.02 µmol/g dry matter. It then gradually increased throughout the ripening process. The TCA-soluble peptide content of Nham control and Nham with starters reached 23.6 to 87.2 µmol/g dry matter and 24.1 %to 65.2 µmol/g dry matter, respectively, at the end of ripening (hour 72). However, there was not significant difference (P<0.05) between the TCA-soluble peptide content of Nham control and samples inoculated with starter culture after 48 hour

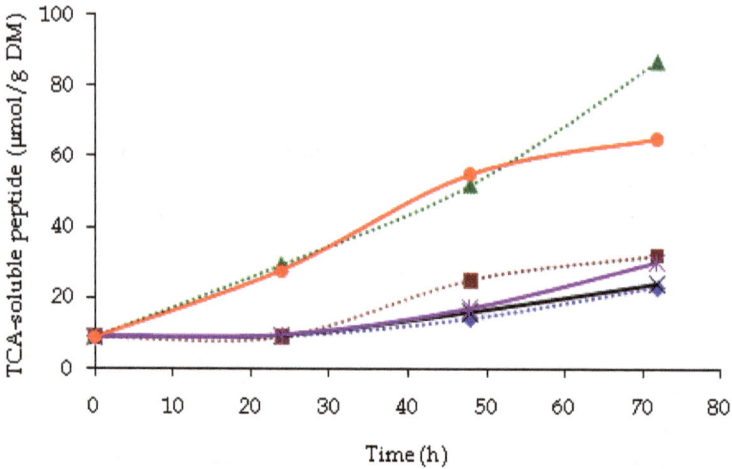

Figure 6. TCA-soluble peptide content during ripening of Nham control at 25°C (♦), 30°C (■), 37°C (▲) and Nham with starters (*L. plantarum* + *L. sake*) at 25°C (×), 30°C (✳), 37°C (●).

of ripening at each ripening temperature, and 72 hour of ripening at 25C and 30C. The results was shown that Nham control fermented at 37°C contained TCA-soluble peptide content higher than the other Nham samples after ripening for 72 hour.

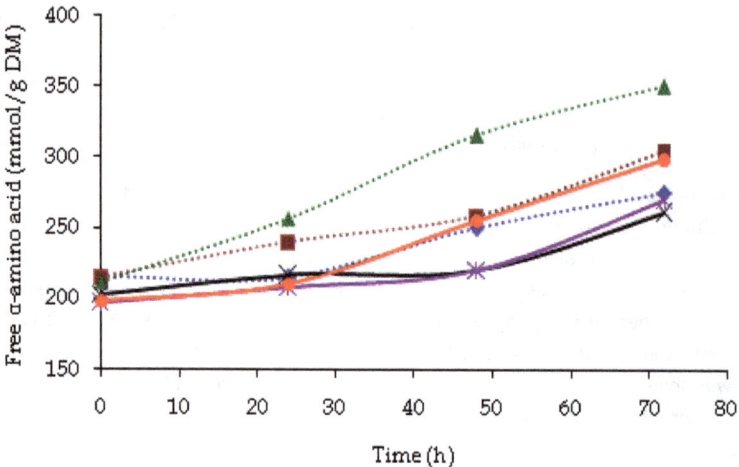

Figure 7. Free α-amino acid content during ripening of Nham control at 25°C (♦), 30°C (■), 37°C (▲) and Nham with starters (*L. plantarum* + *L. sake*) at 25°C (×), 30°C (✳), 37°C (●).

Fig. 7 shows that free α-amino acid content of Nham control samples and Nham with starters, the initial content were 216.2 mmol/g dry matter and 203.7 mmol/g dry matter, respectively. It then gradually increased throughout the ripening process. The free α-amino acid content of Nham control and Nham with starters reached 275.3 to 351.6 mmol/g dry matter and 262.4 to 302.2 mmol/g dry matter, respectively, at the end of ripening (hour 72). However, there was not significant difference ($P<0.05$) between the free α-amino acid content of Nham control and samples inoculated with starter culture during ripening at 25°C. The results was shown that Nham control fermented at 37°C contained free α-amino acid content higher than the other Nham samples throughout the ripening process.

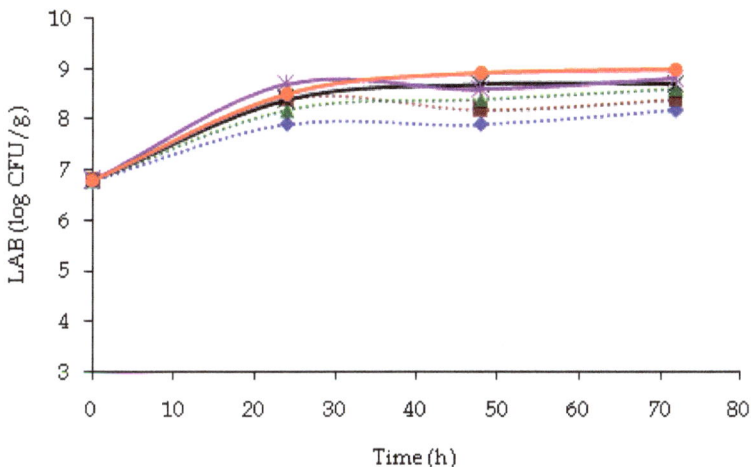

Figure 8. Total count of LAB during ripening of Nham control at 25°C (♦), 30°C (■), 37°C (▲) and Nham with starters (*L. plantarum* + *L. sake*) at 25°C (×), 30°C (✳), 37°C (●).

The differences between Nham in counts of LAB during ripening are shown in Fig. 8. LAB in Nham with starters was increase until the 72 h of ripening. Counts of LAB in Nham with starters (8.7 log CFU/g) were higher ($P < 0.05$) than in Nham control (7.7 log CFU/g).

Fig. 9 shows that cadaverine content of Nham samples, the initial content was 14.89 mg/kg dry matter. It then gradually increased throughout the ripening process. The cadaverine content of Nham control and Nham with starters reached 86.2 to 98.7 mg/kg dry matter and 42.4 to 51.6 mg/kg dry matter, respectively, at 72 hour of ripening. However, there was not significant difference ($P<0.05$) between the cadaverine content of Nham with starters during ripening at 25°C and 30°C. The results was shown that Nham control fermented at 37°C contained cadaverine content higher than the other Nham samples throughout the ripening process.

Fig. 10 shows that putrescine content of Nham samples, the initial content was 23.7 mg/kg dry matter. It then gradually increased throughout the ripening process and there was significant difference at each time of sampling ($P <0.05$). The putrescine content of Nham control and Nham with starters reached 115.4 to 242.6 mg/kg dry matter and 65.2 to 98.4 mg/kg dry matter, respectively, at 72 hour of ripening. However, there was not significant

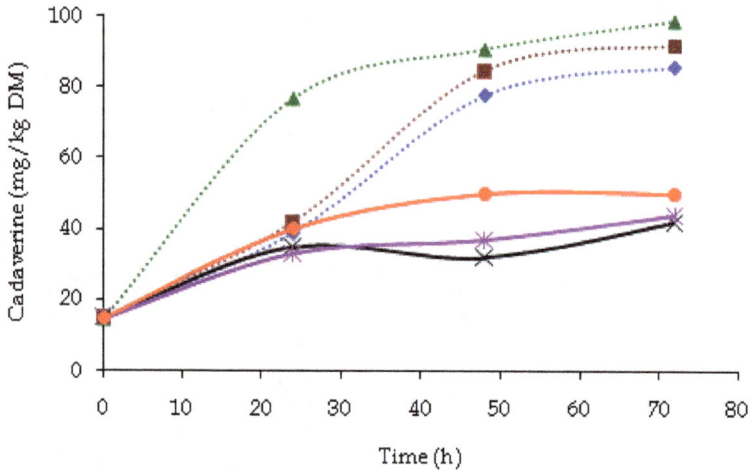

Figure 9. Cadaverine content during ripening of Nham control at 25°C (♦), 30°C (■), 37°C (▲) and Nham with starters (*L. plantarum* + *L. sake*) at 25°C (×), 30°C (＊), 37°C (●).

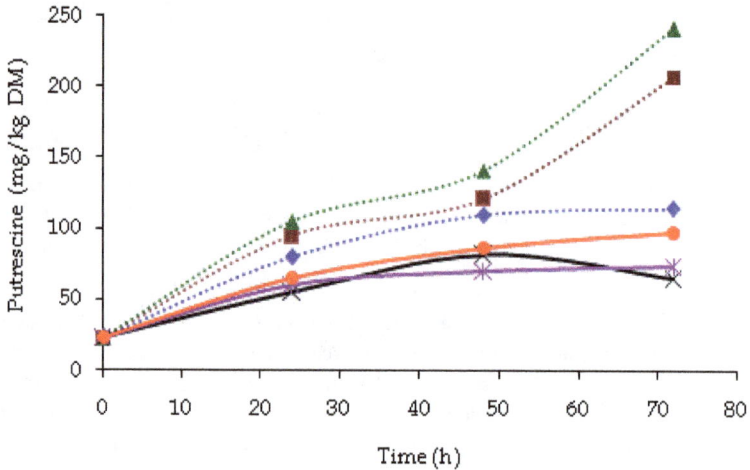

Figure 10. Putrescine content during ripening of Nham control at 25°C (♦), 30°C (■), 37°C (▲) and Nham with starters (*L. plantarum* + *L. sake*) at 25°C (×), 30°C (＊), 37°C (●).

difference (*P*<0.05) between the putrescine content of Nham with starters during ripening at 25°C and 30°C. The results was shown that Nham control fermented at 37°C contained putrescine content higher than the other Nham samples throughout the ripening process.

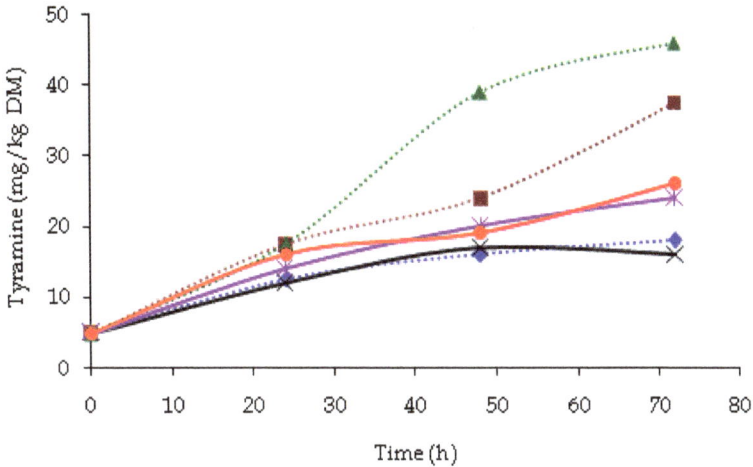

Figure 11. Tyramine content during ripening of Nham control at 25°C (♦), 30°C (■), 37°C (▲) and Nham with starters (*L. plantarum* + *L. sake*) at 25°C (×), 30°C (✳), 37°C (•).

Fig. 11 shows that tyramine content of Nham samples, the initial content was 5.63 mg/kg dry matter. It then gradually increased throughout the ripening process and there was significant difference at each time of sampling (P <0.05). The tyramine content of Nham control and Nham with starters reached 17.6 to 46.4 mg/kg dry matter and 16.3 to 27.8 mg/kg dry matter, respectively, at 72 hour of ripening. However, there was not significant difference (P<0.05) between the tyramine content of Nham with starters during ripening at 30°C and 37°C and Nham control and Nham samples inoculated with starter culture during ripening at 25°C. The results was shown that Nham control fermented at 37°C contained tyramine content higher than the other Nham samples after 48 hour of the ripening process.

The effect of temperature on BA content was evaluated (Fig. 6-9). The storage temperature of Nham with starters at 30°C and 37°C were shown higher BA oxidation comparing Nham control, a low content was observed at 25°C. This suggested that at ripening temperature of 30 °C and 37°C, a strong oxidation of the AO activity of the starters was evident, whereas at 25°C activity was low for amino acid decarboxylase for lysine (precursor of putrescine) and tyrosine (precursor of tyramine) in Nham control.

Fig. 12 showed a* values represent red color of Nham during stored at 15°C, 4°C and 25°C. The initial a* values of Nham control ranged from 7.0 to 8.4. The results showed a* value increased according to 4 week storage for storage temperature at 15°C and 25°C. However, there was significant decrease (P<0.05) between the a* value of 4°C storage of the initial 1 week storage and after 4 week of storage. For Nham with starters, the initial a* values ranged from 7.5 to 8.5. The a* values of Nham with starter decreased according to 4 week storage for storage temperature at 4°C and 15°C. However, there was not significant difference (P<0.05) between the a* value of 25°C storage of the initial 1 week storage and after 4 week of storage.

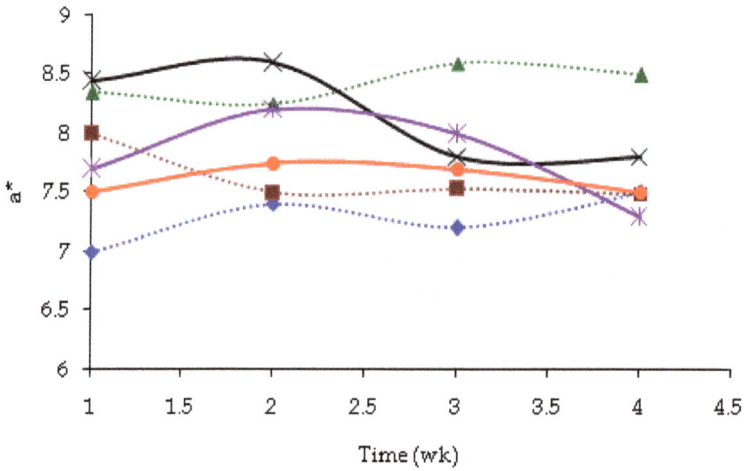

Figure 12. a* Value during storage of Nham control at 25°C (♦), 30°C (■), 37°C (▲) and Nham with starters (*L. plantarum* + *L. sake*) at 25°C (×), 30°C (✳), 37°C (●).

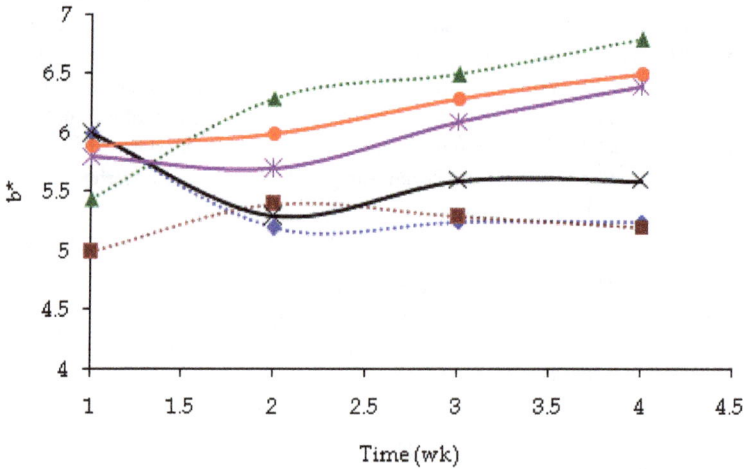

Figure 13. b* Value during storage of Nham control at 4°C (■), 15°C (♦), 25°C (▲) and Nham with starters (*L. plantarum + L. sake*) at 4°C (✳), 15°C (×), 25°C (●).

Fig. 13 showed b* values represent yellow color of Nham during stored at 15°C, 4°C and 25°C. The initial a* values of Nham control ranged from 5.0 to 6.1. The results showed a* value increased according to 4 week storage for storage temperature at 4°C and 25°C. However, there was significant decrease ($P<0.05$) between the b* value of 15°C storage of the initial 1 week storage and after 4 week of storage. For Nham with starters, the initial a* values ranged from 5.5 to 6.2. The b* values of Nham with starter increased according to 4 week storage for storage temperature at 4°C and 25°C. However, there was significant decrease ($P<0.05$) between the b* value of 15°C storage of the initial 1 week storage and after 4 week of storage.

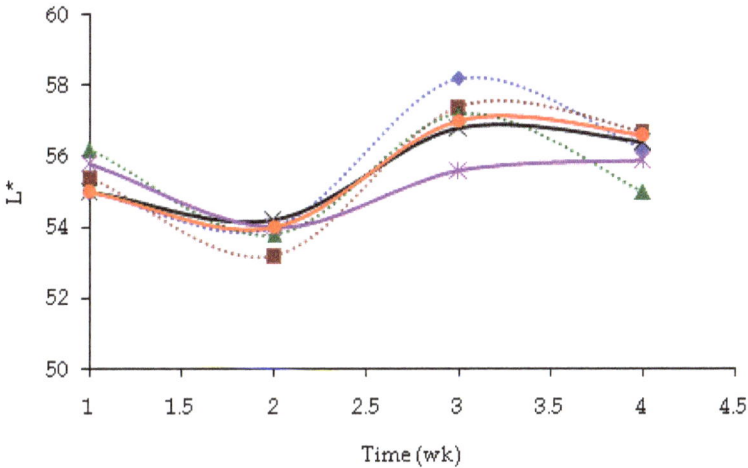

Figure 14. L* Value during storage of Nham control at 4°C (■), 15°C (♦), 25°C (▲) and Nham with starters (*L. plantarum*   + *L. sake*) at 4°C (✳), 15°C (×), 25°C (•).

Fig. 14 showed L* values represent white color of Nham during stored at 15°C, 4°C and 25°C. The initial a* values of Nham control ranged from 55.1 to 56.4. The results showed L* value decreased after 2 week storage and then increased after 3 week storage for each storage temperature. However, there was not significant difference ($P<0.05$) between the L* value of 25°C storage of the initial 1 week storage and after 4 week of storage. For Nham with starters, the initial L* values ranged from 55.0 to 55.6. The L* values of Nham with starter decreased after 2 week storage then the L* values increased after 3 week storage for each storage temperature and after 4 week storage at 15°C and 25°C, the L* value was significant increased. However, there was no significant difference ($P<0.05$) between the L* value of 4°C storage of the initial 1 week storage and after 4 week of storage.

Fig. 15 shows that the initial pH of Nham samples ranged from 4.3 to 4.5. It then gradually decreased throughout the storage. The pH value reached 4.1 to 4.5 at 4 week of storage. The pH values at each storage temperature of Nham with starter were higher than Nham control at each time of sampling. The results was shown that pH value of Nham control stored at 25°C was lower than the other Nham samples throughout the storage process.

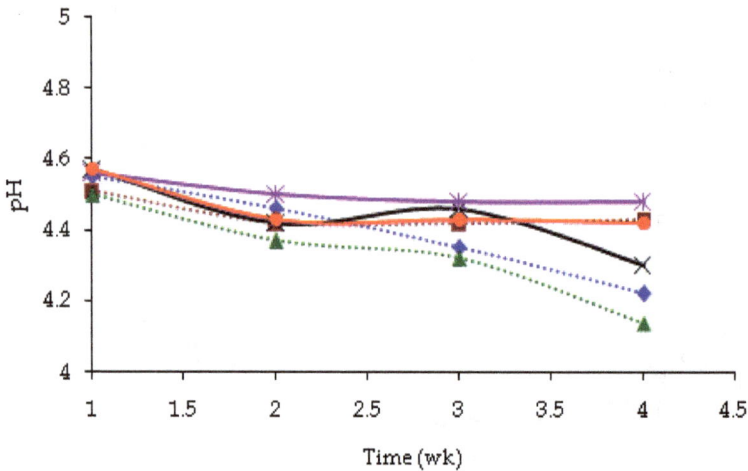

Figure 15. pH during storage of Nham control at 4°C (■), 15°C (♦), 25°C (▲) and Nham with starters (*L. plantarum* + *L. sake*) at 4°C (✳), 15°C (×), 25°C (•).

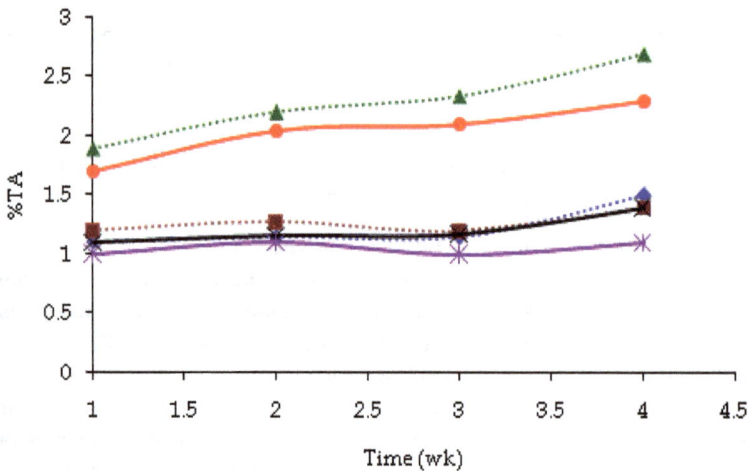

Figure 16. Total acid content during storage of Nham control at 4°C (■), 15°C (♦), 25°C (▲) and Nham with starters (*L. plantarum* + *L. sake*) at 4°C (✳), 15°C (×), 25°C (•).

Fig. 16 shows that the initial total acid content of Nham samples ranged from 1.1 to 1.7. It then gradually increased throughout the ripening process and there was significant difference at each time of sampling (P <0.05). The total acid content of Nham control and Nham with starters reached 1.14% to 2.72% and 1.04 %to 2.32% at 4 week of storage, respectively. However, there was not significant difference (P<0.05) between the total acid content of Nham control stored at 4°C and 15°C and Nham with starters stored at 15°C and from the results, the total acid content of Nham with starters stored at 4°C was not significant difference ($P<0.05$) during storage process. The total acid content of Nham control stored at 25°C was higher than the other Nham samples throughout the storage process.

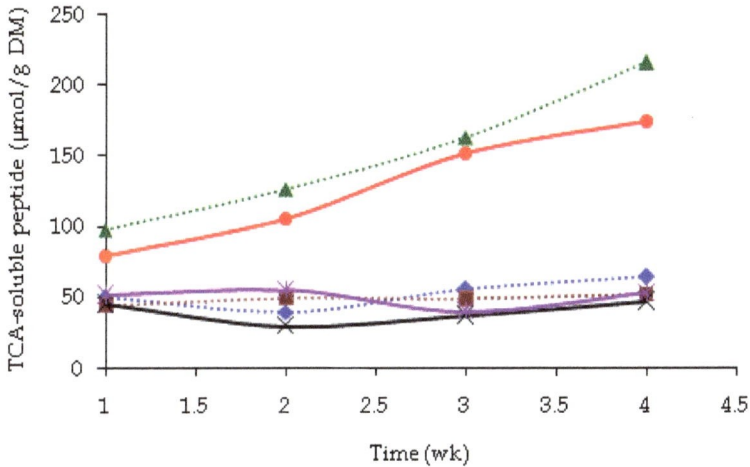

Figure 17. TCA-soluble peptide during storage of Nham control at 4°C (■), 15°C (♦), 25°C (▲) and Nham with starters   (*L. plantarum* + *L. sake*) at 4°C (✳), 15°C (x), 25°C (•).

Fig. 17 shows that the TCA-soluble peptide of Nham control and Nham with starters ranged from 45.2 to 98.4 and 46.3 to 79.6 μmol/g dry matter, respectively. Nham control and Nham with starters stored at 25°C showed gradually increased throughout the storage process and there was significant difference at each time of sampling ($P <0.05$). The TCA-soluble peptide of Nham control and Nham with starters stored at 4°C and 15°C. However, there was not significant difference ($P<0.05$) between the TCA-soluble peptide of Nham control and samples inoculated with starters culture throughout the storage process at 4°C and 15°C. From the results, the TCA-soluble peptide of Nham control stored at 25°C was higher than the other Nham samples throughout the storage process.

Fig. 18 shows that free α-amino acid content of Nham control samples and Nham with starters, the initial ranged from 342.3 to 603.4 and 346.6 to 507.2 mmol/g dry matter, respectively. It then gradually increased throughout the storage process at 15°C and 25°C and there was significant difference at each time of sampling ($P <0.05$). The free α-amino acid content of Nham control and Nham with starters reached 375.2 to 1867.6 mmol/g dry matter and 359.4 to 1252.4 mmol/g dry matter, respectively, at 4 week of storage. However, there was not significant difference ($P <0.05$) between the free α-amino acid content of Nham control and samples inoculated with starters during storage at 4°C. The results was shown that Nham control stored at 25°C contained free α-amino acid content higher than the other Nham samples after 3 week storage.

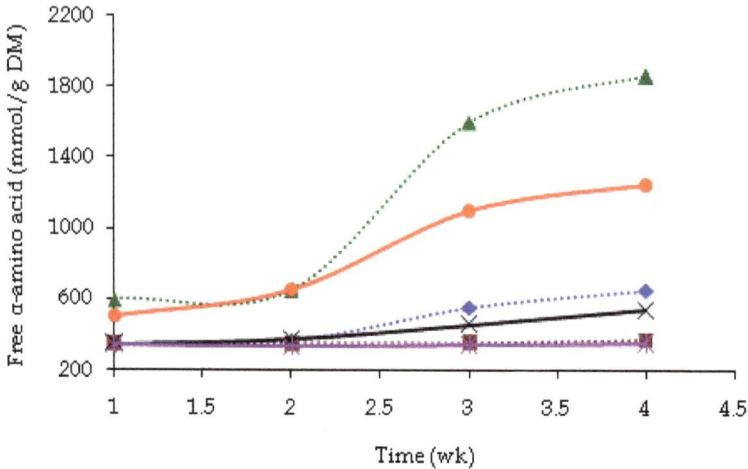

Figure 18. Free α-amino acid content during storage of Nham control at 4°C (■), 15°C (♦), 25°C (▲) and Nham with starters (*L. plantarum* + *L. sake*) at 4°C (✳), 15°C (×), 25°C (●).

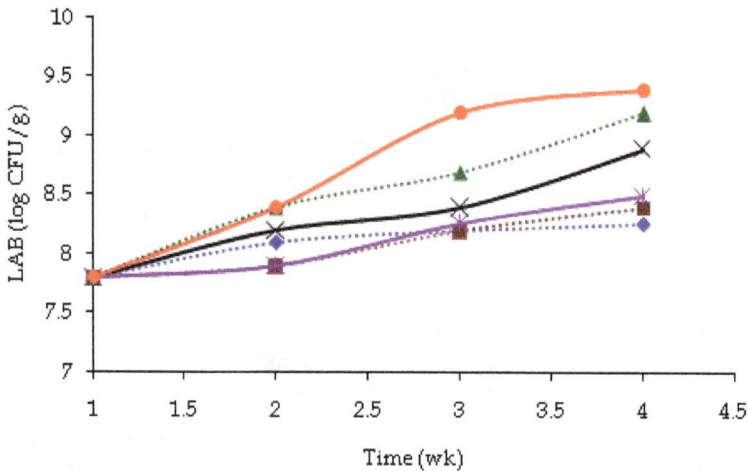

Figure 19. Total count of LAB during storage of Nham control at 4°C (■), 15°C (♦), 25°C (▲) and Nham with starters (*L. plantarum* + *L. sake*) at 4°C (✳), 15°C (×), 25°C (●).

The differences between Nham in counts of LAB during ripening are shown in Fig. 19. LAB in Nham with starters was increase until the 4 week of storage. Counts of LAB in Nham with starters stored at 25°C (9.4 log CFU/g) were higher ($P < 0.05$) than in Nham control stored at 25°C (9.1 log CFU/g). LAB counts in Nham increased steadily during storage, the dependence of the LAB counts of Nham control and Nham with starters on ripening at each storage temperature were significant differences. In the present study in Nham are concerned, total LAB counts in Nham with starters on 3 week of storage were higher ($P <$ 0.05) in comparison with the Nham control produced at the same storage temperature. An increase of LAB in Nham with starters until 3 week of storage and consecutive increase till 4 week of storage was significant. LAB of Nham with starters produced increase steadily during ripening and stored at different temperatures, however, at 4°C storage, LAB counts storage was not different significant.

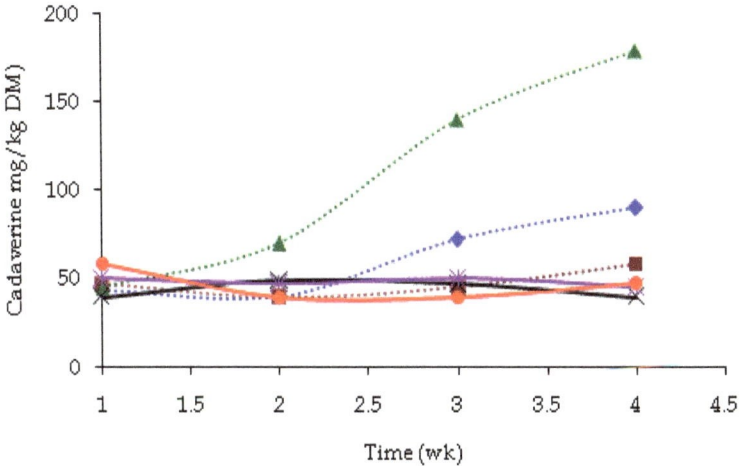

Figure 20. Cadaverine content during storage of Nham control at 4°C (■), 15°C (♦), 25°C (▲) and Nham with starters (*L. plantarum* + *L. sake*) at 4°C (✳), 15°C (×), 25°C (●).

Fig. 20 shows that cadaverine content of Nham samples, the initial ranged from 43.7 to 58.2 mg/kg dry matter. In Nham control stored at 15°C and 25°C, it then gradually increased throughout the storage process. The cadaverine content of Nham control and Nham with starters reached 58.4 to 91.2 mg/kg dry matter and 41.6 to 47.3 mg/kg dry matter, respectively, at 72 hour of storage. However, there was not significant difference ($P<0.05$) between the cadaverine content of Nham with starters during stored at 4°C. The results was shown that Nham control stored at 25°C contained cadaverine content higher than the other Nham samples throughout the storage process and there was significant decreased ($P<0.05$) in the cadaverine content of Nham with starters stored at 25°C for 4 week.

Fig. 21 shows that putrescine content of Nham control and Nham with starters, the initial ranged from 124.6 to 176.3 mg/kg dry matter and 126.2 to 98.3 mg/kg dry matter. Nham control stored at 4°C, 15°C and 25°C gradually increased throughout the storage process. The putrescine content of Nham control and Nham with starters reached 175.3 to 339.4 mg/kg dry matter and 122.6 to 129.3 mg/kg dry matter, respectively, at 4 week of storage. However, there were significant increase ($P<0.05$) between the putrescine content of Nham with starters at each storage temperature for 2 week and then the putrescine content decreased and there was not significant difference ($P<0.05$) between the putrescine content of Nham with starters after 3 week storage. The results were shown that Nham control stored at 25°C contained putrescine higher than the other Nham samples throughout the storage process.

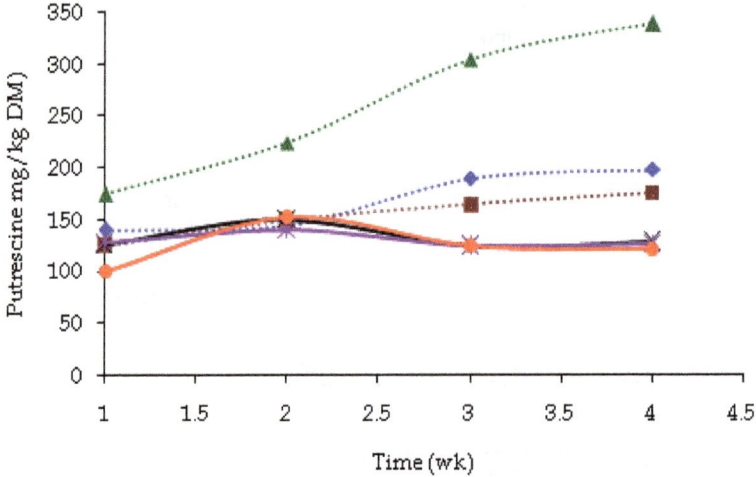

Figure 21. Putrescine content during storage of Nham control at 4°C (■), 15°C (♦), 25°C (▲) and Nham with starters (*L. plantarum* + *L. sake*) at 4°C (✳), 15°C (×), 25°C (●).

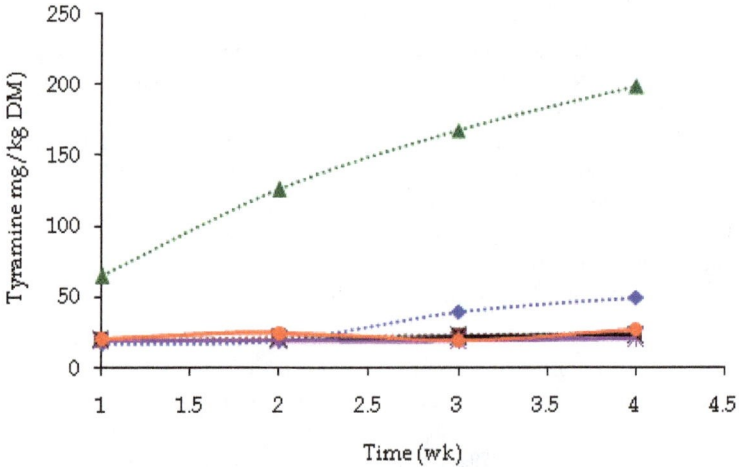

Figure 22. Tyramine content during storage of Nham control at 4°C (■), 15°C (♦), 25°C (▲) and Nham with starters (*L. plantarum* + *L. sake*) at 4°C (✳), 15°C (×), 25°C (●).

Fig. 22 shows that tyramine content of Nham control and Nham with starters, the initial ranged from 18.2 to 65.3 mg/kg dry matter and 19.2 to 22.4 mg/kg dry matter. Nham control stored at 15°C and 25°C gradually increased during storage process. The tyramine content of Nham control and Nham with starters reached 25.2 to 198.6 mg/kg dry matter and 21.4 to 27.6 mg/kg dry matter, respectively, at 4 week of storage. There was not significant difference ($P<0.05$) between the tyramine content of Nham control stored at 4°C and Nham with starters at each storage temperature during storage time. The results was shown that Nham control stored at 25°C contained tyramine higher than the other Nham samples throughout the storage process.

One of the most important factors influencing BA formation in Nham is starter culture (Maijalaet al., 1995). Increase of LAB starters culture in Nham resulted in overgrowth more than the microflora and LAB producing BAs in Nham control and caused decrease in BA contents in Nham during ripening and storage. A higher amount of BAs was formed in the Nham control than in starters culture-ones. However, strains of the starters showed lower decarboxylase activity (lower total free amino acid content in Nham) in comparison with the Nham control. Moreover, from the fact that BA production increased in Nham control after the ripening was finished and Nham was stored at the 15°C and 25°C, which coincided with the temporary increase of total LAB, the presence of spontaneous decarboxylating microflora can be inferred, and the refrigerated storage should be recommended. Simultaneously, higher concentration of BAs was found in Nham fermented at high 30°C and 37°C as compared to 25°C at the end of ripening. However, as regards to the strongly hypothetical effects of some substances in the Nham spicing mixtures in connection with the BA formation, more research is needed. Decarboxylase activities present in microflora in Nham are influenced by pH, temperature (Gardini et al., 2001; Silla-Santos, 1996; Suzzi and Gardini, 2003). The decarboxylation of FAAs to BAs was found to be inhibited by low pH (Gardini et al., 2001). Though amino acid decarboxylase activities usually have acid pH optimum (Gale, 1946), the pH rise could favour the cell yield and growth (Maijala, 1994) of decarboxylase-positive microflora.

3.1. BAs contents of Nham

Occurrence of toxic compounds such as BAs is favoured by a high concentration of substrates (i.e., free amino acids) together with environmental and technological factors (e.g. NaCl content, chemico-physical variables, hygienic procedure adopted during production) promoting microbial growth and the decarboxylase activity of microorganisms (Silla-Santos, 1996). In this study, a high correlation among total BAs and total FAAs content was observed. Temperature markedly influences the formation of BAs, and at 15°C decarboxylases might be still active (Bover-Cid et al., 2001). During storage, the more temperature exceeds 14–15°C the more decarboxylase activities might release BAs from FAAs. In this respect, processing procedures for Nham based on low salt addition, high ripening temperatures (over 20°C), may favour proteolytic and decarboxylase activities. The high values of cadaverine, putrescine and tyramine detected in some Nham, may be ascribed to inadequate microflora and LAB producing BAs reduction occurring in some Nham control (Fig. 9- Fig. 11 and Fig. 20- Fig. 22).

The toxicological level of BAs depends on the individual characteristics and the presence of other amines (Brink et al., 1990; Halasz et al., 1994). Toxic doses of tyramine in foods were reported in the range 100–800 mg/kg, but average amounts of tyramine detected in analysed samples (Fig. 9- Fig. 11 and Fig. 20- Fig. 22) were below this range, even if in case of a few samples, the 100 mg/kg value was exceeded. Putrescine has been regarded as not toxic by itself, but as a potentiator for the toxic effect of tyramine and histamine if present (Hui and Taylor, 1985). However, it was probable to demonstrate significant relationship between the concentration of a specific FAA and its corresponding BA in meat products (Eerola et al., 1998). Fig. 9- Fig. 11 and Fig. 20- Fig. 22 shows the BAs content of Nham evidence the effect of starters on the decrease of the BAs occurrence in Nham after ripening and storage. Histamine was always below the minimum detectable, in spite of the abundance of their precursors (histidine) released during the process; phenylethylamine was also not detected.

The concentration of tyramine was high in Nham control while low concentration, of their precursors (tyrosine) released during the process. Moreover, tryptamine resulted absent in all the investigated samples. The sum of vasoactive BAs, VBA; (tyramine, phenylethylamine, histamine and tryptamine) lower than 200 mg/Kg has been suggested by Eerola et al. (1998) as a quality index (VBA index) for ripened meat products. It is interesting to note that the computed VBA index of Nham with starters with differently processed resulted appreciable samples (3.70 ± 2.46 mg/Kg). These results could be related to the specific characteristics of the product as well as to the process conditions adopted that could, in general, have limited the growth and activity of amino acid decarboxylase positive microorganisms (Suzzi and Gardini, 2003). Cadaverine, putrescine and tyramine were found in high amounts in Nham control. However, the occurrence of BAs in Nham control, and after the storage could be due to the microflora and LAB producing BAs that could have favoured their formation during ripening and storage. During ripening and storage of Nham control, putrescine and cadaverine show a marked increase with high amounts of their precursor, arginine and lysine, respectively, were detected. In fact, arginine may generate putrescine both via arginine deiminase pathway (ADI) leading to ornithine (Montel and Champomier, 1987) and their subsequent decarboxylation to putrescine, and via arginine decarboxylation to agmatine followed by deamination to putrescine and removal of urea (Moreno-Arribas et al., 2003). It seems reasonable to postulate that the large amounts of arginine could be the source of putrescine, which subsequently may be converted in spermine and spermidine by transamination reactions (Lehninger et al., 1999).

3.2. FAAs contents of Nham

FAAs were reported in Table 2- Table 5 as net amounts (mmol/g dry matter) in order to investigate the differences in contents due to starters in Nham during ripening and storage. FAAs were compared to evaluate if the extended storage times gave a similar increase in all of them or different patterns were detectable. Most single FAAs increased during ageing with particular reference to the lipophylic ones; a rise in lypophilic valine, phenylalanine and tryptophan processed following a traditional prolonged way (Ruiz et al., 1999). In the present study, stored Nham showed a FAA pattern enriched with glutamic acid, alanine, arginine, cysteine, serine, threonine and glycine, most FAAs displayed a rise during the extended storage. Arginine found in the most stored Nham was increase, due to changing of its content by proteolysis; and rise in arginine in stored Nham control was higher than stored Nham with starters. Arginine hydrolysis could be hydrolysed via the arginine deiminase pathway (ADI) leading to ammonia and ornithine. It seemed reasonable to postulate that ADI pathway enzymes (arginine deiminase and ornithine transcarbamylase) could be still active during storage times. Arginine catabolism, may be regarded as a source of the BA putrescine both via ADI ornithine generation (Montel and Champomier, 1987) and subsequent decarboxylation to putrescine, and via arginine decarboxylation to agmatine followed by deamidation to putrescine and removal of urea (Moreno-Arribas et al., 2003). The presence in Nham of environmental conditions suitable for decarboxylase activities together with large amounts of arginine may be consistent with the increase in putrescine.

The evolution during incubation/storage of the total free amino acid content, in both the Nham control and after inoculation with either of the two Lactobacillus strains selected, is shown in Table 2- Table 5, and encompassed 17 different amino acids. The control Nham showed the highest concentration of total amino acids at a 5% level of significance. The contents of total amino acids in Nham inoculated with L. plantarum and L. sake, increased throughout time, but at lower rates than the control. The contents of free amino acids and BAs in control and experimental Nham increase significantly throughout incubation/storage. However, specific lactic acid strains of the Lactobacillus genus can effectively prevent BAs from building-up excessively, putrescine (for quantitative reasons, owing to its level). This may lead to a favourable contribution to public health, especially in regions where Nham is frequently included in the diet. To have an overall evolution index of the proteases action in the Nham during processing the TCA-soluble peptide was evaluated (Fig. 6 and Fig. 17) (Toldr, 2005). More intense proteolytic activity occurred in the Nham control. The TCA-soluble peptide values of Nham control are quite high compared to those generally observed in other Nham with starters. This could be attributed to the microflora in Nham control slightly higher proteolytic activity during the process, in comparison with those Nham applied with starters. Proteolysis contributes to texture by breakdown of the muscle structure (Monin et al., 1997).

Table 2- Table 5 show the FAAs content of Nham during ripening and storage arginine and glutamic acid were the FFAs most representative; after ripening and storage a marked increase of alanine was observed. Table 6 shows the effect of the starters treatment on the evolution of the FAAs pattern of the Nham investigated during the ripening and storage: a significant increase in the concentration of all FAAs with respect to their initial occurred in Nham control and Nham with starters, resulting from the aminopeptidases activity of meat (Toldr, 2006) as well as microbial proteases (Dur et al., 2004; Molina and Toldrá, 1992; Rodrguez et al., 1998; Scannell et al., 2004). Moreover, starters in Nham seems to affect the production of some amino acids (Table 6). A lower concentration of lysine, threonine, glycine and proline was detected, after storage, in Nham processed. Arginine was the most abundant amino acid in all the final products, and its level was significantly higher in Nham control than in those subjected with starters to Nham. At the end of the ripening step, cysteine was also present in a relative higher concentration in Nham control, whereas significant larger amounts of proline, lysine, histidine, serine and threonine were reached in Nham control samples. The different profile of FAAs observed in Nham control and Nham with starters may be due to a different evolution of reactions and processes involving both production and consumption of amino acids that occur simultaneously during the various steps of the ripening process and storage and whose combined effects could give rise to an increase or, on the contrary, to a decrease of their concentration. The aminopeptidase activity is considered the main process implied in the FFA release in meat. Moreover, free amino acids concentration could be decreased either by chemical and enzymatic reactions where they act as substrates leading to the formation of secondary products (Ruiz et al., 1999; Ventanas et al., 1992) and/or by microbial amino acid decarboxylase activity with consequent BA production (Virgili et al., 2007).

In Nham control, an effect due to higher concentration of decarboxylase than that of Nham with starters, thus, their reaction with the free amino acids causing an increase of their BA concentration in these samples.

The ripened taste could be related to lysine and glutamic acid, while isoleucine and aspartic acid are implied in acid taste and unpleasant aroma (Buscailhon et al., 1994; Flores et al., 1998). In this study, the increase in concentration of lysine and glutamic acid was observed. The changes in the contents of free amino acids observed in fermented sausages during ripening are given in Table 2. The total free amino acid contents of the Nham control and Nham with starters constituted 212.7–216.4 mmol/g and 197.2–203.4 mmol/g dry matter, respectively (before ripening) on 0 day. An increase in the content of amino acids of Nham control and Nham with starters was observed and ranged between 275.2–349.8 mmol/g and 259.8–300.3 mmol/g dry matter during the ripening on day 3, and a further increase up to the range of 377.6–1851.7 mmol/g and 348.1 nmol/g–1256.0 mmol/g dry matter of total free amino acids was observed during storage at 4°C-25°C of Nham control and Nham with starters (4 weeks). The highest total free amino acid concentration of 1867.2 mmol/g was observed with Nham control stored at 25 °C for 4 week, whereas the lowest total free amino acid concentration of 359.6 mmol/g was observed with Nham with starters stored at 25 °C for 4 week. The hydrolysis of meat proteins generates polypeptides that can be further degraded to smaller peptides and free amino acids. This degradation can be produced by endogenous and microbial enzymes (De Masi et al., 1990; Hughes et al., 2002; Molly et al., 1997). The increase in the total free amino acid concentration was detected in all batches (Hierro et al., 1999, Bruna et al., 2000, Bolumar et al., 2001 and Hughes et al., 2002).

The main differences in the content of total free amino acids among batches were detected during 72 hour of ripening and during 4 week of storage. The amino acids in which differences, which were primarily responsible for the increase in total free amino acids during ripening, were observed were Glu (glutamic acid), Ala (alanine) and Arg (arginine) in Nham control and Nham with starters. Mateo et al. (1996) reported an increase in the total free amino acid content during the ripening. The change occurred during ripening and storage process indicating that the highest enzymatic activity took place during these stages

Amino acid	25°C Control 24 (h)	48 (h)	72 (h)	25°C L. plantarum + L. sake 24 (h)	48 (h)	72 (h)	30°C Control 24 (h)	48 (h)	72 (h)	30°C L. plantarum + L. sake 24 (h)	48 (h)	72 (h)	37°C Control 24 (h)	48 (h)	72 (h)	37°C L. plantarum + L. sake 24 (h)	48 (h)	72 (h)
Ala	12.5a	14.5b	16.3c	13.9d	13.6d	15.4e	14.8b	15.7e	17.9f	13.4d	14.2b	16.6g	16.6g	20.4i	22.1j	12.9a	15.7e	18.6h
Arg	90.9a	105.9b	119.5c	94.2d	95.6e	108.0f	101.0g	107.2h	126.6i	90.3a	95.5e	115.3j	109.8k	134.0m	149.8n	90.7a	109.4k	127.2l
Asp	1.9a	2.3b	2.5b	1.9a	1.9a	2.2b	2.1b	2.3b	2.7b	1.8a	1.9a	2.3b	2.2b	2.7b	3.0b	1.9a	2.2b	2.7b
Cys	10.9a	13.3b	14.6c	11.2d	11.2d	13.2b	12.0e	13.3b	15.5f	10.1a	11.1d	13.1b	13.1b	15.7f	17.4g	10.4a	12.8e	15.5f
Glu	54.5a	65.8b	72.9c	55.1a	54.9a	64.2d	61.6e	67.7f	80.7g	51.4h	56.8i	69.1j	64.2d	80.4m	88.3n	52.5k	65.0b	77.0l
Gly	6.6a	7.9b	8.7c	6.7a	6.7a	8.0b	7.3b	8.1b	9.4d	6.3a	6.9a	8.4bc	7.7b	9.3d	10.3e	6.0a	7.3b	8.8c
His	6.3a	7.5b	8.4c	6.8a	6.7a	7.7b	7.6b	8.2c	9.4d	6.3a	6.9a	8.3c	8.2c	10.1de	10.9e	6.6a	8.2c	9.5d
Leu	1.8a	2.2ab	2.4ab	1.9a	1.9a	2.2ab	2.1ab	2.2ab	2.7bc	1.7a	1.9a	2.3ab	2.1ab	2.7bc	3.0c	1.8a	2.2ab	2.7bc
Lys	4.3a	5.1b	5.6b	4.5a	4.6a	5.2b	5.0b	5.5b	6.5c	4.3a	4.7a	5.7b	5.3a	6.5c	7.4d	4.4a	5.3b	6.2c
Ile	1.2a	1.4a	1.5a	1.2a	1.2a	1.4a	1.3a	1.4a	1.6a	1.1a	1.2a	1.4a	1.4a	1.7ab	1.9b	1.1a	1.4a	1.6a
Met	0.4a	0.4a	0.5a	0.4a	0.4a	0.4a	0.4a	0.5a	0.5a	0.4a	0.4a	0.5a	0.4a	0.5a	0.6a	0.4a	0.4a	0.5a
Phe	0.9a	1.0a	1.2a	0.9a	0.9a	1.0a	0.9a	1.0a	1.2a	0.8a	0.9a	1.0a	1.1a	1.3a	1.4a	0.9a	1.0a	1.3a
Pro	2.6a	3.1ab	3.4b	2.6a	2.7a	3.2b	3.0b	3.3b	3.8b	2.6a	2.8a	3.4b	3.1ab	3.8b	4.3c	2.5a	3.0b	3.6b
Ser	7.6a	9.2bc	10.4c	7.9a	7.7a	8.9b	8.7b	9.4c	11.2d	7.1e	7.9f	9.5c	9.2bc	11.6d	12.8e	7.7e	9.1bc	10.9c
Thr	7.7a	8.4b	9.3c	7.7a	7.5a	8.7b	8.3b	9.2c	10.5d	7.2a	7.8a	9.3c	8.6b	10.3d	11.3f	6.6a	8.1b	9.4c
Tyr	1.1a	1.3a	1.4a	1.1a	1.1a	1.3a	1.2a	1.3a	1.5a	1.0a	1.1a	1.3a	1.3a	1.6a	1.8a	1.1a	1.3a	1.6a
Val	2.2a	2.6a	2.9a	2.3a	2.2a	2.6a	2.5a	2.7ab	3.1b	2.2a	2.3a	2.7ab	2.7a	3.3b	3.6b	2.2a	2.6a	3.2b
Total	214.9a	250.3b	275.2c	217.1a	220.2a	259.8b	239.8c	259.0b	304.8d	208.1a	220.0a	270.1c	257.1b	315.9f	349.8g	209.7a	255.1b	300.3e

Control (without starter culture).

Results are expressed as means of three replicates in mmol/g dry matter.

Means with different letters along rows are significantly different (P<0.05).

Table 2. Amino acid content of Nham without and with starter cultures during ripening at different temperature.

(Verplaetse et al., 1989). A major release of free amino acids at the beginning of the process have been studied in coincidence with the ripening stage (Diaz et al., 1997). This increase has been attributed to the higher temperatures applied during ripening compared to the low temperature. The most significant increases occurred in the content of Arg (arginine) in the sample. The decrease in the content of amino acids may indicate their metabolism by bacteria (Bover-Cid et al., 2000; Ordonez et al., 1999; Sekikawa et al., 2003).

Amino acid	Control				L. plantarum + L. sake			
	Storage time (wk)				Storage time (wk)			
	1	2	3	4	1	2	3	4
Ala	21.4a	22.3a	23.0ab	24.8b	22.1a	20.7a	21.1a	20.2a
Arg	151.9a	154.4a	154.9a	168.3b	149.8a	145.5a	153.8a	147.2a
Asp	3.2a	3.4a	3.5a	3.7a	2.9a	2.9a	3.1a	3.0a
Cys	18.2a	18.5a	18.6a	19.9b	17.8ac	17.0c	17.7ac	17.7ac
Glu	96.0a	98.4a	99.6ab	103.6b	88.3c	86.5c	92.2d	91.4d
Gly	11.2ac	11.4ac	11.6ac	12.4b	10.8c	10.4c	10.8c	10.6c
His	10.5a	10.5a	10.6a	11.3a	10.8a	10.2a	10.7a	10.2a
Leu	2.9a	3.0a	3.1a	3.3a	2.9a	2.8a	2.9a	2.9a
Lys	7.7a	7.8a	7.9a	8.5a	7.4a	7.0a	7.5a	7.3a
Ile	2.0a	2.0a	2.1a	2.2a	1.9a	1.8a	1.9a	1.9a
Met	0.6a	0.6a	0.6a	0.7a	0.6a	0.6a	0.6a	0.6a
Phe	1.5a	1.5a	1.6a	1.7a	1.4a	1.3a	1.4a	1.4a
Pro	4.6a	4.6a	4.7a	4.8a	4.3a	4.2a	4.4a	4.3a
Ser	12.7a	13.2a	13.6ab	14.4b	12.3ac	11.8c	12.4ac	12.1ac
Thr	12.5ac	12.8ac	13.1ab	14.0b	12.2a	11.9c	12.3c	12.3c
Tyr	1.8a	1.8a	1.8a	1.9a	1.7a	1.7a	1.8a	1.7a
Val	3.7a	3.8a	3.9a	4.2a	3.6a	3.4a	3.5a	3.5a
Total	352.9a	358.6a	363.1ab	377.6b	344.9ac	335.1c	357.3a	348.1ac

Control (without starter culture).
Results are expressed as means of three replicates in mmol/g dry matter.
Means with different letters along rows are significantly different (P<0.05).

Table 3. Amino acid content of Nham without and with starter cultures during stored at 4°C.

Amino acid	Control				L. plantarum + L. sake			
	Storage time (wk)				Storage time (wk)			
	1	2	3	4	1	2	3	4
Ala	20.3a	21.7ad	31.2b	37.9c	23.0d	23.1d	27.3e	31.8b
Arg	147.5a	159.4b	237.5c	288.0d	155.9b	162.4b	191.7e	231.4c
Asp	3.2a	3.3a	5.0b	6.3c	3.1a	3.2a	3.9a	4.8b
Cys	18.5a	19.5a	28.5b	36.1c	18.6a	19.0a	23.4d	27.8b
Glu	91.6a	97.3b	146.3c	183.7d	91.1a	93.2a	114.0e	138.8f
Gly	11.0a	11.5a	16.9b	21.1c	11.1a	11.5a	14.1d	16.8b

His	10.4a	11.2a	16.2b	20.5c	11.3a	11.4a	13.8d	16.0b
Leu	3.1a	3.3a	4.9b	6.0c	3.1a	3.2a	3.9a	4.6b
Lys	7.1a	7.5a	11.3b	13.9c	7.5a	7.8a	9.2d	10.9e
Ile	1.9a	2.1a	3.1ab	3.8b	1.9a	2.0a	2.5a	3.0ab
Met	0.6a	0.6a	0.9a	1.2a	0.6a	0.6a	0.8a	0.9a
Phe	1.4a	1.5a	2.3ab	2.9b	1.5a	1.5a	1.8a	2.2ab
Pro	4.3a	4.5a	6.7b	8.3c	4.3a	4.6a	5.6a	6.7b
Ser	12.8a	13.9a	20.7b	26.2c	13.0a	13.1a	15.8d	19.3b
Thr	11.7a	12.5a	18.7b	23.1c	12.7a	12.8a	15.4d	18.4b
Tyr	1.8a	1.9a	2.7b	3.4c	1.8a	1.9a	2.3ab	2.7b
Val	3.7a	3.9a	5.9b	7.5c	3.8a	3.8a	4.6a	5.5b
Total	348.8a	367.2a	542.4b	657.7c	359.1a	374.0a	461.2d	547.1e

Control (without starter culture).
Results are expressed as means of three replicates in mmol/g dry matter.
Means with different letters along rows are significantly different (P<0.05).

Table 4. Amino acid content of Nham without and with starter cultures during stored at 15°C.

Amin o acid	Control				L. plantarum + L. sake			
	Storage time (wk)				Storage time (wk)			
	1	2	3	4	1	2	3	4
Ala	35.6a	40.6b	103.8c	122.6d	31.5e	41.8b	68.2f	76.2g
Arg	232.9a	280.6b	710.9c	832.7d	212.3e	291.3f	475.2g	531.2h
Asp	4.7a	5.8ad	14.7b	17.1c	4.1a	5.6d	9.5e	11.0f
Cys	28.7a	34.6b	86.9c	101.8d	24.0a	32.3b	55.1e	64.4f
Glu	140.9a	174.2b	433.8c	512.6d	118.7e	160.1f	270.5g	313.2h
Gly	16.9a	20.3b	50.6c	59.2d	14.2e	19.3b	32.9f	37.8g
His	17.8a	21.0b	52.8c	62.4d	15.1a	20.4b	34.1e	39.5f
Leu	4.8a	5.9a	14.6b	17.0c	4.1a	5.6a	9.4d	10.9e
Lys	10.9a	13.5b	34.6c	39.8d	9.9a	13.3b	21.8e	25.0f
Ile	3.0a	3.6a	8.9b	10.1c	2.6a	3.4a	5.9d	6.7d
Met	0.9a	1.1ad	2.8bc	3.2c	0.8a	1.1ad	1.9d	2.1d
Phe	2.3ac	2.8a	7.1b	8.2b	1.9c	2.6a	4.4d	5.1d
Pro	6.5a	8.0b	20.5c	23.6d	5.9a	7.9b	13.5e	15.0f
Ser	20.1a	24.2b	60.3c	68.2d	16.8e	22.8e	38.3f	44.7g
Thr	17.7a	21.9b	56.0c	66.2d	16.4a	22.2b	36.8e	41.2f
Tyr	2.8a	3.4a	8.4b	9.8c	2.4a	3.2a	5.5d	6.4d
Val	5.7a	7.0b	17.8c	21.0d	5.0a	6.6b	10.9e	12.7f
Total	541.0a	651.8b	1637.2c	1851.7d	472.0e	647.8b	1103.9f	1256.0g

Control (without starter culture).
Results are expressed as means of three replicates in mmol/g dry matter.
Means with different letters along rows are significantly different (P<0.05).

Table 5. Amino acid content of Nham without and with starter cultures during stored at 25°C.

Two types of fermented sausage differing in starter culture were produced in parallel with two different starter cultures (no starter and *L. plantarum* + *L. sake*). The sausages were ripened 3 days and subsequently stored 7, 14, 21 and 28 days at the 4 °C, 15 °C and 25 °C. Concentration of three most abundant amines, cadaverine, putrescine and tyramine increased significantly ($P < 0.05$) in Nham during ripening and and also during storage. The dominant BAs in the control were cadaverine – and tyramine and putrescine, to a lesser extent; the cadaverine, putrescine and tyramine content were lower if inoculation had added with *L. plantarum* + *L. sake*; whereas they ranked above 300 mg/kg in the control by 3 d. At the end of ripening, cadaverine (98.7 mg/kg dry matter), putrescine (242.6 mg/kg dry matter) and tyramine (46.4 mg/kg dry matter) content in the A-samples-sausage was higher ($P < 0.05$) than in Nham with starters (51.6, 98.4 and 27.8 mg/kg dry matter, respectively). Starter culture influenced significantly in decrease of ($P < 0.05$) cadaverine, putrescine and tyramine content in the sausage. Due to the significant ($P < 0.05$) increase of total aerobic counts in the Nham control between the end of ripening and during storage, followed by the significant ($P < 0.05$) increase of the sum of total BAs between the 72 hour of ripening (387.7 mg/kg dry matter) and the 4th week of storage at 25°C (629.2 mg/kg dry matter).

The main rate of BAs production was during the first two days, when a sharp pH decrease and the development of LAB occurred. Sausages fermented with starters had lower amounts of cadaverine, putrescine and tyramine than naturally fermented sausages (control) during storage at 15°C and 25°C. However, phenylethylamine, histamine and tryptamine were not detected.

Nham control showed proteolysis that was correlated with pH values higher than those with starters. However, no positive correlation was found between the proteolysis index and BAs production. Since proteolysis was stronger during the second half of the ripening process, the FAAs occurred later than the early amine production. No effect on pH development in the fermented sausage was observed when non-amine forming strain of *L. plantarum* + *L. sake* were present during 4 week of 4°C storage period. A study on the evolution of FAAs and BAs in Nham during 4 week at different temperatures of storage (4°C, 15°C and 25°C) was performed. FAAs and BAs were determined by RP-HPLC. Storage temperature of 15°C and 25°C promoted a significant increase of the contents of arginine, glutamic acid, cadaverine, putrescine and tyramine, expressed as g/kg of dry matter while storage temperature of 4°C decreased a significant of the contents of arginine, glutamic acid, cadaverine, putrescine and tyramine, expressed as g/kg of dry matter. These two amino acids and three BAs may serve as indicators of temperatures changes in stored fermented sausage.

4. Conclusions

The aim of this study was to investigate the effect of non-amine forming LAB as starter culture during ripening and storage time and temperature on the evolution of FAAs of Nham during processing. The correlation between FAAs and BAs content was also investigated. Larger increases of FAAs occurred in Nham without starter in the ripening and storage step. Total FAAs content was highly correlated with total BAs amount. Sausage ripening was further carried out with non-amine forming strain of *L. plantarum* + *L. sake* after ripening and stored at different temperature. The amount of amine in the product was significantly less than the control. The results obtained for BAs degradation by bacteria in a synthetic medium suggest that AO activity is strain dependent rather than being related to specific species. In all batchs, the total amino acid contents increased with time – and the predominant ones were arginine and glutamic acid. However, upon inoculation with non-amine forming strain, the total BAs contents remained considerably lower than those of the control. Hence, an efficient food-grade biological tool was made available that constrains buildup of BAs in fermented sausage during storage.

Author details

Jirasak Kongkiattikajorn
School of Bioresources and Technology, King Mongkut's University of Technology Thonburi, Thailand

Acknowledgement

This study was supported by Office of the Higher Education Commission (OHEC), Thailand.

5. References

AOAC. (2000). *Official methods of analysis* (17th ed.). Gaithersburg, Maryland: Association of Official Analytical Chemists.

Benjakul, S., & Morrissey, M. T. (1997). Protein hydrolysates from Pacific whiting solid wastes. *Journal of Agricultural and Food Chemistry*, Vol.45, pp. 3423–3430.

Beutling, D. (1996). Biogenic Amines in Nutrition (Biogene Amine in der Ernaehrung). Springer-Verlag, Berlin, Germany, pp. 59–67.

Bover-Cid, S., Hugas, M., Izquierdo-Pulido, M. & Vidal-Carou, M.C. (2000). Reduction of biogenic amine formation using a negative amino acid-decarboxylase starter culture for ripening of *fuet* sausage. *Journal of Food Protection*, Vol. 63, No.2, pp. 237–243.

Bover-Cid, S., Izquierdo-Pulido, M. & Vidal-Carou, M. C. (2000). Influence of hygienic quality of raw materials on biogenic amine production during ripening and storage of dry fermented sausage. *Journal of Food Protection*, Vol. 63, pp. 1544–1550.

Bover-Cid, S., Hugas, M., Izquierdo-Pulido, M. & Vidal-Carou, M. C. (2001). Amino acid-decarboxylase activity of bacteria isolated from fermented pork sausages. International *Journal of Food Microbiology*, Vol. 66, pp. 185–189.

Bolumar, T., Nieto, P. & Flores, J. (2001). Acidity, proteolysis and lipolysis changes in rapid cured fermented sausage dried at different temperatures. *Food Science and Technology International*, Vol.7, pp. 269–276.

Brink, B. ten., Damink, C., Joosten, H. M. L. J. & Huis in't Veld, J. H. J. (1990). Occurrence and formation of biologically active amines in foods. *International Journal of Food Microbiology*, Vol.11, pp. 73–84.

Bruna, J. M., Fernandez, M., Hierro, E. M., Ordonez, J. A. & De la Hoz, L. (2000). Combined use of pronase E and a fungal extract (*Penicillium aurantiogriseum*) to potentiate the sensory characteristics of dry fermented sausages. *Meat Science*, Vol.54, pp. 135–145.

Buscailhon, S., Berdagué, J. L., Bousset, J., Gandemer, G., Cornet, M. & Touraille, C. (1994). Relations between compositional traits and sensory qualities of French dry-cured ham. *Meat Science*, Vol.37, pp. 229–243.

Cooper, R.A., 1997. On the amine oxidases of *Klebsiella aerogenes* strain W70. *FEMS Microbiology Letters* 146, 85– 89.

Hernández-Jover, T., Izquierdo-Pulido, M.,Veciana-Nogués, M.T. & Vidal-Carou, M.C. (1996). Ion-pair high-performance liquid chromatographic determination of biogenic amines in meat and meat products. *Journal of Agricultural and Food Chemistry*, Vol. 44, No.9, pp. 2710–2715.

De Masi, T. W., Wardlaw, F. B., Dick, R. L. & Acton, J. C. (1990). Non protein nitrogen (NPN) and free amino acid contents of dry fermented and non fermented sausages. *Meat Science*, Vol.27, pp. 1–12.

Durà, M. A., Flores, M. & Tolrà, F. (2004). Effect of *Debaryomyces* spp. On the proteolysis of dry-fermented sausages. *Meat Science*, Vol.68, pp. 319–328.

Dapkevicius, M.L.N.E., Nout, M.J.R., Rombouts, F.M., Houben, J.H. & Wymenga, W., 2000. Biogenic amine formation and degradation by potential fish silage starter microorganisms. International *Journal of Food Microbiology*, Vol.57, pp. 107– 114.

Eerola, S., Roig-Saguèz, A. X. & Hirvi, T. K. (1998). Biogenic amines in Finnish dry sausages. *Journal of Food Safety*, Vol.18, pp. 127–138.

Flores, M., Spanier, A. M. & Toldrá, F. (1998). Flavour analysis of dry-cured ham. In Shahidi, F. (Ed.), Flavour of meat product and seafood pp. 320–341. London: Blackie.

Gale, E. F. (1946). The bacterial amino acid decarboxylases. In Nord, F. F. (Ed.), Advances in enzymology, Vol. 6, New York: Interscience Publisher.

Gardini, F., Martuscelli, M., Caruso, M. C., Galgano, F., Crudele, M. A. & Favati, F. (2001). Effects of pH, temperature and NaCl concentration on the growth kinetics, proteolytic activity and biogenic amine production of *Enterococcus faecali*. *International Journal of Food Microbiology*, Vol.64, pp. 105–117.

Gardini, F., Martuscelli, M., Crudele, M.A., Paparella, A. & Suzzi, G., 2002. Use of *Staphylococcus xylosus* as a starter culture in dried sausages: effect on the biogenic amine content. *Meat Science*, Vol. 61, pp. 275– 281.

Greene, D. H., & Babbitt, J. K. (1990). Control of muscle softening and protease–parasite interactions in arrowtooth flounder, Ateresthes stomias. *Journal of Food Science*, Vol.55, pp. 579–580.

Hierro, E., De la Hoz, L. & Ordonez, J. A. (1999). Contribution of the microbial and meat endogenous enzymes to the free amino acid and amine contents of dry fermented sausages. *Journal of Agriculture and Food Chemistry*, Vol.47, pp. 1156–1161.

Hughes, M. C., Kerry, J. P., Arendt, E. K., Kenneally, P. M., McSweeney, P. L. H. & O'Neill, E. E. (2002). Characterization of proteolysis during the ripening of semidry fermented sausages. *Meat Science*, Vol.62, pp. 205–216.

Ishizuka, H., Horinouchi, S. & Beppu, T., 1993. Putrescine oxidase of *Micrococcus rubens*: primary structure and *Escherichia coli*. *General Microbiology*, Vol.139, pp. 425–432.

Joosten, H.M.L.J. & Northolt, M.D. (1989). Detection, growth and amine producing capacity of lactobacilli in cheese. *Applied and Environmental Microbiology*, Vol. 55, No.9, pp. 2356–2359.

Kalac, P., Spicka, J., Krizek, M. and Pelikanova, T. (2000). The effect of lactic acid bacteria inoculants on biogenic amines formation in sauerkraut. *Food Chemistry*, Vol.70, pp. 355–359.

Lehane, L. & Olley, J., 2000. Histamine fish poisoning revisited. International *Journal of Food Microbiology*, Vol.58, pp. 1 –37.

Lehninger, A. L., Nelson, D. L. & Cox, M. M. (1999). Principi di biochimica, ed. Zanichelli, Bologna.

Leuschner, R.G.K. & Hammes, W.P., 1998. Tyramine degradation by micrococci during ripening of fermented sausages. *Meat Science*, Vol.49, pp. 289–296.

Lowry, Q. H., Rosebrough, N. J., Farr, L. A., & Randall, R. J. (1951). Protein measurement with the Folin phenol reagent. *Journal of Biological Chemistry*, Vol.193, pp. 256–275.

Mah, J.H. & Hwang, H.J. (2009). Inhibition of biogenic amine formation in a salted and fermented anchovy by *Staphylococcus xylosus* as a protective culture. *Food Control*, Vol. 20, pp. 796–801.

Maijala, R. (1994). Histamine and tyramine production by a Lactobacillus strain subjected to external pH decrease. *Journal of Food Protection*, Vol.57, pp. 259–262.

Maijala, R., Eerola, S., Lievonen, S., Hill, P. & Hirvi, T. (1995). Formation of biogenic amines during ripening of dry sausages as affected by starter culture and thawing time of raw materials. *Journal of Food Science*, Vol.60, pp. 1187–1190.

Martuscelli, M., Crudele, M.A., Gardini, F. & Suzzi, G., 2000. Biogenic amine formation and oxidation by *Staphylococcus xylosus* strains from artisanal fermented sausages. *Letters in Applied Microbiology*, Vol.31, pp. 228– 232.

Molina, I. & Toldrà, F. (1992). Detection of proteolytic activity in microorganisms isolated from dry-cured ham. *Journal of Food Science*, Vol. 57, pp. 1308–1310.

Montel, M. C. & Champomier, M. C. (1987). Arginine catabolism in *Lactobacillus sake* from meat. *Applied and Environmental Microbiology*, Vol.53, pp. 2683–2685.

Molly, K., Demeyer, D., Johansson, G., Raemaekers, M., Ghistelinck, M. & Geenen, I. (1997). The importance of meat enzymes in ripening and flavour generation in dry fermented sausages. First results of a European project. *Food Chemistry*, Vol.59, pp. 539–545.

Monin, G., Marinova, P., Talmant, A., Martin, J. F., Cornet, M. & Lanore, D. (1997). Chemical and structural changes in dry-cured hams (Bayonne hams) during processing and effects of the deharing technique. *Meat Science*, Vol.47, pp. 29–47.

Montel, M. C. & Champomier, M. C. (1987). Arginine catabolism in Lactobacillus sake from meat. *Applied and Environmental Microbiology*, Vol.53, pp. 2683–2685.

Moreno-Arribas, M. V., Polo, M. C., Jorganes, F., & Munoz, R. (2003). Screening of biogenic amine production by acid lactic bacteria isolated from grape must and wine. *International Journal of food Microbiology*, Vol.84, pp. 117–123.

Murooka, Y., Doi, N. & Harada, T., 1979. Distribution of membrane bound monoamine oxidase in bacteria. *Applied and Environmental Microbiology*, Vol.38, pp. 565– 569.

Murooka, Y., Higashiura, T. & Harada, T., 1976. Regulation of tyramine oxidase synthesis in *Klebsiella aerogenes*. *Journal of Bacteriology*, Vol.127, pp. 24– 31.

Ordonez, J. A., Hierro, E. M., Bruna, J. M. & De la Hoz, L. (1999). Changes in the components of dry-fermented sausages during ripening. *Critical Reviews in Food Science and Nutrition*, Vol.39, pp. 329–367.

Rodrìguez, M., Núñez, J. J., Códoba, M. E., Bermúdez, M. E. & Asensio, M. A. (1998). Evolution of proteolytic activity of micro-organisms isolated from dry-cured ham. *Journal of Applied Microbiology*, Vol.85, pp. 905–912.

Rozan, P., Kuo, Y.H. & Lambein, F. (2000). Free amino acids present in commercially available seedlings sold for human consumption: a potential hazard for consumers. *Journal of Agricultural and Food Chemistry*, Vol. 48, No.3, pp. 716–723.

Ruiz, J., Garcia, C., Diaz, M. C., Cava, R., Tejeda, F. J. & Ventanas, J. (1999). Dry cured Iberian ham non-volatile components as affected by the length of the curing process. *Food Research International*, Vol.32, pp. 643–651.

Ruiz-Capillas, C. & Moral, A. (2001). Production of biogenic amines and their potential use as quality control index foer hake (*Merluccius merluccius* L.) stored in ice. *Journal of Food Science*, Vol.66, pp. 1030–1032.

Scannell, A. G. M., Kenneally, P. M. & Arendt, E. K. (2004). Contribution of starter cultures to the proteolytic process of a fermented non-dried whole muscle ham product. *International Journal of Food Microbiology*, Vol.93, pp. 219–230.

Schomburg, D. & Stephan, D., 1993. Enzyme Handbook. Springer-Verlag, Berlin, Germany.

Sekikawa, M., Kawamura, T., Fujii, H., Shimada, K., Fukushima, M. & Mikami, M.(2003). Effect of arginine on growth of lactic acid bacteria for fermented sausage under high concentration of salt. *Hokkaido Animal Science and Agriculture Society*, Vol.45, pp. 17–23.

Silla-Santos, M. H. (1996). Biogenic amines: Their importance in food. *International Journal of Food Microbiology*, 29, pp. 213–221.

Suzzi, G. & Gardini, F. (2003). Biogenic amines in dry fermented sausages: A review. *International Journal of Food Microbiology*, Vol.88, pp. 41–54.

Toldrá, F. (2005). Dry-cured ham. In Hui, Y. H., Culbertson, J. D., Duncan, S., Guerrero-Legarreta, I., Li-Chan, E. C. Y., Ma, C. Y., Manley, C. H., McMeekin, T. A., Nip, W. K., Nollet, L. M. L., Rahman, M. S., Toldrá, F., &. Xiong, Y. L (Eds.). Handbook of food science, technology and engineering (Vol. 1). Boca Raton, FL: USA, Marcel-Dekker Inc./CRC Press.

Toldrá, F. (2006). The role of muscle enzymes in dry-cured meat products with different drying conditions. *Trends in Food Science and Technology*, Vol.17, pp. 164–168

Ventanas, J., Cordoba, J. J., Antequera, T., Garcìa, C., López-Bote, C. & Asensio, M. A. (1992). Hydrolysis and Maillard reactions during ripening of Iberian ham. *Journal of Food Science*, Vol.57, No.4, pp. 813–815.

Verges, C.M.C., Zuniga, M., Morel-Deville, F., Perez-Martinez, G., Zagorec, M. & Ehrlich, S.D. (1999). Relationships between arginine degradation, pH and survival in *Lactobacillus sake*. *FEMS Microbiology Letter*, Vol. 180, pp. 297–304.

Verplaetse, A., De Bosschere, M. & Demeyer, D. (1989). Proteolysis during dry sausage ripening. In Proceedings of the 35th international congress on meat science and technology. Copenhaegen, Denmark.

Virgili, R., Saccani, G., Gabba, L., Tanzi, E. & Soresi Bordini, C. (2007). Changes of free amino acids and biogenic amines during estende ageing of Italian dry-cured ham. *LWT – Food Science and Technology*, Vol.40, No.5, pp. 871–878.

Yamashita, M., Sakaue, M., Iwata, M., Sugino, H. & Murooka, Y. (1993). Purification and characterization of monoamine oxidase from *Klebsiella aerogenes*. *Journal of Ripening and Bioengineering*, Vol. 76, pp. 289–295.

Section 3

Vegetable & Cereal Products

Chapter 6

Fermentation of Vegetable Juices by *Lactobacillus Acidophilus* LA-5

Lavinia Claudia Buruleanu, Magda Gabriela Bratu,
Iuliana Manea, Daniela Avram and Carmen Leane Nicolescu

Additional information is available at the end of the chapter

1. Introduction

Probiotics foods represent one of the largest sectors in functional food markets. Most of the available probiotic products are some form of dairy, despite the continuous growth of the non-dairy probiotic sector, with products like soy-based drinks, fruit-based foods, and other cereal-based products. Both non-dairy (in general) and soy-based probiotic products represent a huge growth potential for the food industry, and may be widely explored through the development of new ingredients, processes, and products. For this purpose, new studies must be carried out to: test ingredients, explore more options of media that have not yet been industrially utilized, reengineer products and processes, towards potentially meet the demands of lactose-intolerant and vegetarian consumers for new nourishing and palatable probiotic products [1].

Lactic acid bacteria are among the most important probiotic microorganisms typically associated with the human gastrointestinal tract. Traditionally, lactic acid bacteria have been classified on the basis of phenotypic properties, e.g. morphology, mode of glucose fermentation, growth at different temperatures, lactic acid configuration, and fermentation of various carbohydrates. However some species, like the so-called *Lactobacillus acidophilus* group and some bifidobacteria, are not readily distinguishable by phenotypic characteristics [2]. From the physiological point of view, *Lactobacillus acidophilus* strains were characterized as lactic acid bacteria with strictly homofermentative metabolism (> 85% lactic acid). The hexoses are preferential fermented via Embden – Meyerhof – Parnas (EMP), (as the strains produce aldolase and phosphoketolase), and only then the pentoses and gluconate are fermented. LAB of the *Lactobacillus acidophilus* group *as* well as of the *Bifidobacterium* group isolated from the human faeces or intestine are thought to have beneficial effects on health being thus considered to be probiotic bacteria [3].

For use in food, important criteria for probiotics must be met, in particular that they should not only be capable of surviving passage through the digestive tract, by exhibiting acid and bile tolerance, but also have the capability to proliferate in the gut.

Probiotics must be able to exert their benefits on the host through growth and/or activity in the human body. Although generally recognised as safe a probiotic strains must be characterized by a set of tests that assure its safety to consumer (1, 2, 3, 5, 6).

Inclusion of probiotic bacteria in fermented dairy products enhances their value as better therapeutic functional foods. However, insufficient viability and survival of these bacteria remain a problem in commercial food products. By selecting better functional probiotic strains and adopting improved methods to enhance survival, including the use of appropriate prebiotics and the optimal combination of probiotics and prebiotics (synbiotics), an increased delivery of viable bacteria in fermented products to the consumers can be achieved [5].

The fermentation of vegetable products, applied as a preservation method for the production of finished and half-finished food products, is considered as an important technology, though requiring more research, as a growing number of raw materials are being processed in this way by the food industry. The main reasons for this interest are nutritional, physiological and hygienic aspects of the process [6]. Thus, according to Kelwicka, (2010) [7], the fermentation of beetroot juice requires selected starter cultures made of LAB, naturally present in this vegetable although their number is usually very small. This makes them un-appropriate to, alone, conducting a fermentation that ensures satisfying sensory properties of the fermented juice, with improved health promoting activity.

Thus, probiotic juices represent an alternative to dairy products that suits consumers who don't want to eat dairy foods or are lactose intolerant. Adding probiotics to juices is more complex than formulating in dairy products where the bacteria can be easily added to other cultures.

Despite its potential for healthy products development, there is very little research activity addressing the fermentation of vegetable juices using probiotic bacteria.

2. Materials and methods

2.1. Vegetables treatments

Fresh vegetables (carrots, cucumbers, beetroot, white cabbage, red cabbage) were purchased from a retail market and specifically processed by removing the non-edible pieces. The raw material processing was made faster, because the possibility of contamination and proliferation of microorganisms in the products is very high in comparison with their intact counterparts (Lee, 2011). Using a domestic extractor the vegetables were turned into juice. The heating treatment of the juice, applied at 80°C with a view to destroy the undesirable microorganisms under the limit of detection, was followed by cooling at 40°C.

2.2. Microorganisms and fermentation conditions

The strain *Lactobacillus acidophilus* LA-5 from Christian Hansen (Romania) was used in this study.

The lyophilized culture was aseptically inoculated into the vegetable juices and vigorously homogenized for 15 min, according to the producer's specification. The fermentation experiments were carried out using Erlenmeyer flasks containing 50ml of juice, without pH adjustment. The flasks were incubated statically in an incubator chamber at 37±0.2⁰C. Sampling was taken at regular interval of times for physico-chemical and microbiological analysis.

The tested supplements were: L-cysteine hydrochloride monohydrate (Merck, Darmstadt, Germany), L-lysine hydrochloride (Merck), L-valine (Merck), L-leucine (Calbiochem, San Diego, CA, USA) and yeast extract (Merck). Cysteine, lysine, valine and leucine were separately added in quantity by 0.1% (w/v) into carrot juice, while amounts by 0.2% (w/v) were tested, also individual, in the case of the yeast extract and cysteine. A control sample without supplements was carried out for each experiment.

2.3. Physico – Chemical analysis

Metabolic activity of the strain LA-5 in the conditions mentioned above was evaluated based on the dynamics of pH, respectively end products of fermentation. The pH values were measured with a HACH pH-meter. Lactic acid was determined using commercial kits (K-DLATE from Megazyme International). The calculations were made with Megazyme Mega-Calc™ and expressed as g lactic acid/l. Reducing sugars were analyzed applying the spectrophotometric method with 3.5-dinitrosalicilic acid (DNS) after the removing of other substances with reducing character using basic lead acetate and expressed as g glucose/l. Ascorbic acid was determined applying the 2,6-dichloroindophenol titrimetic method, based on the reduction of the sodium salt of the dye by ascorbic acid (AOAC method). It was expressed as mg/100ml. The amino acids content, expressed as g glycine/100ml, was determined through the Sörensen method.

2.4. Microbiological analysis

The amount of viable cells of *Lactobacillus* sp. was determined by serial tenfold dilution with sterile peptone water. Aliquots of 1ml were plated, in duplicate, in plates with Man-Rogosa-Sharpe agar, enriched with L-cysteine HCl. The Petri plates were incubated for 48-72h at 37°C and the results were expressed as log colony forming units (CFU)/ml juice.

The optical density of biomass was measured with the UV-Visible spectrophotometer at 610nm. In the preparation of the calibration curve for optical density vs. dry cell weight several dilutions of the juices were made. According Altiok [8], for each dilution 2 ml of sample was used to obtain optical densities at 610 nm wavelength and 15 ml of sample was filtered with a pre-weighed cellulose acetate membrane filter having a pore size of 0.45 μm using a vacuum pump. The biomass collected on the filters was washed with 15 ml of water and the filters were dried at 100⁰C for approximately 24 h until constant weight was observed. The results were expressed as g.

2.5. Statistical analysis

Statistical analysis was carried out using the software SPSS (Statistical Package for the Social Science 17.0 trial version).

3. Results and discussions

3.1. Effect of inoculum size on the lactic acid accumulation and biomass growth

A comparative study of the dynamics of lactic acid fermentation of carrot juice using three different concentrations of lyophilized pure culture was realized (Figure 1).

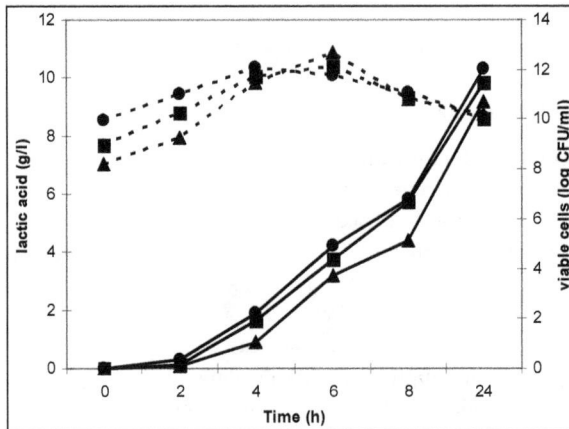

Figure 1. Correlation between lactic acid production by *Lactobacillus acidophilus* LA-5 and number of viable cells during fermentation of carrot juice with different inoculum size ▲ 0.2g/l; ■ 0.3g/l; ● 0.4 g/l (smooth lines - lactate, dashed lines - viable cells count)

Relative higher differences concerning the lactate increasing were observed between the variant with 0.2g/l pure culture initial added and the other two within 24 hours of fermentation. Thus, at the end of this interval, the excess was by 7.06% in the juice with 0.3g/l inoculum and 12.06% in the juice with 0.4g/l inoculum respectively. However, in all the batches the lactic acid accumulation, higher than 9g/l, could be considered satisfactory for the shelf life of the final products. From the other part, the number of viable cells is decisive for the probiotic feature of these ones. A direct proportionality between the amount of the lyophilised culture initial added and the viable cells was observed only in the first 4h of the fermentation. As a general characteristic, in the interval 6 - 24h pH values less than 4.5 have become inhibitory for the useful microbiota in all the experimental samples.

The initial concentration of reducing sugars of the carrot juices, by 25.2g/l, was favourable for the growth of *Lactobacillus acidophilus* LA-5. Testing two strains of Lactobacillus (one genetically selected Mont4+ and the other genetically altered, Mont4+pxyAB-mod). Kiouss [9] established that the Mont4+ had the highest yield of lactic acid fermenting with six percent concentration of glucose, whereas the L strain utilized the sugar best at the four percent concentration. In the same time temperature and pH seemed to play the largest role in the organisms ability to grow and thus affecting its production of lactic acid.

Concluding, higher inoculum densities of *Lactobacillus acidophilus* LA-5 were not significantly influenced the survival yield of the useful microbiota in the lactic acid fermented juices after 24h. In the same time, no parallel relationships between lactic acid concentration and the inoculum size were determined. The result agrees to those obtained by Agarwal, Dutt, Meghwanshi and Saxena [10] using *Enterococcus flavescens* for production of lactic acid. In their opinion, beyond a certain concentration lactic acid yield dropped due to high cell density resulting in fast depletion of essential nutrients, limiting further growth and reducing the yield. Referring to bifidobacteria, Dave and Shah [11] reported also that a higher inoculum did not always improve their viability to a satisfactory level. No data referring to *Lactobacillus acidophilus* were found in the literature.

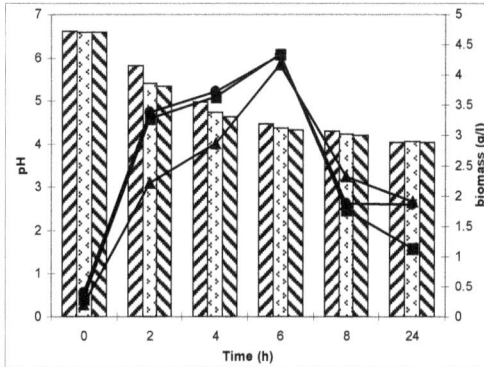

Figure 2. pH and biomass evolution during lactic acid fermentation of carrot juice with different inoculum of *Lactobacillus acidophilus* LA-5: 0.2g/l (▨ and ▲); 0.3g/l (▨ and ■); 0.4 g/l (▨ and ●); columns - pH values, lines - biomass

Although the pH dynamics was quite different in the first 6h of the process, the initial amount of the pure culture did not affect the subsequent evolution or the final value of this parameter (Figure 2).

The sharp decrease in biomass from 6 to 8h has been correlated with the viable cells tendency, as result of reaching pH values by 4.34 to 4.47. Being known that *Lb. acidophilus* is more sensitive in acidic environment, this result underlines the necessity to manage the size of inoculum in order to obtain a balance between the lactic acid accumulation and the survival of the probiotic microorganisms.

The maximum rate of acidification v_{max} was calculated as the time variation of pH (dpH/dt) and expressed as pH units/min (Table 1). Other kinetic parameters were also calculated: time to reach v_{max} (t_{max}, hours), time to reach pH 5.0 ($t_{pH\ 5.0}$, hours), time to complete the fermentation ($t_{pH\ 4.2}$, hours).

Inoculum, g/l	$v_{max} \cdot 10^{-3}$ (units/min.)	t_{max} (h)	$t_{pH\ 5.0}$ (h)	$t_{pH\ 4.2}$ (h)
0.2	7.08	4	2.95	8.4
0.3	9.83	2	2.88	8.2
0.4	10.41	2	2.67	8.05

Table 1. 1. Acidification kinetic parameters of fermentation of carrot juices by *Lactobacillus acidophilus* LA-5

A double amount of inoculum had an insignificant influence on the time to reach pH 5.0, important parameter from the shelf life of the fermented juices. Thus, $t_{pH\ 5.0}$ (h) was 1.1-fold higher in the case of the batch with 0.2g/l lyophilized pure culture initial added to juice than that one with 0.4g/l. A different situation was registered concerning the maximum rate of acidification (v_{max}) and the time to reach this rate (t_{max}). Thus, a polynomial equation of the form $y = -108.5x^2 + 81.75x - 4.93$ correlated the size of inoculum with the corresponding values of v_{max} at R squared = 1. Although at the initial moment of fermentation seems to be advantageous to use a higher amount of pure culture, this aspect lessen in time, from the economic point of view being important to obtain a balance between the quantity of inoculum and the targeted parameters which ensure the preservation of the final product.

The values of the biomass content became close after about 6h of fermentation. No parallel relationship between lactic acid concentration and biomass was observed, result that agrees to those obtained by Amrane [12] and Kotzamanidis [13].

However, taking into account the lactic acid accumulation and the dynamics of the number of viable cells, it was obvious that the utilization of higher amount of inoculum is not justified.

3.2. Effect of temperature on the dynamics of fermentation

According to the information provided by the producer of the lactic culture, respectively to the data found in literature, two different incubation temperatures were tested: 37°C and 41°C respectively.

The dynamics of both pH and lactic acid (Figure 3) emphasizes the influence of the higher temperature on the rate of acidification. After 24h no significant differences between the pH values were determined, while the lactic acid content of the samples fermented at 41°C was 1.24-fold higher comparatively with those fermented at 37°C. This situation may be due to the higher amino acids content in the samples fermented at 41°C, that act as buffer. Thus, expressed as glycin, the total amount was by 0.165g/100ml at the end of the analyzed interval, which represented an increase by 10% comparatively with the batch fermented at lower temperature.

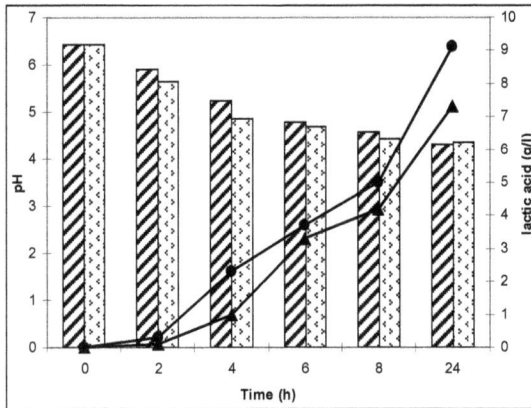

Figure 3. pH and lactic acid dynamics during the lactic acid fermentation of carrot juice at different temperatures: 37°C (▨▨▨ and ▲) and 41°C (▨▨▨ and ●); columns - pH values, lines - lactic acid content

The rate of acidification has been correlated with the glucose consumption: 38.9% in the case of the juice fermented at 37°C, respectively 53.89% in the case of the juice fermented at 41°C. The different tendency of this parameter became obviously after 4h of fermentation (Figure 4), being the consequence of the different rate of growth of *Lactobacillus acidophilus*, expressed as optical density at 610nm.

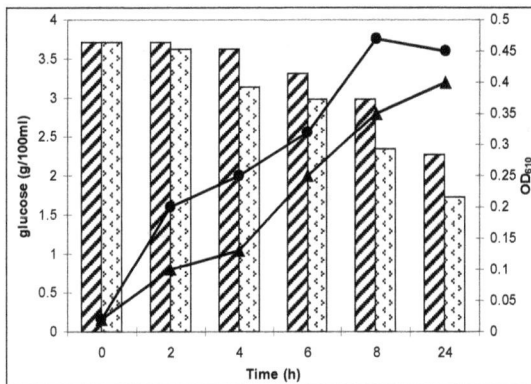

Figure 4. Glucose consumption and microbial evolution during the lactic acid fermentation of the carrot juices at different temperatures: 37°C (▨▨▨ and ▲) and 41°C (▨▨▨ and ●); columns - glucose, lines - optical density at wavelength by 610nm

Although close, the yields of glucose conversion to lactic acid have inclined the balance in favour of the juices fermented at 37°C, the corresponded value being by 0.5, unlike 0.45 in the case of the juices incubated at 41°C.

The faster consumption of the carbon source, correlated with the growth of the useful microbiota at higher temperature, respectively with the increase of the lactic acid content until the value by 9.1g/l, was followed at 24h by the decline of the viability of *Lactobacillus acidophilus*. Taking into account the dynamics of all the above mentioned parameters, the incubation temperature applied in the further studies was by 37±0.1⁰C.

3.3. The behaviour of different raw materials during the lactic acid fermentation by Lactobacillus acidophilus LA-5

Fresh white cabbage (*Brassica oleracea* L.), red cabbage (*Brassica oleracea* var. *capitata* f. *rubra*), red beet (*Beta vulgaris* var. *vulgaris*), cucumbers (*Cucumis sativus*) and red onion (*Allium cepa* var. *ascalonicum*) were chosen in order to perform different experimental batches, as follows: Cb - cabbage juice, RCb - red cabbage juice, Rb - red beet juice, Cc - cucumber juice, CcO - cucumber juice with 0.1% (v/v) onion juice added after the heating and cooling of the batches.

pH and lactic acid dynamics during the lactic acid fermentation of vegetable juices with *Lb. acidophilus* are shown in Figure 5 and Figure 6 respectively. The pH values ranged from 6.29 to 3.74, no significant differences between the analyzed batches being observed, excepting the red beet juice. Thus, after one day a higher value by 4.28 was determined, the prolongation of the time of fermentation with other 24h hadn't a positive influence on this parameter.

After 24h, the highest decrease of pH was determined in the case of the cucumber juice (2.51 units), correlated with the increase of the lactic acid amount until 9.36g/l. Although the pH values of the samples Cc and Cb were close during the process development, the maximum rate of acidification v_{max} registered a better value of $9.33 \cdot 10^{-3}$ units/min. in the case of the cucumber juice. This could explain the fermentation slowdown in the batch Cb the interval 6 - 8 hours. Correlated with the results of the microbiological analysis, it seems that this time the process was directed towards the growth of the useful microbiota. A minimum value of the maximum rate of acidification, by $6.66 \cdot 10^{-3}$ units/min., was determined in the case of CcO, while the time to reach pH 5.0 ($t_{pH\ 5.0}$, hours) ranged between 1.9 (Cb) to 3.5 (CcO).

A relative distinct behaviour was observed in the case of red cabbage juice, red beet juice and cucumber juice with onion juice added, in the sense of the slowdown of the metabolism objectified in the dynamics of the parameters that describe the process unfolding. The differences could be explained through the presence of some chemical constituents which can act as inhibitors on useful bacteria, like anthocyanins in the red cabbage, betacyanins in red beet, respectively constituent sulfides in the onion juice. According [14], sulfides, especially those with three or more sulfur atoms, apparently possess potent antimicrobial activity. However, concerning the batch with onion juice added the initial trend was attenuated after 6 hours of fermentation, the oils and their sulfides constituent showing weak antimicrobial activity ([15]).

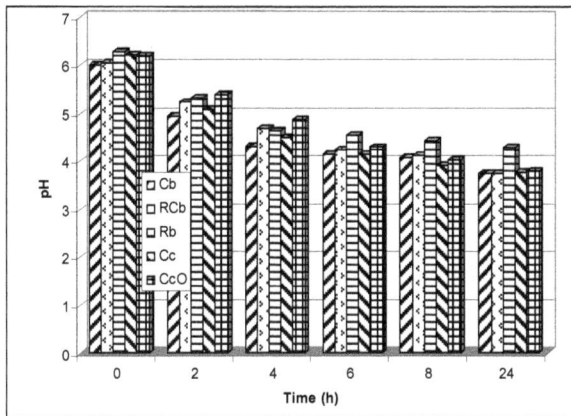

Figure 5. pH dynamics in vegetable juices obtained from different raw materials, during fermentation with *Lactobacillus acidophilus* LA-5

Referring to the red cabbage juice, although after 24 hours of fermentation the pH values were similar, the lactic acid content was lesser with about 1.5g/l compared with the white cabbage juice. This can be due to the amphoteric nature of the anthocyanins.

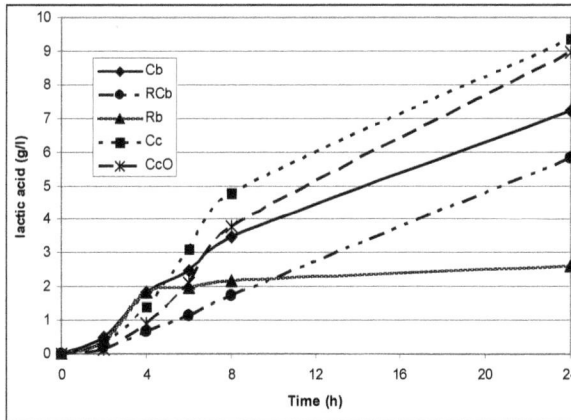

Figure 6. Lactic acid accumulation in vegetable juices obtained from different raw materials, during fermentation with *Lactobacillus acidophilus* LA-5

[16] studied the fermentation of cucumber juices with a 0.5%, 1% and 2% additions of the onion juices by *Lb. plantarum* CCM 7039. It was found that in the initial stages of fermentation, the presence of onion in the juices positively influenced lactic and acetic acid production. However, in further course of fermentation, slight inhibition effects of onion in the fermented juices were observed, especially at elevated onion/cucumber ratio.

The correlation between the biomass amount and the production of lactic acid (Figure 7) in the case of lactic acid fermentation of red beet juices with *Lactobacillus acidophilus* in the first 24 hours, was described using the Luedeking & Piret model [17]. According to this model, the instantaneous rate of lactic acid formation (dP/dt) can be related to the instantaneous rate of bacterial growth (dN/dt), and to the bacterial density (N), throughout fermentation at a given pH, by the expression:

$$dP/dt = \alpha \, dN/dt + \beta \, N$$

where the constants α and β are determined by the pH of the fermentation.

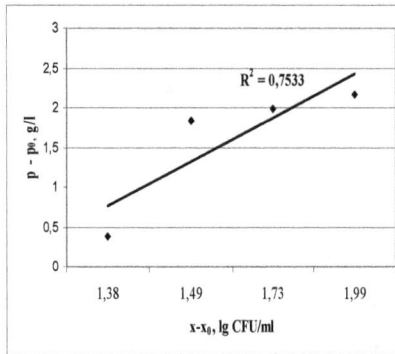

Figure 7. The correlation between the lactic acid production and viable cells count of *Lactobacillus acidophilus* LA-5 growing on red beet juices

A simplified presentation of the above model relates to the linear part of the equation which is presented as:

$$\left(p - p_0\right) = \alpha \left(x - x_0\right)$$

where p_0 and p are the concentrations of lactic acid (g/l) initially and at time t, respectively, and x_0 and x are the increases of the biomass (log CFU/mL) initially and at time t, respectively.

The R squared coefficient closed by the ideal value "1" ($R^2 = 0.9989$) in the case of the carrot juices fermented with *Lactobacillus acidophilus* LA-5 (data not shown) highlights a better linear correlation, respectively a strong connection between the lactic acid production and the lactic acid bacteria growth. Not the same situation has registered in the lactic acid fermentation of the red beet juices with the same strain. The highest value of the coefficient $(1 - R^2)$ it is caused by the increase of the lactic acid amount in the first 4 hours, followed by a steady interval of evolution of this parameter. From the other hand, according [18], the deviations from the linear dependence are mostly caused by nutritive limitations of the substrates, and are related to the specific bacterial species. Not at least, the initial content of reducing sugars of the red beet, by 21.2g/l, could be limiting. However, taking into account the fact that the cucumber juice underwent a tumultuous fermentation although its content was only with 15.09% higher, it seems that other chemical constituents of the raw materials are responsible for the above mentioned differences.

The initial content of sugars in cucumber juice was situated at the maximum limit determined by [19], while in the case of the white cabbage juice was close to that one determined by [20].

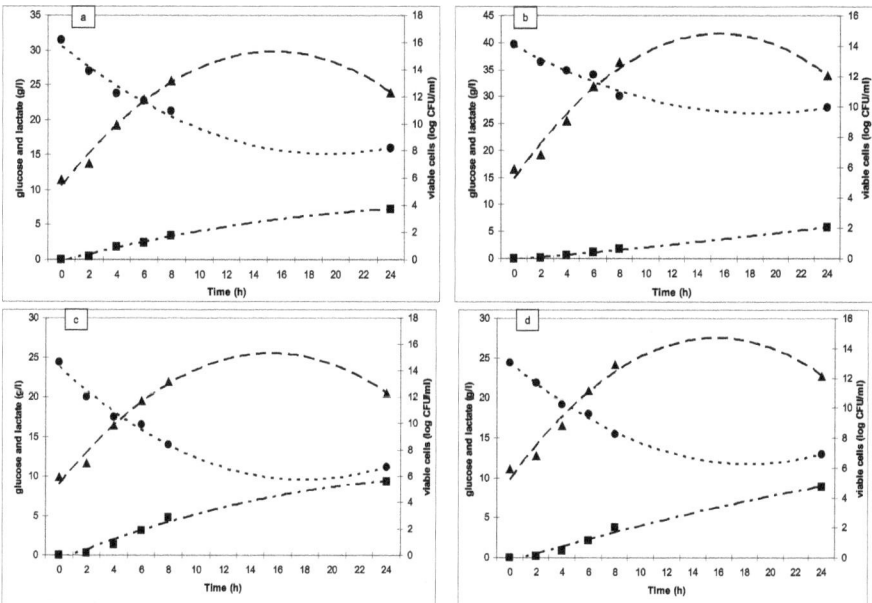

Figure 8. Correlation between the substrate consumption, lactate production and viable cells Cb (a), RCb (b), Cc (c) and CcO (d)
● - glucose, ■ - lactate, ▲ - viable cells (points - experimental data, smooth lines - predicted values)

The metabolization of the reducing sugars after 24h of lactic acid fermentation of vegetable juices with $Lb.$ $acidophilus$ LA-5 ranged between 26.66% (Rb) to 54.09% (Cc). Relative close values were obtained by other authors in lactic acid fermentation of vegetable juices. Thus, the utilization of sugar during fermentation in a mixture of beetroot juice and carrot juice and different content of brewer's yeast autolysate with $Lb.$ $plantarum$ A112 and with $Lb.$ $acidophilus$ NCDO 1748 varied from 19.4 to 24.1% ([21]).

The tested pure culture, routinely used for dairy products, was found to be capable of growing on pure vegetable juices without nutrients added. In the batches obtained from cabbage, respectively cucumber, the maximum volumetric productivity was determined after 8 hours as follows: 19.25×10^{14} CFU/(l·h) for Cb, 11.9×10^{14} CFU/(l·h) for RCb, 18.6×10^{14} CFU/(l·h) for Cc and 10.25×10^{14} CFU/(l·h) for CcO respectively.

The relationship between the growth of *Lactobacillus acidophilus*, the substrate metabolization and the lactic acid accumulation is shown in Figure 8. The prediction functions of the values of the analyzed parameters in all the samples were defined as polynomial, the R squared being very close to unit.

Correlating the number of viable cells with the dynamics of the lactic acid, the values were lower until 6 hours in the red cabbage juice and cucumber juice with onion juice added respectively. The differences were lessened in the next period of the process. However, the final yield of the lactic acid production was better in the sample CcO, by 0.78, comparatively with 0.7 in the sample Cc.

3.4. Effect of growth factors on the dynamics of the lactic acid fermentation of the carrot juices by Lactobacillus acidophilus LA-5

Kinetic parameters such as the time to reach pH 5.0 and the maximum rate of acidification are important in terms of the shelf life of the fermented vegetable juices. These ones were differently modified by the presence of the amino acids or of the yeast extract at the initial moment of fermentation. From Table 2 we deduced that a highest influence on both $t_{pH\ 5.0}$ and v_{max} was exerted by cysteine, added to the juice in amount by 0.2% (w/v). Compared with the other supplements, the yeast extract had a relative good effect on the analyzed parameters. At the used concentrations, the behavior of valine and lysine seems to be unobservable from this point of view, excepting the poor effect of lysine on the maximum rate of acidification. Time to complete the fermentation ($t_{pH\ 4.2}$, h) ranged between 7.4 (YE) and 10.42 (Leu), trend that underline the statement that in the above mentioned experimental conditions *Lactobacillus acidophilus* growing faster.

Kinetic parameter	Supplements[1]					
	Cys_1	Leu	Val	Lys	Cys_2	YE
Time-decreasing of $t_{pH\ 5.0}$[2]	1.28	0.85	0	0	1.69	1.1
Time-increasing of v_{max}[3]	0.82	0.84	0.98	1.05	1.1	1.05

[1]The notations used for the samples are in agreement with the nutrients added, as follows: L-Cysteine (Cys_1 sample with 0.1% cysteine and Cys_2 sample with 0.2% cysteine), L-Leucine (Leu), L-Valine (Val), L-Lysine (Lys) and yeast extract (YE) respectively
[2]The data were obtained by dividing the kinetic parameters of the control to the corresponding values of the samples
[3]The data were obtained by dividing the kinetic parameters of the samples to the corresponding values of the control Subunit or null values should be considered as lack of effect on the analyzed parameters.

Table 2. Effect of supplements on the kinetic parameters

MRS broth used for lactobacilli enumeration often incorporates L-cysteine to improve the recovery of these ones, especially due to the fact that *Lactobacillus acidophilus* LA-5 is micro-aerophilic. Cysteine, a sulfur containing amino acid, could provide amino nitrogen as a growth factor while reducing the redox potential. [22] reported that the incubation time to reach a pH of 4.5 was greatly affected by the addition of cysteine in yogurts made with different commercial cultures, although their viability was adversely affected in function of the amount of supplement and the type of the starter culture.

Lactic acid is the major metabolite of *Lactobacillus acidophilus*, influencing both the preservation of the fermented products and the sensorial characteristics of these ones. The effect of the amino acids and of the yeast extract on the dynamics of the lactic acid, assessed against the control, is underlined through the data from Table 3. The buffering capacity of the amino acids prevented a direct proportionality between the pH values and the lactic acid content.

Time, h	Cys_1	Leu	Val	Lys	Cys_2	YE
2	8.737864	-12.6214	21.52778	-29.8611	107.6923	15.38462
4	17.66784	-23.6749	-5.55556	-2.77778	28.125	12.5
6	16.98113	-1.50943	11.71717	3.636364	1.818182	5.454545
8	20.63492	-1.5873	8.571429	1.428571	-11.1111	15.87302
24	0.925926	-0.92593	5.076142	4.568528	-14.433	11.34021

Table 3. Time-increasing of lactic acid during 24h of lactic acid fermentation of carrot juices by *Lactobacillus acidophilus* LA-5

The values were expressed in percents by reporting the difference between sample and control to the control, at the same moment of time

Negative values shows that for the corresponding interval of time the supplements had not influence on the lactic acid production at the used levels.

Analyzing the whole process, only the samples with a minimum amount of cysteine added and those with yeast extract have been a great effect on the time-increasing of lactic acid. At the other opposite were found the samples with leucine added, this amino acid with non-polar hydrophobic chains clumsying the fermentation. From the viewpoint of increase the lactic acid content in the final stages of the process, the supplementation of the carrot juices with 0.2% (w/v) cysteine seems to be undesirable.

The beneficial effect of cysteine on the lactic acid accumulation in vegetable juices can occur due to its buffering capacity, which may diminish the toxic effects of organic acids on lactobacilli. Referring to the yeast extract, which contains more cell growth factors, being used generally as a source of assimilable nitrogen, vitamins and minerals, its influence at the level of 0.2%(w/v) on the time-increasing of lactic acid could be characterized as moderate. If some authors reported different maximum lactic acid concentration in media supplemented with yeast extract, several possible explanations include the strain of microorganism, the chemical composition of the substrate, the fermentation system, and generally the conditions employed during fermentation ([12]).

Effect of supplements on the performance of lactic acid production was evaluated based on lactic acid productivity and lactic acid yield, respectively on glucose ratio (Table 4).

The previous conclusion referring to the positive influence of the yeast extract and cysteine (in minimum amount) on the development of the lactic acid fermentation of vegetable juices is confirmed by the data from Table 4. Good values of lactic acid productivity were obtained after 24 h of fermentation in the samples with valine and lysine added, although in these ones the substrate consumption seems to be directed to the increasing of biomass, aspect emphasized by the average values of the lactic acid yield.

Parameter	Cys_1	Leu	Val	Lys	Cys_2	YE
Lactic acid yield[2]	1.1	0.85	0.88	0.79	0.85	1.15
Lactic acid productivity[3]	1.01	0.99	1.06	1.05	0.7	1.13
Glucose conversion ratio[4]	1.1	0.9	1.2	1.05	0.92	1.25

[1]The data from the table were obtained by dividing the corresponding values for the samples to those of the control
[2]Lactic acid yield was calculated by dividing the amount of lactic acid produced to the amount of glucose consumed
[3]Lactic acid productivity was defined as the amount of lactic acid produced per hour per liter
[4]Glucose conversion ratio was calculated by dividing the amount of glucose consumed to the initial amount of glucose.

Table 4. Effect of supplements on lactic/acetic acid production after 48 h of fermentation[1]

The effect of supplements (amino acids and yeast extract) on the ascorbic acid dynamics is shown in Figure 9. L-Ascorbic acid (AA), also known as vitamin C, is a representative water-soluble vitamin possessing a variety of biological, pharmaceutical, and dermatological functions; it promotes collagen biosynthesis, provides photoprotection, causes melanin reduction, scavenges free radicals, and enhances immunity ([23]).

Due to the heat treatment applied with a view to destroy the epiphytic microbiota of the fresh vegetable juices, the losses occurred in the ascorbic acid content represented about 65%.

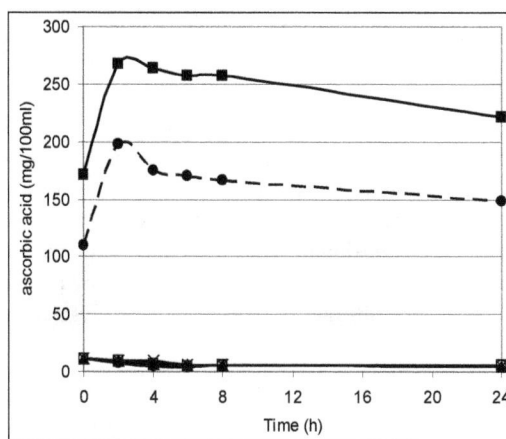

Figure 9. Time-course (0-24h) of the relative levels of ascorbic acid
(•Cys_1, ■Cys_2, ○YE, □Leu, ▲Val, x Lis). The data shown are average values of two independent replicate experiments

The presence of ascorbic acid into vegetable juices submitted to fermentation by probiotic bacteria, especially by *Lactobacillus acidophilus* strains, is desired not only from the nutritional point of view, but also due to the fact that it could promote anaerobic conditions, acting as an oxygen scavenger. [24] have shown also that the fruit juices may be an alternative vehicle for the incorporation of probiotics because they are rich in nutrients and do not contain starter cultures that compete for nutrients with probiotics. Furthermore, fruit juices are often supplemented with oxygen scavenging ingredients such as ascorbic acid, thus promoting anaerobic conditions.

L-cysteine, a sulfur-containing amino acid known as a powerful reducing agent, caused the reduction of dehydroascorbic acid to ascorbic acid, which led a different behavior of the samples Cys_1 and Cys_2 by the others. The increase of this parameter was by 80% and 56.4% respectively, after 2h from the initial moment of fermentation. Subsequently, the analyzed parameter had the same diminishing tendency as in the other batches.

The losses occurred after 24h of lactic acid fermentation of carrot juices with *Lactobacillus acidophilus* LA-5 ranged from 48.39% (YE) to 61.9% (control). The possible reason could be the oxygen traces that cause the chemical oxidation of the vitamin C.

In order to evaluate the probiotic feature of the vegetable juices, the study of the effect of supplements on *Lactobacillus acidophilus* growth is from overwhelming importance, both during the lactic acid fermentation and during the storage of the final products.

Between the analyzed samples, those with yeast extract and 0.1% (w/v) cysteine added registered a higher increase of the number of viable cells till 14.4 - 14.5 log CFU/ml in the first 8h of the process. Concerning the yeast extract, the most possible explanation is due to an enhanced availability of minerals, which are growth promoters for *L. acidophilus* ([25]), while discussing the factors that affect the activity of endogenous probiotics, (26) mentioned that some of the growth promoters in cow milk were apparently cysteine-containing peptides.

Referring to the juices with leucine, lower values were determined comparative with the control during 24h, while in the samples with 0.2% (w/v) cysteine added the trend of the survival of lactobacilli was slow down in the period 6 - 8h, the level being by 13.5 and 13.6 log CFU/ml respectively. The last observation agrees with this one of [27], which have shown that the increasing of cysteine concentration improved the viability of *B. bifidum* in bio-yogurt, although it had no important effect on the viability of *Lactobacillus acidophilus.*

The batches supplemented with valine and lysine had occupied an intermediate position, the growth until 14.2 log CFU/ml after 8h of fermentation making from the utilization of these amino acids a promising variant in the future, with a view to optimize the conditions of the process unfolding. In the period 8 - 24h the number of viable cells decreased, as result of the lack of tolerance at lower pH of the analyzed strain.

The correlation between the most important parameters of the lactic acid fermentation of the carrot juices with *Lactobacillus acidophilus* LA-5 were evaluated using Pearson correlation analysis (significance level $p < 0.01$; confidence level of 99%).

Analytical variables	pH	lactic acid	glucose	viable cells	glycine	ascorbic acid
pH	1	-0.889**	0.829**	-0.940**	0.099*	-0.184*
lactic acid		1	-0.891**	0.843**	-0.201*	0.016*
glucose			1	-0.789**	0.093*	0.084*
viable cells				1	-0.061*	0.066*
glycine					1	-0.103*
ascorbic acid						1

** Correlation is significant at the 0.01 level (2-tailed)
* Not significant

Table 5. The Pearson coefficients for the experimental batches

The correlations are strong between pH and lactic acid, respectively pH and glucose, while a very strong relationship pH - viable cells could be considered (Table 5). A non-existent relationship between ascorbic acid / amino acids content (expressed as glycine) and the other analyzed parameters was determined.

A firm correlation between glucose and lactic acid was expected, but on the one hand it is known that the practical yield of sugars conversion to lactic acid of the strains of the group *Lb. acidophilus* is about 85%, while on the other hand the analysis does not include supplementary data referring to other factors that might be involved in the dynamics of the lactic acid fermentation of vegetable juices.

Factor Analysis (FA) is a multidimensional statistic method whose purpose is the analysis of the structure of mutual dependences of variables. The method is similar to the Principal Component Analysis (PCA) with the exception of the factor weights that are scaled ([28]).

Applying FA to the experimental data, the analytical variables were reduced to two principal components, which accounted for 59.72% (PC1) and respectively 18.95% (PC2) from the total variance. According to the component matrix, respectively to the values of the component loadings expressed by the first second principal components (rotation method: Varimax with Kaiser normalization), the most notable variables were pH and lactic acid (equal loading values by 0.954). Higher values were obtained also for viable cells (loading 0.939) and glucose (loading 0.933).

The combination of PC1 and PC2 (Figure 10) underlined the lack of correlation between amino acids content / ascorbic acid and all the other parameters taking into account both control and supplemented samples. While PC1 affected the dependent and independent variables involved in the progress of the lactic acid fermentation of vegetable juices, respectively in their probiotic feature, PC2 separated the variables which contribute to the nutritional characteristics of the final products.

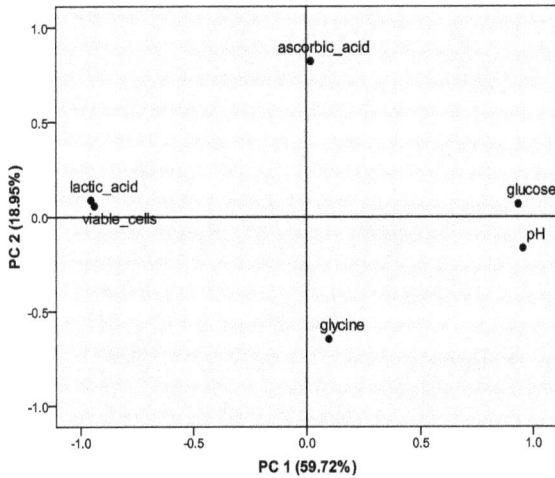

Figure 10. Component plot in rotated space

Applying PCA to the lactic acid fermentation of cabbage juices with various microorganisms, [29] established that the original 7 analytical variables were reduced also to 2 independent components that explained 88.2% from total variance of input data (PC1 66.9% and PC2 21.3%).

Cluster Analysis (CA) is a statistic method whose purpose is to join data into clusters with a view to increase their withingroup homogeneity. Usually, the FA is considered the first step of CA, with a view to reduce the data dimensionality. In order to better distinguish among experimental samples, the cluster method of the nearest neighbour was used. The distances between objects were measured as squared Euclidean distance. K-Means Cluster Analysis divided the experimental data into three groups, characterized by similar analytical properties, as follows:

- cluster 1: all the carrot juices (control samples and the batches with amino acids and yeast extract added) at the initial moment of fermentation, respectively at 2th h of fermentation. Supplementary, this cluster included the control and the sample with leucine at 4th h of fermentation (C_4 and Leu_4);
- cluster 2: all the carrot juices at 24th h of fermentation and the sample with lysine added at 8th h of fermentation ;
- cluster 3: the carrot juices with leucine and lysine added, respectively the control, at 6th and 8th h of fermentation (Leu_6, Leu_8, Lys_6, Lys_8, C_6, C_8); the carrot juices with cysteine, valine, respectively yeast extract from 4th to 8th h of fermentation (Cys_4 - Cys_8, (Val_4 - Val_8), (YE_4 - YE_8).

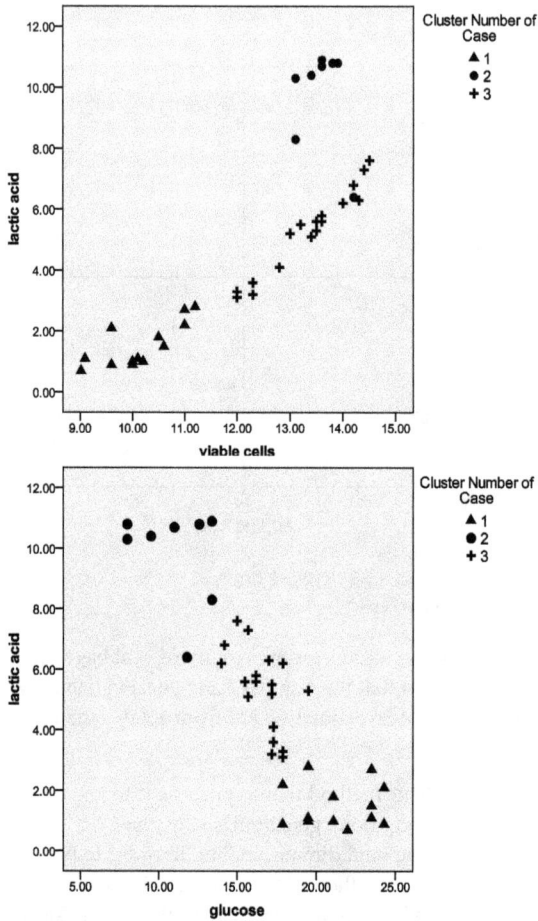

Figure 11. Clusters plotting in coordinate of two selected variables: lactic acid - viable cells and lactic acid - glucose

The clusters in axes of two selected variables (Figure 11) denote that the samples from the first cluster were marked with a higher content of substrate, null or very lower lactate amount and pH values more than 5. The corresponding time was both 0 and 2h (C, Val, Leu, and YE) or the entire interval 0 - 4h (Leu and control).

The samples at the final moment of fermentation and those with lysine after 8h of the process were included in the second cluster, characterized through lower or average values of glucose content, higher lactic acid amount and pH values close to 4.2. This cluster marks the achievement of the optimum characteristics of the lactic acid fermented products.

The samples included in the third cluster best describes a vigorous process, being characterized through average values of the main parameters involved in the dynamics of the lactic acid fermentation of vegetable juices.

The usefulness of the methods of statistical analysis is underlined by a lot of applications of CA that could be reported: in evaluation of analytical and sensory characteristics of vegetable juices ([28], [29]), in distinguishing between wines aged a different number of months ([30]).

4. Further research

The importance of consuming probiotic foods for the improvement of the quality of life increasingly more in the last years, being underlined by the scientific literature. The diversification of the market from this point of view could be strong correlated with the increasing of the life expectancy worldwide.

Our further researches are needed in order to optimize the level of nutrients (individually and in combination) and in the same time their influence on growth and viability of probiotics (in particular of *Lactobacillus acidophilus*, single strain or in combination with other probiotics), not only during fermentation but especially during the storage of the final products.

5. Conclusions

Different vegetable juices are suitable and alternative food matrices for the production of functional foods with *Lactobacillus acidophilus* LA-5, a probiotic strain which is not present in the epiphytic microbiota. Although some differences between the growths trends were determined, all the analyzed vegetables could be considered proper in order to obtain lactic acid fermented juices with a higher self-life. Application of Principal Component Analysis selected the most important parameters from analytical point of view: pH, lactic acid, biomass and viable cells, while the Cluster Analysis divided the experimental variables into three groups.

Author details

Lavinia Claudia Buruleanu, Magda Gabriela Bratu,
Iuliana Manea, Daniela Avram and Carmen Leane Nicolescu
Department of Food Engineering, Faculty of Environmental Engineering and Biotechnology,
Valahia University of Targoviste, Romania

Acknowledgement

The research was funded by Executive Unit for Financing Higher Education, Research, Development and Innovation (UEFISCDI) in the frame of the Project PN-II-ID-PCE-2008-2 (ID_1359). The microorganisms were kindly provided by Chr. Hansen, Romania.

6. References

[1] Granato D, Branco GF, Nazzaro F, Cruz AG, Faria JAF (2010) Functional Foods and Nondairy Probiotic Food Development: Trends, Concepts and Products. Compr. Rev. Food Sci. Food Saf. 9: 292-302.

[2] Holzapfel WH, Haberer P, Geisen R, Björkroth J, Schillinger U (2001)Taxonomy and Important Features of Probiotic Microorganisms in Food and Nutrition. Am. J. Clin. Nutr. 73(suppl): 365S-73S.

[3] Saito T (2004) Selection of Useful Probiotic Lactic Acid Bacteria from the *Lactobacillus acidophilus* Group and their Applications to Functional Foods. Animal Sci. J. 75:1-13.

[4] Pineiro M, Stanton C (2007) Probiotic Bacteria: Legislative Framework - Requirements to Evidence Basis. J. Nutr. 850S-853S.

[5] Kailasapathy K, Chin J (2000) Survival and Therapeutic Potential of Probiotic Organisms with Reference to *Lactobacillus acidophilus* and *Bifidobacterium* spp. Immun. Cell Biol. 78: 80–88.

[6] Karovičová J, Kohajdová Z (2003) Lactic Acid Fermented Vegetable Juices. Hort Sci.(Prague), 30 (4): 152-158.

[7] Klewicka E (2010) Antimutational Activity of Beetroot Juice. Food Technol. Biotechnol. 48(2): 229–233

[8] Altiok D (2004) Kinetic Modelling of Lactic Acid Production from Whey, Ph D Thesis, Izmir Institute of Technology, Turkey

[9] Kious, Jessica J. Lactobacillus and Lactic Acid Production [online]. Energy Research Undergraduate Laboratory Fellowship Program (ERULF), National Renewable Energy Laboratory, Colorado, 2000. Portable Document Format. Available from Internet: http://www.nrel.gov/docs/gen/fy01/NN0017.pdf.

[10] Agarwal L, Dutt K, Meghwanshi GK, Saxena RK (2008) Anaerobic Fermentative Production of Lactic Acid using Cheese Whey and Corn Steep Liquor. Biotechnol. Lett. 30: 631–635.

[11] Dave RI, Shah NP (1997) Viability of Yoghurt and Probiotic Bacteria in Yoghurts Made from Commercial Starter Cultures. Int. Dairy J. 7: 31-41.

[12] Amrane A (2005) Analysis of The Kinetics of Growth and Lactic Acid Production for *Lactobacillus Helveticus* Growing on Supplemented Whey Permeate. J. chem. technol. biotechnol. 80: 345-352.

[13] Kotzamanidis C, Roukas T, Skaracis G (2002) Optimization of Lactic Acid Production from Beet Molasses by Lactobacillus delbrueckii NCIMB 8130. World j. microbiol. biotechnol. 18: 441-448.

[14] Kim JW, Kim YS, Kyung KH (2004) Inhibitory Activity of Essential Oils of Garlic and Onion against Bacteria and Yeasts. J. food prot. 67(3): 499-504.

[15] Kim JW, Huhi JE, Kyung SH, Kyung KH (2004) Antimicrobial Activity of Alk(En)Yl Sulfides Found in Essential Oils of Garlic and Onion. Food sci. biotechnol. 13(2): 235-239.

[16] Kohajdová Y, Karovičová J, Greifová M (2007) Analytical and Organoleptic Profiles of Lactic Acid-Fermented Cucumber Juice with Addition of Onion Juice. J. food nutr. res. 46(3): 105-111.

[17] Rakin M, Baras J, Vukašinović M, Maksimović M (2004) The Examination of Parameters for Lactic Acid Fermentation and Nutritive Value of Fermented Juice of Beetroot, Carrot an Brewer's Yeast Autolysate, J. serbian chem. soc. 69 (8-9): 625-634.

[18] Luedeking R, Piret EL (1959) A Kinetic Study of the Lactic Acid Fermentation. Batch Process at Controlled pH, J. biochem. microbiol. technol. eng., 1: 393-412.

[19] Amrane A, Prigent Y (1999) Analysis of Growth and Production Coupling for Batch Cultures of Lactobacillus Helveticus with Help of Anunstructurated Model. Process biochemistry, 34: 1-10.

[20] Lu Z, Fleming HP, McFeeters RF (2002) Effects of Fruit Size on Fresh Cucumber Composition and the Chemical and Physical Consequences of Fermentation. J. food sci. 67(8): 2934-2939.

[21] Yoon KY, Woodams EE, Hang YD (2005) Production of Probiotic Cabbage Juice by Lactic Acid Bacteria. Bioresour. technol. 97(12): 1427-1430.

[22] Rakin MB, Baras JK, Vukašinović MS (2005) Lactic Acid Fermentation in Vegetable Juices Supplemented with Different Content of Brewer's Yeast Autolysate. Acta periodica technol. 36: 71-80.

[23] Dave RI (1998) Factors Affecting Viability of Yoghurt and Prbiotic Bacteria in Commercial Starter Cultures. PhD thesis, Victoria University of Technology, Werribee Campus, Victoria, Australia.

[24] Jang KI, Lee HG (2008) Influence of Acetic Acid Solution on Heat Stability of L-Ascorbic Acid. Food sci. biotechnol. 17: 637-641.

[25] Ding WK, Shah NP (2008) Survival of Free and Microencapsulated Probiotic Bacteria in Orange and Apple Juices. Int. food res. J. 15: 219-232.

[26] Lourens-Hattingh A, Viljoen BC (2001) Yogurt as Probiotic Carrier Food. Int. dairy j. 11: 1-17.

[27] Bezkorovainy A. (2001) Probiotics: Determinant of Survival and Growth in the Gut. Am. j. clin. nutr. 73(suppl): 399S-405S.

[28] Bari MR, Ashrafi R, Alizade M, Rofegarineghad L. (2009) Effects of Different Contents of Yogurt Starter/Probiotic Bacteria, Storage Time and Different Concentration of Cysteine on the Microflora Characteristics of Bio-yogurt. Res. j. biol. sci. 4: 137-142.

[29] Karovičová J, Kohajdová Z (2002) The Use of PCA, FA, CA for Evaluation of Vegetable Juices Processed by Lactic Acid Fermentation. Czech j. food sci. 20: 135-143.

[30] Kohajdová Z, Karovičová J. (2004) Optimisation of Method of Fermentation of Cabbage Juice. Czech j. food sci. 22: 39-50.

[31] Li Z, Pan QH, Jin ZM, He JJ, Liang NN, Duan CQ (2009) Evolution of 49 Phenolic Compounds in Shortly-Aged Red Wines Made from Cabernet Gernischt (*Vitis vinifera* L. cv.). Food sci. biotechnol. 18: 1001-1012.

Chapter 7

Lactic Acid Bacteria in Biopreservation and the Enhancement of the Functional Quality of Bread

Belal J. Muhialdin, Zaiton Hassan and Nazamid Saari

Additional information is available at the end of the chapter

1. Introduction

LAB have a long history in preserving foods from spoilage microorganisms - they are commonly used in food fermentation, may produce several metabolites with beneficial health effects and, thus, are generally recognized as safe (GRAS). The increasing resistance of food spoilage microorganisms to current preservatives, the consumer's high demand for safe, minimally processed foods and the hazards associated with the use of high doses of chemical preservatives has led to the need for finding safer alternatives in food preservation. The application of LAB with the simultaneous control of factors that affect fungal growth can help to minimize food spoilage. The selection and addition of novel isolates of LAB may be the key to reducing the use of chemicals, enhancing nutrients and extend the shelf life of bakery products. In this chapter, the focus will be on the use of LAB as biopreservative agents to extend the shelf life of bakery products and the inhibition of the common spoilage fungi of bread.

2. Sources of LAB

LAB are found in many habitats and occur naturally in a variety of food products, such as dairy, vegetables and meat products (Carr et al., 2002), all of which are rich in the nutrients required for the fastidious metabolism of LAB (Björkroth & Holzapfel, 2003; Hammes & Hertel, 2003). Some LAB are associated with the mouth flora, intestine and vagina of mammals (Whittenbury, 1964), while others are present in fermented seafood, such as *Lactobacillus plantarum* (IFRPD P15) and *L. reuteri* (IFRPD P17), which are reported to be associated with plaa-som fermented Thai fish (Saithong et al., 2010). LAB are the most important bacteria used in the fermentation industry of dairy products, such as yogurt, cheese, sour milk and butter, and in combination with yeast are commonly used to ferment cereal products such as dough (Lavermicocca et al., 2000; Muhialdin et al., 2011a; Ryan et al., 2008).

3. Spoilage fungi in food

The economic losses and the health hazards of the mycotoxins produced by spoilage fungi are the main concerns of the food industry (Gray & Bemiller, 2003). According to Gerez et al., (2009) the spoilage of bakery products by fungi is more common in countries with a high humidity and temperature. Pitt and Hocking (1999) estimated that about 5-10% of food production is spoiled by the growth of yeast and fungi in food materials. Similarly, in Western Europe, the growth of the spoilage fungi of bread is estimated to reach more than 200 million Euros per year (Legan, 1993; Schnürer & Magnusson, 2005). The history conditions of the food can be a major factor in determining any fungal spoilage - for example, stored and processed foods are more sensitive to spoilage when compared with fresh and prepared foods. *Aspergillus* and *Penicillium* species are the most common spoilage fungi for many foods and feeds while *Fusarium* species are reported to attack the cereal grains in the field (Samson et al., 2000).

The most widespread species of fungi that contaminate bakery products belong to the genera *Aspergillus, Penicillium, Eurotium* (Abellana et al., 1997; Guynot et al., 2005), *Monilia, Mucor, Endomyces, Cladosporium, Fusarium* and *Rhizopus* (Lavermicocca et al., 2000, 2003). In addition, fungi may be responsible for off-flavours, the production of mycotoxins and allergenic compounds. There are more than 400 known mycotoxins produced by different fungi (Filtenborg et al., 1996). Mycotoxigenic fungi such as *Aspergillus, Fusarium* and *Penicillium* are serious hazards for human health. The six classes of mycotoxins frequently encountered in different food systems are: aflatoxins, fumonisins, ochratoxins, patulin, trichothecenes and zearalenone (Dalié et al., 2009).

4. Common techniques to control spoilage fungi in bakery products

Two types of techniques/factors are commonly used to control spoilage fungi: physical ones such as drying, freeze drying, cold storage, modified atmosphere storage, irradiation, the pasteurization of packaged bread and heat treatment; and chemical ones, in general based on the use of organic acids such as propionic acid and its salts (Farkas, 2001; Legan, 1993). Heat treatment is one of the most important physical factors in controlling fungi growth and mycotoxin production, as mycotoxins are destroyed by heat, although the effectiveness of destruction is affected by the food matrix and the composition of the mycotoxin (Scott, 1984). Mycotoxins have different heat stability - for example, ochratoxin A is highly stable even at 200 °C (Trivedi et al., 1992), aflatoxins are destroyed only at temperatures of approximately 250 °C (Levi, 1980), while zearalenone and fumonisin require high temperatures between 150-200 °C to be efficiently destroyed (Bennett et al., 1980). Microwaves are effective in destroying mycotoxins - the aflatoxin in peanuts is reported to be destroyed using microwaves at a power level of 1.6 kW for 16 min and at 3.2 kW for 5 min (Luter et al., 1982). Among the physical methods, a modified atmosphere and gamma irradiation are preferred to the chemical methods and they have been used successfully in grain storage (Shapira & Paster, 2000).

Chemical methods that use weak acids and salts such as propionic, sorbic and benzoic acids, are usually applied only to inhibit the growth of spoilage microorganisms. The allowable concentrations of sorbate, propionate and ethanol have a limit up to 0.2% (wt/wt), 0.3% (wt/wt) and 2% (wt/wt) respectively. The use of such low concentration may not be sufficient to prevent the growth of spoilage fungi (Dantigny et al., 2005; European Union, 1995). Propionic acid is inhibitory to fungi and *Bacillus* spores and has commonly been used to preserve bakery products. Its activity relies on the un-dissociated form which, at low pH, has optimum activity (Coda et al., 2008; Pattison et al., 2004). The use of propionic acid at a concentration of 4% led to the appearance of cancer-like tumours in rats and eventually led to the prohibition of the use of calcium propionate in some European countries (Pattison et al., 2004). There is a major concern with microorganisms that can develop resistance to chemical preservatives, namely food spoilage and human pathogen fungi resistant to antibiotics and chemicals additives, such as sorbic and benzoic acids (Brul & Coote, 1999; Lourens-Hattingh & Viljoen, 2001). Calcium propionate has been reported to inhibit the growth of many fungi but, after a lag phase, it stimulated the growth of resistant strains of *Penicillium roqueforti* (Suhr & Nielsen, 2004). Interest in natural bio-preservation from LAB has been on the rise as an alternative to chemical preservatives.

5. Significance of the metabolites of LAB

LAB are well known for their antifungal activity, which is related to the production of a variety of compounds including acids, alcohols, carbon dioxide, diacetyl, hydrogen peroxide, phenyllactic acid, bacteriocins and cycle peptides (Gerez et al., 2009; Lavermicocca et al., 2000; Magnusson et al., 2003; Prema et al., 2008). These compounds were added to several foods in order to conserve them from food-borne and spoilage microorganisms. Organic acids are the main product of LAB in the fermentation systems of the raw materials. The main acids produced by LAB are lactic acid and acetic acid, besides certain other acids depending upon the strain of LAB (El-Ziney, 1998). These acids will be diffused through the membrane of the target organisms in their hydrophobic un-dissociated form and then used to reduce the cytoplasmic pH and stop metabolic activities (Piard & Desmazeaud, 1991). Other factors that contribute to the preservative action of the acids are the sole effect of pH, the extent of the dissociation of the acid and the specific effect of the molecule itself on the microorganisms (Axelsson, 1998).

Bacteriocins exhibit good potential for use in the food industry and as bio-preservation agents (Ennahar et al., 1999). Bacteriocins are small, ribosomally synthesized, antimicrobial peptides or proteins that display inhibition activity toward related species, with no reports about fungal inhibition (Cotter & Ross, 2005). The notable property of LAB supernatant is the heat stability of the antifungal compounds present in it. This will promote the use of LAB supernatant and/or antifungal compounds in heat-treated foods. The supernatant of certain LAB observed to be active within a wide range of pH, starting from as low as 3 and up to 9 depending upon the strain (Muhialdin et al., 2011b). This could be considered as a major factor whereby LAB are used in food preservation when compared with the chemical preservative which are usually active at low pH between 3 and 4.5. Additionally, LAB have a broad spectrum of antifungal activity against several food spoilage and mycotoxin-producing fungi while commercial preservatives are usually used to control only one or few fungi.

6. Bioactive compounds as antifungal agents

Several lactobacilli species are reported to have antifungal activity (Gerez et al., 2009; Muhialdin et al., 2011b; Plockova et al., 2001; Stiles et al., 1999). The antifungal compounds consist of organic acids, reuterin, hydrogen peroxide and other peptides (Table 1). The organic acids are active at low pH and the activity relies on the un-dissociated form of the acids. Recently, interest has dramatically increased in the use of bioactive peptides produced by LAB as an antifungal agent. The use of protein-like compounds are preferred over the use of acids because their activity is present over a wide range of pH and they are heat stable compounds which are ideal for use in heat processed foods (Muhialdin et al., 2011a). Cyclic dipeptides cyclo (Phe-Pro) and cyclo (Phe-OH-Pro) were produced by the *L. corniformis* subsp. *corniformis* Si3 strain and were inhibitory to *Aspergillus* sp. (Magnusson, 2003; Ström et al., 2002). Ryan et al. (2011) observed that sourdough made with *L. amylovorus* DSM 19280 had a longer shelf life compared with bread produced with calcium propionate. The selected strain inhibited the growth of *Fusarium culmorum* FST4.05, *Aspergillus niger* FST4.21, *Penicillium expansum* FST4.22, *Penicillium roqueforti* FST4.11 and *L. amylovorus* DSM 19280 and produced seventeen antifungal compounds.

Compound	Producer	Inhibited fungi	References
Possibly proteinaceous	*Pediococcus acidilactici*	*Saccharomyces cerevisiae*	Vandenbergh & Kanka (1989)
Possibly proteinaceous	*L. lactis* subsp. *Lactis* CHD 28.3	*A. flavus, A. parasiticus, Fusarium* spp.	Roy et al. (1996)
Caproic acid, propionic acid, butyric acid, valeric acid	*L. sanfranciscencis* CB1	*Fusarium* spp., *Penicillium* spp., *Aspergillus* spp., *Monilia* spp.	Corsetti et al. (1998)
Benzoic acid, methylhydantoin, mevalonolactone,	*L. plantarum* VTT E78076	*F. avenaceum*	Niku-Paavola et al. (1999)
phenyllactic and 4-hydroxy-phenyllactic acids	*L. plantarum* 21B	Broad spectrum against bakery spoilage fungi	Lavermicocca et al. (200)
3-Phenyllactic acid, cyclo (Phe-OH-Pro), cyclo (Phe-Pro).	*L. plantarum* MiLAB 393	*F. sporotrichioides* and *A. fumigatus*	Ström et al. (2002)
Hydroxy fatty acids, phenyllactic acid, cyclo(Phe-Pro), cyclo(Phe-OH-Pro),	*L. plantarum* MiLAB14	Broad spectrum	Magnusson et al. (2003)

Compound	Producer	Inhibited fungi	References
Possibly cyclic dipeptide	*P. pentosaceus*	*P. expansum*	Rouse et al. (2008)
diacetyl and hydrogen peroxide	*L. fermentum* and *Leuconostoc mesenteroides*	*Rhizopus oryzae, A. niger, A. flavus, Penicillium* sp and *F. oxysporum*	Ogunbanwo et al. (2008)
Acetic acid, phenyllactic acid	*L. reuteri* 1100	*F. graminearum*	Gerez et al. (2009)
(cyclo(Leu–Leu))	*L. plantarum* AF1	*Aspergillus flavus* ATCC 22546	Yang & Chang (2010)
Four peptides and organic acid mixture	*L. plantarum* LB1 and *L. rossiae* LB5	*Penicillium roqueforti* DPPMAF1	Rizzello et al. (2011)
Mixture of peptides	*L. plantarum* 1A7 (S1A7)	Broad spectrum	Coda et al. (2011)
Possibly protein-like	*L. fermentum* Te007, *P. pentosaceus* Te010, *L. pentosus* G004, and *L. paracasi* D5	*A. niger* and *A. oryzae*	Muhialdin et al. (2011a)
nine carboxylic acids, two nucleosides, sodium decanoate and five cyclic dipeptides	*L. amylovorus* DSM 19280	*A. niger* FST 4.21, *A. fumigatus* J9, *F. culmorum* TMW 4.0754 *P. expansum* FST 4.22 and *P. roqueforti* FST 4.11	Ryan et al. (2011)
3-phenyllactic acid and Benzene acetic acid, 2- propenyl ester	*L. plantarum* IMAU10014	*Botrytis cinerea, Glomerella cingulate, Phytophthora drechsleri Tucker, P. citrinum, P. digitatum* and *F. oxysporum*	Wang et al. (2012)

Table 1. Antifungal compounds produced by lactic acid bacteria and their target fungi

7. Method for determining antifungal activity

Rapid, reliable and sensitive methods for the detection of the antifungal activity of LAB becomes essential in the search for new replacements for chemical preservatives with potential industrial applications.

7.1. Dual agar overlay method

This method has been described by several authors (Magnusson & Schnürer, 2001; Ström et al., 2002; Hassan & Bullerman, 2008) and it is accurate and simple for determining the antifungal activity of LAB isolates. The method consists of inoculating the LAB cells in two 2-cm-long lines and/or small circle spots on a MRS agar surface then incubating the plates at 30 °C for 24-48 h in anaerobic jars. The plates are overlaid with 10 ml of malt extract soft agar (2% malt extract, 0.7% agar; Oxoid) containing different concentrations of the spore inoculant of 10^4 and 10^5 spore/ml. The plates are then incubated aerobically at 30 °C for 48-72 h. The inhibition activity is indicated by the clear zones around the bacterial streaks. The scale for measuring the activity can be recorded as follows: -, no activity; +, no fungal growth on 0.1 to 3% of the plate area; ++, no fungal growth on 3 to 8% of the plate area; and+++, no fungal growth on 8% of plate area. Another way to measure the activity is by recording the clear zone diameter around the isolates streak, which refers to the inhibition of the fungi growth. The dual agar overlay method is also a good method for the screening of the antifungal activity of the supernatant of LAB isolates. The supernatant can be mixed with the de Man, Rogosa and Sharpe (MRS) agar or potato dextrose agar (PDA) and poured into Petri dishes followed by a similar step, mentioned previously. The supernatant can be added to the agar before it is autoclaved in order to determine the heat stability of the antifungal compounds present in the supernatants, which is a good indicator of whether the supernatant is used in heat processed foods.

7.2. Agar well diffusion method

The well diffusion method is another approach for determining the antifungal activity of LAB, described as a simple, accurate and flexible method. It is suitable to determine the inhibition activity of LAB supernatant. A fungi numbering 10^4-10^5 spore/ml are mixed with the selected agar and allowed to solidify. The wells can be made on a variety of agar surfaces - for example, wells are made on potato dextrose agar if the target is a fungi or on a nutrient agar if the target is a bacteria; the wells are made by using a sterilized cork borer with a diameter of 3 or 5 mm. 50 µl of the same agar is added to each well in order to seal the base so as to avoid leakage. The cell-free supernatants are then added to wells in amounts of 30-80 µl and incubated at room temperature for 3-6 h in order to allow the supernatant to be diffused through the agar. The antifungal activity is recorded by measuring the clear zones' diameters around the wells.

7.3. Dry weight of biomass

The reduction of the biomass of the fungi can be a tool for determining the growth inhibition activity of the supernatant. 50 ml of the supernatant is inoculated into a 250 ml flask containing the growth medium for the target fungi and then the suspension of the fungi spores is added at a concentration of 10^5. The fungal mass is harvested on filter paper and dried in an oven at 50 °C for 2 days. The average of the fungal biomass inhibition can be calculated by comparing the weight of treated fungi with the positive control which contains the fungi and the growth medium with no supernatant.

7.4. Micro-titter 96 well plate

The method is simple, inexpensive and practical for determining antibacterial and antifungal activity. The supernatant of LAB is placed into the wells of 190 µl and inoculated with 10 µl of a conidial suspension containing about 10^4-10^5 spore/ml. The plates are then incubated at 25-30 °C. The control is a conidial suspension placed in the wells in equal amounts without the addition of the LAB supernatant. Fungal growth is observed by the naked eye and determined by measuring the optical density at 560-580 nm, starting from 0 h and repeated every 24 h with a spectrophotometer. The result can be obtained by comparing the OD readings of the control with the treated wells. The method is appropriate for evaluating the MIC, heat stability, enzyme activity and effects of pH for the LAB supernatant.

8. Effect of the addition of LAB on bread quality

8.1. Shelf life

Traditionally, chemical preservatives and fungicides are used to inhibit fungal growth but concerns about environmental pollution and consumer health, along with problems of microbial resistance, favour the demand for alternative methods in controlling the growth of fungi (Druvefors et al., 2005). The shelf life of bread has been reported to be extended when certain LAB strains were added to bread formulations (Muhialdin et al., 2011a; Ogunbanwo et al., 2008; Rizzello et al., 2010; Ryan et al., 2011) (Table 2). The use of safe microbes in bread to extend the shelf life of the product is a great research area. Since LAB isolates are safe for use in foods, they are a significant alternative to chemical preservatives. Several researchers in the area of the bakery industry have successfully added LAB to dough and these strains grew well, producing the desired antifungal compounds in the dough.

Various fungi isolated from bakeries were inhibited by *L. plantarum* (LB1) and *L. rossiae* (LB5) isolated from raw wheat germ. Organic acids and peptides synthesized during fermentation were responsible for the antifungal activity; formic acid had the highest inhibition activity (Rizzello et al., 2011). However, the inhibitory compounds characterized were different, depending upon the LAB strains and flour type used. Dal Bello et al., (2007) characterized lactic acid, phenyllactic acid (PLA), cyclic dipeptides cyclo (L-Leu–L-Pro) and cyclo (L-Phe–L-Pro) produced by *L. plantarum* FST 1.7 and found them to inhibit the growth of *Fusarium* spp. in wheat bread. Ryan et al., (2008) reduced the use of calcium propionate from 3000 ppm to 1000 ppm when using sourdough fermented with *L. plantarum* FST 1.7 (LP 1.7) and *L. plantarum* FST 1.9 (LP 1.9), in which the growth of *A. niger*, *F. culmorum* and *P. expansum* was delayed for over six days while the growth of *P. roqueforti* appeared after three days of incubation at 30 °C. *L. plantarum* VTT E-78076. *Pediococcus pentosaceus* VTT E-90390 was reported to inhibit the growth of rope-forming *Bacillus subtilis* and *Bacillus licheniformis* in laboratory conditions and in the bread when the selected strains were inoculated to sourdough and subsequently 20-30 g of the inoculated sourdough was added to 100 g of wheat dough (Katina et al., 2002). Lavermicocca et al. (2000) found that *L.*

Strains	No. of days	Target fungi	Storage temperature °C	Reference
L. plantarum 21B	7	Broad spectrum	20	Lavermicocca et al., (2000)
L. plantarum	12	*Rhizopus oryzae A. niger A. flavus Penicillium sp. F. oxysporum*	27	Ogunbanwo et al., (2008)
L. brevis AM7	21	*P. roqueforti* DPPMAF1	25	Coda et al., (2008)
L. plantarum	10	*A. niger, F. culmorum,* and *P. expansum*	25	Ryan et al., (2008)
L. plantarum CRL 778, *L. reuteri* CRL 1100, and *L. brevis* CRL 772 and CRL 796	8	*Aspergillus, Fusarium,* and *Penicillium*	30	Gerez et al., (2009)
L. plantarum 1A7 (S1A7)	28	*P. roqueforti* DPPMAF1	25	Coda et al., (2011)
L. amylovorus DSM 19280	14	*F. culmorum* FST 4.05, *A. niger* FST4.21, *P. expansum* FST 4.22, *P. roqueforti* FST 4.11	25	Ryan et al., (2011)
L. fermentum Te007, *P. pentosaceus* Te010, *L. pentosus* G004, and *L. paracasi* D5	9-12	*A. niger* and *A. oryzae*	30	Muhialdin et al., (2011)

Table 2. Delay of the appearance of fungal growth on bread with added lactic acid bacteria cells

plantarum 21B inhibited the bread spoilage fungi *Aspergillus, Fusarium, Penicillium* and *Eurotium;* the active compounds were phenyllactic and 4-hydroxyphenyllactic acids. The growth of *Aspergillus niger* appeared after two days in the control sample while *L. plantarum* 21B delayed the growth of the stated fungi for seven days at 20 °C.

8.2. Flavour

Flavour is one of the most valued sensory attributes in bread - volatile and non-volatile compounds produced during the fermentation of dough contribute to bread's flavour. Reports show that the fermentation of dough with LAB can enhance the aroma and flavour

(Ryan et al., 2011; Muhialdin et al., 2011a). The growth of fungi is responsible for the formation of off-flavours and the production of mycotoxins; adding LAB to dough can prevent the growth of fungi and enhance the flavour of bread. The produced compound plays an important role for any technological application to enhance the flavour, such as diacetyl which gives a buttery flavour. Sourness in white bread indicates spoilage in contrast to the sourness of sourdough bread; for this reason, the search for new LAB for application in white bread becomes essential. Finding a new LAB strain that produces less acid and does not drop the pH below 4 will mark a good strategy for resolving such an issue. The addition of *L. paracasi* D5 and *L. fermentum* Te007 in the production of white bread resulted in an improved aroma and a pleasant caramel-like flavour in the baked bread itself (Muhialdin et al., 2011a).

8.3. Quality and acceptability

The quality of bread produced with LAB as a starter culture was reported to improve the texture and the quality of bread by increasing the air cells (Coda et al., 2008; Katina et al., 2002; Lavermicocca et al., 2000). Baker's yeast - also referred to as 'baking yeast' (*Saccharomyces cerevisiae*) - has the ability to ferment different carbohydrates and produce CO_2; the most important factor involving baking yeast in bread manufacturing is to leaven the dough during the bread's preparation. The presence of antimicrobials in the dough is used to inhibit the growth of spoilage microorganisms that can affect the growth of the baker's yeast and delay the fermentation of dough, thereby resulting in economic losses to the bakery industry (Pattison & von Holy, 2001). Baking yeast is a excellent producer of the necessary flavour and aroma compounds from the products of secondary metabolism (Evans 1990).

Pattison & von Holy (2001) found that the presence of propionic salts reduced the baking yeast activity by up to 34.4% in an *in vitro* study carried out using several natural antimicrobials with positive control calcium propionate. In comparison, lactic acid and acetic acid displayed slight effects on the activity reduction of the yeast compared with the positive control. Baking yeast and lactic acid bacteria commonly have live symbiotically in the natural ecosystem of fermenting food and beverages (Kenns et al., 1991). The volume of the dough was increased by adding sourdough containing *L. amylovorus* DSM 19280 when compared with chemical acidification (Ryan et al., 2011). Rizzello et al. (2010) reported the improvement of bread texture properties and the delaying of the staling of the bread because of the anti-staling effect produced by LAB and the synthesis of antifungal compounds. As mentioned previously, *S. cerevisiae* is responsible of leaving the dough and giving the most desirable texture to the bread.

The key role in achieving the optimum growth and activity of the bakery yeast is played by selecting a LAB that does not exhibit inhibition activity against the bakery yeast. Before choosing the LAB to be added to the dough as a co-starter, a simple experiment can be conducted in order to examine the tolerance of the bread yeast to the selected LAB strain. In a test tube mix of 10 ml water, 5 g of white flour, the LAB strain and baking yeast, we incubate and observe the production of gas at the top of the tube, which is a good indicator of the yeast activity. Ogunbanwo et al. (2008) isolated LAB from retted cassava and studied the effects of lactic acid bacteria as a starter co-culture in combination with *S. cerevisiae* in order to produce cassava-wheat bread. The improvement in the nutritional contents, physical properties and the extension of the shelf life were reported. Bread produced using *L. acidophilus* and *L. brevis* had the highest acceptability on average in relation to the bread produced with other strains of LAB. The use of LAB in bread in terms of improving the quality of wheat bread, bread volume and crumb structure has been reported (Clarke et al., 2002; Zannini et al., 2009).

8.4. Enhancement of a specific nutrient

LAB fermentation in dough has been approved for enhancing the nutritional value and digestibility of bread. Vitamin B, organic acids and the free amino acids produced through the fermentation of LAB can enhance the nutrients' presence in bread. The human body cannot synthesize B-group vitamins and this is why the body needs an external source of the vitamins. Certain LAB has been proven to synthesize B-group vitamins during the fermentation of foods; at the same time, LAB are considered to be the perfect vehicle for delivering the vitamins to the human body.

There are reports about the production of B-group vitamins by LAB isolates. Keuth and Bisping (1993) described the production of Riboflavin (Vitamin B 2) by *Streptococcus* and *Enterococcus* isolated from tempeh (Indonesian fermented food). Folates were observed to be produced by *L. plantarum* in low amounts (Sybesma et al., 2003). Vitamin B 12 (Cobalamin) was also produced by *L. reuteri* as well as the other groups of vitamin B (Santos et al., 2008). LAB enzymatic activity by proteases that take place during dough fermentation will release small peptides and free amino acids, which are considered to be important nutrients that should be present in bread in high quantities (Thiele et al., 2002). Essential amino acids, including lysine, threonine, phenylalanine and valine were reported to be produced by LAB (Gerez et al., 2006). The enzymes produced by LAB including amylases, proteases, phytases and lipases improve the food quality through the hydrolysis of polysaccharides, proteins, phytates and lipids. Anti-nutrients such as phytic acid and tannins can be reduced by LAB fermentation in food, leading to increased sensory properties of the bread (Chelule et al., 2010). The growth of fungi in food materials can cause the synthesis of allergenic spores and hazardous mycotoxins, which will lead to the reduction of the nutritional value of food stuffs. Adding 4% of fermented sourdough to the white wheat flour improved the texture and physical sensation of the bread. Furthermore, it enhanced the free amino acids, protein digestibility, phytase and antioxidant activities (Rizzello et al., 2010).

9. Starter cultures for the bread industry

Lactic acid bacteria were reported as being used as a starter culture or co-culture in the bread industry with success in terms of survivability in dough (Lavermicocca et al., 2000;

Rezzillo et al., 2011). The use of lactic acid bacteria as an antifungal agent or as a starter culture for bakery and processed foods can solve two global issues; firstly, it can extend the shelf life of the food products, which will reduce their cost and the need for low temperatures, secondly, it will satisfy the high demand of modern consumers for high quality food that is free of chemicals. Above all, the product must be safe with an extended shelf life and good sensory properties.

10. Production of LAB cells and inhibitory compounds

10.1. Growth medium

The growth of LAB and the production of antifungal compounds are largely affected by the food matrix itself (Helander, 1997). Most of the studies regarding the antifungal activity of LAB were done using the universal MRS agar. As demonstrated earlier, there are few studies that evaluate the ability of LAB isolates to produce the active compounds in non-defined media as well as few *in situ* studies. The challenge for the food industry is the need for the high production of biomass and the bioactive compounds using an inexpensive fermentation growth medium. A defined medium is all well and necessary for laboratory screening purposes but it is not suitable for heavy industrial plant. The question here is whether the selected LAB can produce the biomass and maintain the antifungal activity. In our laboratory, *L. fermentum* Te007, *Pediococcus pentosaceus* Te010, *L. pentosus* G004 and *L. paracasi* D5 were used to ferment white bread dough and they maintained the antifungal activity, as detected using MRS agar, indicating that these isolates produced the antifungal compounds in the bread dough (Muhialdin et al., 2011a). *Pediococcus pentosaceus* Te010 was further investigated for its ability to grow in formulated media from plant extracts supplemented with the basic growth needs of LAB, such as vitamins, carbohydrates, nitrogen sources and salts. The results indicated that the selected isolate was able to grow in the formulated media and maintain the production of the antifungal activity but, unfortunately, the compounds have not yet been characterized (unpublished data).

10.2. Growth conditions

The growth conditions of any microbe are the key to success during the fermentation process. As for LAB, the generally optimum temperature for growth is 37 °C for 48 h in anaerobic conditions. This is not exactly what can be applied for the production of antagonistic fungal inhibitor compounds. Some of the LAB are psychrophilic and prefer low temperatures for their growth while others are thermophilic and prefer high temperatures for their growth. This should be considered as a significant factor because the optimum growth temperature has a significant impact on the production of antifungal compounds. As well as temperature, the incubation time has a significant effect on the production of antifungal compounds with respect to the availability of nutrients in the growth medium and the production of primary or secondary metabolites.

11. Future research

The high demand by consumers for foods free of chemical preservatives has led to increasing amounts of research to provide alternatives for these chemicals. LAB provides technologically practicable alternatives for the replacement of chemical preservatives. The achievement of selecting LAB as starter cultures or co-cultures in fermentation processes can improve the desired properties of bread, at the same time providing consumers with new chemical-free foods. There is a need to study the interaction between the food matrix and the kinetics of the starter culture of LAB in bread; such studies will contribute to the bread industry by increasing the yield of the antifungal and nutritional compounds produced by LAB. Besides using the LAB cells in bread formulations, the use of the supernatant of LAB should be considered, especially the supernatant of LAB that are grown in non-conventional media such as plant extract and other cheap materials. Additional studies on the contribution of bioactive molecules to the quality and shelf life of foods will surely widen the use of LAB strains as a novel bio-control strategy in bakery products.

12. Conclusion

LAB can be used as a starter culture or a co-culture in the bread industry to enhance the sensory properties of bread and extend the shelf life. The nutritional value of the bread is enhanced due to the production of free amino acids, organic acids and a variety of Group-B vitamins. The antifungal compounds produced by LAB are important for the food industry for replacing or reducing the use of chemical preservatives. Several methods have been developed to determine the antifungal activity of the cells and the free cell supernatant. Natural sources of food preservatives - especially LAB - are important and reflect one possibility for fulfilling the needs of modern consumers of bakery products that are free of chemicals. Challenges are evident in finding new and novel isolates of LAB that can be applied in bread and which do not affect the activity of the yeast or inhibit their growth. Future works should consider the use of the LAB supernatant as well as the cells because the active compounds can be present in the supernatant. Inexpensive media are also important for high-scale industry, especially the use of plant extracts that are rich in carbohydrates and which can be supplied in bulk over the course of the year.

Author details

Belal J. Muhialdin and Nazamid Saari
Universiti Putra Malaysia (UPM),
Malaysia

Zaiton Hassan
Universiti Sains Islam Malaysia (USIM),
Malaysia

13. References

Abellana, M. L., Torres, V. S. and A. J. Ramos. 1997. Caracterizacio´n de diferentesproductos de bollería industrial. II. *Estudio de la Microflora. Alimentaria*, 287: 51-56.

Axelsson, L. 1998. Lactic acid bacteria: Classification and physiology. In *Lactic Acid Bacteria: Microbiology and functional aspects*, 2nd Edition, Revised and Expanded. Edited by S. Salminen & A. von Wright. pp. 1-72. Marcel Dekker, Inc. New York.

Bennett, G. A., Shotwell, O. L. and Hesseltine, C. W. 1980. Destruction of zearalenone in contaminated corn. *Journal of the American Oil Chemists' Society*, 57: 245-7.

Björkroth, J. and Holzapfel, W. 2003. Genera *Leuconostoc, Oenococcus* and *Weissella*, In M. Dworkin et al., eds., *The Prokaryotes: An Evolving Electronic Resource for the Microbiological Community*, 3rd edition, Springer-Verlag, New York, http://link.springer-ny.com/link/service/books/10125/

Brul, S. and Coote, P. 1999. Preservative agents in foods-mode of action and microbial resistance mechanisms. International *Journal of Food Microbiology*, 50: 1-17.

Carr, F. J., Chill, D. and Maida, N. 2002. The lactic acid bacteria: a literature survey. *Critical Reviews in Microbiology*, 28: 281-370.

Chelule, P. K., Mokoena, M. P. and Gqaleni, N. 2010. Advantages of traditional lactic acid bacteria fermentation of food in Africa. *RORMATEX*, pp. 1164-1167, http://www.formatex.info/microbiology2/1160-1167.pdf

Clarke, C., Schober, T. J. and Arendt, E. K. 2002. Effect of single strain and traditional mixed strain starter cultures in rheological properties of wheat dough and bread quality. *Cereal Chemistry*, 79: 640-647.

Coda, R., Rizzello, C. G., Nigro, F., De Angelis, M., Arnault, P., and Gobbetti, M. 2008. Long-term fungi inhibitory activity of water-soluble extract from Phaseolus vulgaris cv Pinto and sourdough lactic acid bacteria during bread storage. *Applied and Environmental Microbiology*, 74: 7391-7398.

Coda, R., Cassone, A., Rizzello, C. G., Nionelli, L., Cardinali, G. and Gobbetti, M. 2011. Antifungal Activity of *Wickerhamomyces anomalus* and *Lactobacillus plantarum* during Sourdough Fermentation: Identification of Novel Compounds and Long-Term Effect during Storage of Wheat Bread. *Applied and Environmental Microbiology*, 77: 3484-3492.

Corsetti, A., Gobetti, M., Rossi, J. and Damiani, P. 1998. Antimould activity of sourdough lactic acid bacteria: identification of a mixture of organic acids produced by *Lactobacillus sanfrancisco* CB1. *Applied Microbiology and Biotechnology*, 50: 253-256.

Cotter, P. D., Hill, C. and Ross, R. P. 2005.Bacteriocins: developing innate immunity for food. *Nature Reviews Microbiology*, 3: 777-788.

Dal Bello, F., Clarke, C. I., Ryan, L. A. M., Ulmer, H., Schober, T. J., Strom, K., Sjorgren, J., Van Sinderen, D., Schnurer, J. and Arendt, E. K. 2007. Improvement of the quality and shelf life of wheat bread by fermentation with the antifungal strain *Lactobacillus plantarum* FST 1.7. *Journal of Cereal Science*, 45: 309-318.

Dalié, D. K. D., Deschamps, A. M. and Richard-Forget, F. 2009. Lactic acid bacteria - Potential for control of mould growth and mycotoxins.A review. *Food Control*, 21: 370-380.

Dantigny, P., Guilmart, A., Radoi, F., Bensoussan, M., and Zwietering, M. 2005. Modelling the effect of ethanol on growth rate of food spoilage moulds. *International Journal of Food Microbiology*, 98: 261-269.

Druvefors, U. A., Passoth, V. and Schnurer, J. 2005. Nutrient effects on biocontrol of *Penicillium roqueforti* by *Pichia anomala* J121 during airtight storage of wheat. *Applied Environmental Microbiology*, 71: 1865-1869.

El-Ziney, M. G. and Debevere, J. M. 1998. The effect of reuterin on *Listeria monocytogenes* and *Escherichia coli* O157:H7 in milk and cottage cheese. *Journal of Food Protection*, 61: 1275-1280.

Ennahar, S., Sonomoto, K. and Ishizaki, A. 1999. Class IIa bacteriocins from lactic acid bacteria: antibacterial activity and food preservation. *Journal of Bioscience and Bioengineering*, 87: 705-716.

European Union. 1995. European Parliament and Council directive no. 95/2/EC of 20 February 1995 on food additives other than colours and sweeteners, 53. Office for Official Publications of the European Communities, Luxembourg.
http://europa.eu.int/eur-lex/en/consleg/pdf/1995 /en_1995L0002_do_001.pdf.

Evans, I. H. (1990). Yeast strains for baking. In: *Yeast Technology*, eds., Spencer, J. F. T. and Spencer, D. M. pp. 13±45. Berlin, Germany: Springer-Verlag.

Farkas, J. 2001. Physical methods for food preservation. In *Food microbiology: Fundamentals and frontiers*. Edited by M. P. Doyle, L. R. Beuchat & T. J. Montville. pp. 567-592. ASM press. Washington, USA.

Filtenborg, O., Frisvad, J. C. and Thrane, U. 1996. Moulds in food spoilage. International *Journal of Food Microbiology*, 33: 85-102.

Hammes, W. P. and Hertel, C. 2003. The Genera *Lactobacillus* and *Carnobacterium*. In *The Prokaryotes: An Evolving Electronic Resource for the Microbiological Community*, Edited by M. Dworkin. Springer-Verlag, New York.
http://link.springer-ny.com/link/service/books/10125/.

Hassan, Y. I. and Bullerman, L. B. 2008. Antifungal activity of *Lactobacillus paracasei* ssp. *tolerans* isolated from a sourdough bread culture. *International Journal of Food Microbiology*, 121: 112-115.

Helander, I. M., von Wright, A. and Mattila-Sandholm, T-M. 1997. Potential of lactic acid bacteria and novel antimicrobials against Gram-negative bacteria. *Trends in food Science and Technology*, 8: 146-150.

Gerez, C. L., Rollán, G. and Font de Valdez, G. 2006. Gluten breakdown by lactobacilli and pediococci strains isolated from sourdough. *Letters in Applied Microbiology*, 42: 459-464.

Gerez, C. L., Torino, M. I., Rollan, G. and Font de Valdez, G. 2009. Prevention of bread mould spoilage by using lactic acid bacteria with antifungal properties. *Food Control*, 20: 144-148.

Gray, J. and Bemiller, J. 2003. Bread staling: Molecular basis and control. *Comprehensive Reviews in Food Science and Safety*, 2: 1-21.

Guynot, M. E., Marín, S., Setu, L., Sanchis, V. and Ramos, A. J. 2005. Screening for antifungal activity of some essential oils against common spoilage fungi of bakery products. *Food Science and Technology International*, 11: 25-32.

James, S. and Stratford, M. 2003. Spoilage yeast with emphasis on the genus *Zygosaccharomyces*. In *Yeasts in food*. Edited by B. T & R. V. Behrs, Verlag, pp. 171-191.

Katina, K., Sauri, M., Alakomi, H. L. and Mattila-Sandholm T. 2002. Potential of Lactic Acid Bacteria to Inhibit Rope Spoilage in Wheat Sourdough Bread. *Lebensm.-Wiss. u.-Technology*, 35: 38-45.

Kenns C, Veiga, M. C., Dubourgular, H. C., Touzel, J. P., Albengae, G., Navean, H. and Nyns, E. J. 1991. Tropic relationship between *Saccharomyces cerevisiae* and *Lactobacillus plantarum* and their metabolism of glucose and citrate. *Journal of Applied Environmental Microbiology*, 57:1047-1051.

Keuth, S., and Bisping, B. 1993. Formation of vitamins by pure cultures of tempe moulds and bacteria during the tempe solid substrate fermentation. *Journal of Applied Bacteriology*, 75: 427-434.

Lavermicocca, P., Valerio, F., Evidente, A., Lazzaroni, S., Corsetti, A. and Gobbetti, M. 2000. Purification and characterization of novel antifungal compounds by sourdough *Lactobacillus plantarum* 21B. *Applied and Environmental Microbiology*, 66: 4084-4090.

Lavermicocca, P., Valerio, F. and Visconti, A. 2003. Antifungal activity of phenyllactic acid against molds isolated from bakery products. *Applied Environmental Microbiology*, 69: 634-640.

Legan, J. D. 1993. Mould spoilage of bread: the problem and some solutions. *International Biodeterioration and Biodegradation*, 32: 33-53.

Levi, C. 1980. Mycotoxins in coffee. *Journal - Association of Official Analytical Chemists*, 63: 1282-1285.

Lourens-Hattingh, A. and Viljoen, B. C. 2001.Yoghurt as probiotic carrier food. *International Dairy Journal*, 11: 1-17.

Luter, L., Wyslouzil, W. and Kashyap, S. C. 1982. The destruction of aflatoxin in peanuts by microwave roasting. *Canadian Institute of Food Science and Technology Journal*, 15: 236-238.

Magnusson, J. and Schnurer, J. 2001. *Lactobacillus coryniformis* subsp. *coryniformis* strain Si3 produces a broad-spectrum proteinaceous antifungal compound. *Applied and Environmental Microbiology*, 67: 1-5.

Magnusson, J. 2003. Antifungal activity of lactic acid bacteria. Ph.D. Thesis, Agraria 397, Swedish University of Agricultural Sciences, Uppsala, Sweden.

Magnusson, J., Ström, K., Roos, S., Sjögren, J. and Schnürer, J. 2003. Broad and complex antifungal activity among environmental isolates of lactic acid bacteria. *FEMS Microbiology Letters*, 219: 129-135.

Muhialdin, B. J., Hassan, Z. and Sadon, S. K. 2011a. Antifungal Activity of *Lactobacillus fermentum* Te007, *Pediococcus pentosaceus* Te010, *Lactobacillus pentosus* G004 and *L. paracasi* D5 on Selected Foods. *Journal of Food Science*, 76: 493-499.

Muhialdin, B. J., Hassan, Z. Sadon, S. K. NurAqilah, Z. and AZFAR, A. A. 2011b. Effect of pH and Heat Treatment on Antifungal Activity of *Lactobacillus fermentum* Te007, *Lactobacillus pentosus* G004 and *Pediococcuspentosaceus* Te010. *Innovative Romanian Food Biotechnology*, 8: 41-53.

Muhialdin, B. J. and Hassan, Z. 2011. Screening of Lactic Acid Bacteria for Antifungal Activity against *Aspergillus oryzae*. *American Journal of Applied Science*, 8: 447-451.

Niku-Paavola, ML., Laitila, A. Mattila-Sandholm, T. and Haikara, A. 1999. New types of antimicrobial compounds produced by *Lactobacillus plantarum*. *Journal of Applied Microbiology*, 86: 29-35.

Ogunbanwo, S. T., Adebayo, A. A., Ayodele, M. A., Okanlawon, B. M. and Edema, M. O. 2008. Effects of lactic acid bacteria and *Saccharomyces cerevisiae* co-cultures used as starters on the nutritional contents and shelf life of cassava-wheat bread. *Journal of Applied Biosciences*, 12: 612-622.

Pattison, T. L. and von Holy, A. 2001. Effect of selected natural antimicrobials on Baker's yeast activity. *Letters in Applied Microbiology*, 33: 211-215.

Pattison, T. L., Lindsay, D. and von Holy, A. 2004. Natural antimicrobial as potential replacements for calcium propionate in bread. *South Africa Journal of Science*, 100: 339-342.

Piard, J. C. and Desmazeaud, M. 1991. Inhibiting factors produced by lactic acid bacteria. 1. Oxygen metabolites and catabolism end-products. *Le Lait*, 71: 525-541.

Pitt, J. J. & Hocking, A. D. 1999. Fungi and food spoilage Second ed. Aspen Publications. pp.

Plockova, M., Stiles, J. and Chumchalova, J. 2001 Control of mould growth by Lactobacillus rhamnosus VT1 and Lactobacillus reuteri CCM 3625 on milk agar plates. *Czech Journal of Food Science*, 19: 46-50.

Prema, P., Smila, D., Palavesam, A. and Immanuel, G. 2008. Production and Characterization of an Antifungal Compound (3-Phenyllactic Acid) Produced by *Lactobacillus plantarum* Strain. *Food Bioprocess Technology*, 3: 379-386.

Rizzello, C. G., Nionelli, L., Coda, R., Di Cagno, R. and Gobbetti, M. 2010. Use of sourdough fermented wheat germ for enhancing the nutritional, texture and sensory characteristics of the white bread. *European Food Research and Technology*, 230: 645-654.

Rizzello, C. G., Cassone, A., Coda, R. and Gobbetti, M. 2011. Antifungal activity of sourdough fermented wheat germ used as an ingredient for bread making. *Food Chemistry*, 127: 952-959.

Rouse, S., Harnett, D., Vaughan, A. and van Sinderen, D. 2008. Lactic acid bacteria with potential to eliminate fungal spoilage in foods. *Journal of Applied Microbiology*, 104: 915-923.

Roy, U., Batish, V. K., Grover, S. and Neelakantan, S. 1996. Production of antifungal substance by *Lactococcus lactis* subsp. *lactis* CHD-28.3.*International Journal Food Microbiology*, 32: 27-34.

Ryan, L. A. M., Dal Bello, F. and Arendt, E. K. 2008. The use of sourdough fermented by antifungal LAB to reduce the amount of calcium propionate in bread. *International Journal of Food Microbiology*, 125: 274-278.

Ryan, L. A. M., Zannini, E., Dal Bello, F., Pawlowska, A., Koehler, P. and Arendt, E. K. 2011. *Lactobacillus amylovorus* DSM 19280 as a novel food-grade antifungal agent for bakery products. *International Journal of Food Microbiology*, 146: 276-283.

Saithong, P., Panthavee, W., Boonyaratanakornkit, M. and Sikkhamondhol, C. 2010. Use of a starter culture of lactic acid bacteria in plaa-som, a Thai fermented fish. *Journal of Bioscience and Bioengineering*, 110: 553-557.

Samson, R. A., Seifert, K. A., Kuijpers, A. F. A., Houbraken, J. A. M. P. and Frisvad, J. C. 2004. Phylogenetic analysis of *Penicillium* subgenus *Penicillium* using partial b-tubulin sequences. *Studies in Mycology*, 49: 175-200.

Santos, F., Vera, J. L., van der Heijden, R., Valdez, G., de Vos, W. M., Sesma, F. and Hugenholtz, J. 2008. The complete coenzyme B 12 biosynthesis gene cluster of *Lactobacillus reuteri* CRL1098. *Microbiology*, 154: 81-93.

Scott, P. M. 1984. Effect of food processing on mycotoxins. Journal of Food Protection, 47: 489-499.

Schnürer, J. and Magnusson, J. 2005. Antifungal lactic acid bacteria as biopreservatives. *Trends in Food Science and Technology*, 16: 70-78.

Shapira, R. and Paster, N. 2004.Control of mycotoxins in storage and techniques for their decontamination. Mycotoxins in food. Woodhead Publishing Limited, Abington Hall, Abington, Cambridge CB1 6AH, England. pp. 190-223.

Stiles, J., Plockova, M., Toth, V. and Chumchalova, J. 1999. Inhibition of *Fusarium sp.* DMF 0101 by *Lactobacillus* strains grown in MRS and Elliker broths. *Advances in Food Scien*ce, 21: 117-121.

Ström, K., Sjörgen, J., Broberg, A. and Schnürer, J. 2002. *Lactobacillus plantarum* MiLAB 393 produces the antifungal cyclic dipeptides cyclo(L-Phe-L-Pro) and cyclo(L-Phe-trans-4-OH-L-Pro) and 3 phenyllactic acid. *Applied and Environment Microbiology*, 68: 4322-4327.

Suhr, K. I. and Nielsen, P. V. 2004. Effect of weak acid preservatives on growth of bakery product spoilage fungi at different water activities and pH values. *International Journal of Food Microbiology*, 95: 67-78.

Sybesma, W., Starrenburg, M., Tijsseling, L., Hoefnagel, M. H. and Hugenholtz, J. 2003. Effects of cultivation conditions on folate production by lactic acid bacteria. *Applied and Environmental Microbiology*, 69: 4542-4548.

Thiele, C., Gänzle, M. G. and Vogel, R. F. 2002. Contribution of sourdough lactobacilli, yeast, and cereal enzymes to the generation of amino acids in dough relevant for bread flavor. *Cereal Chemistry*, 79: 45-51.

Trivedi, A. B., Doi, E. and Kitabatake, N. 1992. Detoxification of ochratoxin A on heating under acidic and alkaline conditions. *Bioscience, Biotechnology and Biochemistry*, 56: 741-755.

Vandenbergh, P. A. and Kanka, B. S. 1989. Antifungal product. *United States Patent.* 4,877,615.

Wang H., Yan Y., Wang J., Zhang H., Qi W. (2012). Production and Characterization of Antifungal Compounds Produced by *Lactobacillus plantarum* IMAU10014. *PLoS ONE*, 7(1): e29452. doi:10.1371/journal.pone.0029452

Whittenbury, R. 1964. Hydrogen peroxide formation and catalase activity in the lactic acid bacteria. *Journal of General Microbiology*, 35: 13-26.

Yang, E. J. and Chang, H. C. 2010. Purification of a new antifungal compound produced by *Lactobacillus plantarum* AF1 isolated from kimchi. *International Journal of Food Microbiology*, 139: 56-63.

Zannini, E., Garofalo, C., Aquilanti, L., Santarelli, S., Silvestri, G. and Clementi, F. 2009. Microbiological and technological characterization of sourdoughs destined for bread-making with barley flour. *Food Microbiology*, 26: 744-753.

Health Applications Purposes

Chapter 8

Highlights in Probiotic Research

Gülden Başyiğit Kılıç

Additional information is available at the end of the chapter

1. Introduction

For centuries, lactic acid bacteria (LAB) have been used for the preservation of food for human consumption. LAB are a large group of fermentative, anaerobe facultative, aerotolerant microorganisms which are usually present in the gut of humans and other animals, raw vegetables, meat and meat products, and cereals (Carr et al., 2002). In animals, their numbers may vary with the species, the age of the host, or the location within the gut (De Vries et al., 2006). In the food industry, lactic acid bacterial strains are widely employed either as starter cultures or as non-starter lactic acid bacteria. Furthermore, owing to their probiotic properties, several LAB strains are used as adjunctive cultures in foods and feed (Sanders, 2000; Leroy & de Vuyst, 2004).

The term "probiotic" originated from the Greek word "probios" meaning "for life" (as opposed to "antibiotic," which means "against life") (Longdet et al., 2011). Probiotics are microbial food supplements which, when administered in adequate amounts, confer health benefits to consumers by maintaining or improving their intestinal microbial flora (Salminen et al., 1998; Reid et al., 2003). The US Food and Drug Administration uses other terms for live microbes for regulatory purposes (Sanders, 2008); live microbes used in animal feeds are called "direct-fed microbials" (FDA, 1995), and, when intended for use as human drugs, they are classified as "live biotherapeutics" (Vaillancourt, 2006). Probiotics are mainly members of the genera *Lactobacillus* and *Bifidobacterium* and are normal residents of the complex ecosystem of the gastrointestinal tract (GIT) of humans.

The research of novel probiotic strains is important in order to satisfy the increasing request of the market and to obtain functional products in which the probiotic cultures are more active and with better probiotic characteristics than those already present on the market (Verdenelli, et al., 2009). According to a recent market research report 'Probiotics Market (2009-2014)', the global probiotics market generated US $15.9 billion in 2008 and is expected to be worth US $ 32.6 billion by 2014 with a compound annual growth rate of 12.6 percent from 2009 to 2014 (FB 1046, 2009).

Several aspects, including general, functional and technological characteristics, have to be taken into consideration while selecting probiotic strains (Sanders & Huis in't Veld 1999; Šušković et al., 2001). This chapter includes selection criteria of bacteria as probiotics, technological usage of probiotics, new approaches for enhancing the performance of probiotics, and health effects of probiotic bacteria.

2. Selection of probiotic bacteria

Probiotics are living, health-promoting microorganisms that are incorporated into various kinds of foods. Although there has been a growing interest in using LAB isolated both from naturally fermented products and humans for health benefits (Lim & Im, 2009), the strains should preferably be of human origin and possess a Generally-Recognized-As-Safe status (Rönkä et al., 2003).

In order to exhibit their beneficial effects, probiotic bacteria need to survive during the food-manufacturing process and in human ecosystem conditions; therefore it is important to investigate bacterial behavior under conditions which mimic the GIT (Zago et al., 2011; Lo Curto et al., 2011). Stresses to microorganisms begin in the mouth, with the lysozyme-containing saliva; continue in the stomach, which has a pH between 1.5 and 3.0; and go on to the upper intestine, which contains bile (Corzo & Gilliland, 1999). Acid and bile tolerances are two fundamental properties that indicate the ability of a probiotic microorganism to survive the passage through the GIT, resisting the acidic conditions in the stomach and the bile acids at the beginning of the small intestine (Prasad et al., 1998; Park et al., 2002). To evaluate the probiotic survival in the GIT, several *in vitro* static models of digestion have been developed (Kitazawa et al., 1991; Charteris et al., 1998). One of them is the gastric–small intestinal system TIM-1 (Minekus et al., 1995), which consists of four serial compartments simulating the stomach and the three segments of the small intestine: the duodenum, jejunum, and ileum. Another one, the TIM-2 model, is a more sophisticated *in vitro* model of fermentation in the proximal large intestine. It consists of a series of linked glass vessels containing flexible walls which allow simulation of peristalsis (De Preter et al., 2011). The simulator of the human intestinal microbial ecosystem (SHIME) was developed to simulate the entire human gastrointestinal system (Molly et al., 1993). SHIME consists of a series of five temperature- and pH-controlled vessels that simulate the stomach; small intestine; and ascending, transverse and descending colon, respectively. The SHIME harbors a microbial community resembling that from the human colon both in fermentation activity and in composition (De Preter et al., 2011). Yet another model of the digestive system has been developed by such as TNO to mimic human physiological conditions in the stomach and small intestine (Blanquet et al., 2001). The major limitations of those systems is that digestion products are not removed during the incubation, and they may have a potential inhibitory effect on enzyme activities and on probiotic survival (Pitino et al., 2010). Furthermore, such systems ignore key GIT physical processes, including the temporal nature of gastric and duodenal processing, structure of food, pattern of mixing, particle size reduction and shear, which all affect the digestion rate (Shah 2000; Sumeri et al., 2008).

Effects of probiotics are strain specific. Strain identity is important in order to link a strain with a specific health effect, as well as to enable accurate surveillance and epidemiological studies (Ganguly et al., 2011). It is very important to be able to identify specifically and unambiguously the particular probiotic LAB strains from clinical fecal and intestinal biopsy specimens and from food samples (Tilsala-Timisjärvi & Tapanialtossava, 1998). Identification of bacterial species and strains from commercialized probiotics has been conducted mostly using molecular methods (Holzapfel et al., 2001; Schillinger et al., 2003; Huys et al., 2006; Sheu et al., 2009).

Verdenelli et al. (2009) investigated the probiotic potential of 11 *Lactobacillus* strains isolated from the faeces of elderly Italians. For this purpose, the researchers identified the *Lactobacillus* strains and examined them for resistance to gastric acidity and bile toxicity, adhesion to HT-29 cells, antimicrobial activities, antibiotic susceptibility and plasmid profile. They also examined the survival of the strains as they moved through the human intestine in a 3-month human feeding trial. According to the results, *L. rhamnosus* IMC 501 and *L. paracasei* IMC 502 present favourable strain-specific properties for their utilisation as probiotics in functional foods. Both *in vitro* and *in vivo* studies confirm the high adhesion ability of *L. rhamnosus* IMC 501 and *L. paracasei* IMC 502, used in combination, indicating that the two bacterial strains could be used as health-promoting bacteria.

Başyiğit Kılıç & Karahan (2010) isolated one hundred seven strains of human originated LAB identified by 16S rRNA analysis and examined them for resistance to acidic pH, bile salts and antibiotic susceptibility. They found that *L. plantarum* (AA1–2, AA17–73, AC18–88, AK4–11, and AK7–28), *L. fermentum* (AB5–18, BB16–75, and AK4–180), *Enterococcus faecium* (AB20–98 and BK11–50) and *E. durans* (AK4–14 and BK9–40) are potentially good probiotic candidates for use as health-promoting bacteria. In another study, the *L. plantarum* strains were examined for resistance to gastric acidity in simulated gastric juice at pH 2.0, 2.5, 3.0 and 3.5; 0.4% phenol; production of H_2O_2; adhesion to Caco-2 cell line; and antimicrobial activities. The researchers determined that the artificial gastric juice, even at pH 2.0, did not significantly change the viability of the cultures, and all *L. plantarum* strains showed good resistance to 0.4% phenol. They also reported antimicrobial activity and good adhesion of *L. plantarum* strains to Caco-2 cells. The researchers concluded that all of the strains showed probiotic properties, but *L. plantarum* AB6-25, AB7-35, AA13-59, AB16-65, BC18-81 and AK4-11 were the best potential probiotic strains for human use, given their ability to survive in gastric conditions, strong resistance to phenol, and the ability to adhere to the Caco-2 cell line (Başyiğit Kılıç et al., 2011a).

Lo Curto et al. (2011) investigated the survival of three commercial probiotic strains (*L. casei* subsp. *shirota*, *L. casei* subsp. *immunitas*, *L. acidophilus* subsp. *johnsonii*) in the human upper GIT. They used a dynamic gastric model (DGM) of digestion followed by incubation under duodenal conditions. The DGM is a computer-controlled gastric model which incorporates the chemical, biochemical, physical environment and processes of the human stomach; the model is based on kinetic data derived from the Echo planar-MRI and data on the rates of GI digestion obtained from human studies (Marciani et al., 2001; 2003; 2005; 2006). The researchers used water and milk as food matrices, and survival was evaluated in both logarithmic and stationary phases. The researchers found that the % of recovery in the logarithmic phase ranged from 1.0% to 43.8% in water for all tested strains, and from 80.5% to 197% in milk. They observed higher survival rates in the stationary phase for all strains. *L. acidophilus* subsp. *johnsonii* showed the highest survival rate in both water (93.9%) and milk (202.4%).

The safety of probiotic bacteria must be carefully assessed, with particular attention to transferable antibiotic resistance (Mathur & Singh, 2005). In the last decade, increasing concern has arisen about the safe use of LAB cultures for food and feed applications, in light of the latest knowledge about their possible role as an antibiotic-resistant gene reservoir. Particular concern is due to evidence of widespread occurrence in this bacterial group of conjugative plasmids and transposons (Clementi & Aquilanti, 2011). It is known that lactobacilli have a high natural resistance to bacitracin, cefoxitin, ciprofloxacin, fusidic acid, kanamycin, gentamicin, metronidazole, nitrofurantoin, norfloxacin, streptomycin, sulphadiazine, teicoplanin, trimethoprim/sulphamethoxazole, and vancomycin (Danielsen & Wind, 2003).

One of the primary benefits associated with probiotic bacterial cultures is that they can exclude pathogenic bacteria from the small and large intestine (Kos et al., 2008). Another benefit is that in food products, antimicrobial activity of probiotic bacteria may contribute to an improvement in the quality of fermented foods. This may result from control of spoilage and pathogenic bacteria, extension of shelf life, and improvement of sensory quality (Wei et al., 2006; Siripatrawan & Harte, 2007). Kos et al. (2008) used overnight cultures and cell-free supernatants of the three probiotic strains *L. acidophilus* M92, *L. plantarum* L4, and *E. faecium* L3 for determining the antagonistic effect against *Listeria monocytogenes*, *Salmonella typhimurium*, *Yersinia enterocolitica*, and *Acinetobacter calcoaceticus*. The researchers determined that probiotic strains *L. acidophilus* M92, *L. plantarum* L4, and *E. faecium* L3 demonstrated anti-*Salmonella* activity. *L. acidophilus* M92 was also shown to have antilisterial activity, as demonstrated by *in vitro* competition test.

Production of antimicrobial compounds, which may take part in the inhibition of intestinal pathogens, is another criterion for classifying a potentially probiotic bacteria (Hutt et al., 2006). The inhibition of pathogenic microorganisms by selected probiotic strains may occur via a) production of antibiotic-like substances, b) bacteriocins and bacteriocin-like inhibitory substances such as acidophilin and reuterin, c) lowering of pH by producing organic acids such as acetic, lactic and phenyllactic acid, d) production of hydrogen peroxide and short chain fatty acids, e) decreasing the redox potential, and f) consumption of available nutrients (Holzapfel et al., 1995; Ouwehand, 1998; Tharmaraj & Shah, 2009).

The ability of LAB to adhere to epithelial cells and mucosal surfaces is thought to be an important property of many bacterial strains used as probiotics (FAO/WHO, 2001). Cell adhesion is a complex process involving contact between the bacterial cell membrane and interacting surfaces. Difficulties experienced in studying bacterial adhesion *in vivo*, especially in humans, have stimulated interest in the development of *in vitro* models for preliminary screening of potentially adherent strains (Duary et al., 2011). Attachment and colonization of the gut epithelium prolongs the time for microorganisms to influence the immune system and microbiota of the host (Forestier et al., 2001). HT-29 and Caco 2 cells, the two colonic adenocarcinomas, are derived from human intestinal epithelium. Because they have structural and functional features of normal human enterocytes, they have been extensively used as *in vitro* models in the study of human enterocytic function (Moussavi & Adams, 2009).

The ability of probiotic bacteria to adhere to Caco-2 cells can be determined by plate counting or real time PCR (Matijasic et al., 2003; Candela et al., 2005). Nawaz et al. (2011) used both of these methods and did not find a statistically significant difference. Gaudana et al. (2010) investigated the ability of four different isolates (*L. plantarum* CS23, *L. rhamnosus* CS25, *L. delbrueckii* M and *L. fermentum* ASt1) and two standard strains (*L. plantarum* ATCC 8014 and *L. rhamnosus* GG) to stimulate three types of cells (Caco-2 cells, human peripheral blood mononuclear cells [PBMC] and THP-1 cells). The researchers reported that child faecal isolate CS23 showed high binding ability, high tolerance to acidic pH and bile salts, and significant immunomodulation; therefore they concluded that CS23 can be a good potential probiotic candidate. Duary et al. (2011) determined the colonization potentials of five human faecal *L. plantarum* isolates to the Caco-2 cells. Based on direct adhesion to epithelial cells, *L. plantarum* Lp91 was the most adhesive strain to the Caco-2 cell lines, with adhesion values of approximately 10.2%. They also mentioned that the percentage of adhesion to Caco-2 and HT-29 cell lines was higher among the strains isolated from the human faecal samples and buffalo milk than that which had been isolated from cheese.

3. Technological usage of probiotics

The use of starter cultures in the production of fermented food is necessary for guaranteeing safety and standardizing properties. LAB functions primarily to drop the pH of the batter; lower pH a) promotes product safety by inactivating pathogens, b) creates the biochemical conditions to attain the final sensory properties through modification of the raw materials, and c) improves the product stability and shelf life by inhibiting undesirable changes brought about by spoilage microorganisms or abiotic reactions (Ammor & Mayo, 2007).

Functional starter cultures are defined as microbes that possess at least one inherently functional property aimed at improving the quality of the end product (De Vuyst, 2000). The use of probiotics in food has reinforced the acclaimed healthy properties and given rise to an increased consumption of these products in Europe and the USA (Kristo et al., 2003). Probiotics have been evaluated as functional starter cultures in various types of fermented food products such as yoghurt, cheese, dry sausage, salami, and sourdough. They have also been studied in therapeutic preparations to assess their positive effects on physico-chemical properties of foods and their impact on the nutritional quality and functional performance of the raw material (Knorr, 1998; Rodgers, 2008).

Fermented dairy products are widely-accepted, healthy food products and valued components of diets. The incorporation of probiotic bacteria as adjuncts in various fermented milk products is currently an important topic with industrial and commercial consequences. A number of dairy products containing probiotic bacteria are currently on the market. Fermented milk and cheeses have been described as the most suitable carriers, because they enhance the transit tolerance of bacteria (Saarela et al., 2000; Lourens-Hattingh & Viljoen, 2001). Some strains of *Lactobacillus* and *Bifidobacterium* have been shown to tolerate acidic stress when ingested with milk products (Mater et al., 2005). Lactobacilli (e.g. *L. acidophilus, L. casei* subsp. *casei, L. gasseri, L. paracasei, L. reuteri* and *L. rhamnosus*) and bifidobacteria (e.g. *Bifidobacterium adolescentis, B. bifidum, B. breve, B. infantis* and *B. longum*) constitute a significant proportion of probiotic lactic acid bacterium cultures used in the dairy industry (Wood & Holzapfel, 1995; Klein et al., 1998). It is also important to determine the technological features of the strains because they could greatly affect food quality. Further, probiotic starter cultures need to be tested for large-scale production feasibility in regard to acidification, proteolysis, and aroma formation. They must accomplish this without losing viability and functionality or creating unpleasant flavor or texture (De Vuyst, 2000; Lacroix & Yildirim, 2007).

Although the number of cells required to produce therapeutic benefits is not known and might vary as a function of the strain and the health effect desired, in general a minimum level of more than 10^6 viable probiotic bacteria per mililitre or gram of food product is accepted (Ouwehand & Salminen, 1998). The study of new probiotic strains for their technological relevance and use in food products is important for trade and industry. The search for strains which show resistance to biological barriers of the human GIT, and which possess physiological characteristics compatible with probiotic properties among LAB isolated from food, may eventually lead to the discovery of new probiotic strains for functional food products (Bude-Ugarte et al., 2006).

Studies of fermented food products as a source of new isolates are rapidly accumulating. For example, a mixture of human-derived probiotic strains was tested in the manufacture of ice cream; some of the ice cream was sweetened with sucrose and some was sweetened with aspartame (Başyiğit et al. 2006). The results showed that neither frozen conditions during the storage period nor the type of sweeteners used had any undesired effect on the survival of the probiotic cultures. Georgieva et al. (2009) studied technologically relevant properties of eight candidate probiotic *L. plantarum* strains isolated from cheeses. Researchers tested their capacity to survive over extended shelf-times at refrigerated temperatures and their growth viability in the presence of preservatives widely used in food processing. The researchers determined that the cultures' acidifying and coagulating abilities and enzyme activity make them appropriate for diverse food applications, but especially for dairy products. In another study, the survival of the probiotic strains *L. fermentum* (AB5-18 and AK4-120) and *L. plantarum* (AB16-65 and AC18-82), all derived from human faeces, was investigated in Turkish Beyaz cheese production (Başyiğit Kılıç et al., 2009). The researchers determined the viability of probiotic bacteria in Turkish Beyaz cheese during 4 months of ripening and the bacteria's effect on chemical properties of the cheese. The results of the study revealed that the test probiotic culture mix was successful for cheese production and did not adversely affect cheese quality during ripening.

Essid et al. (2009) characterized 17 strains of *L. plantarum* isolated from traditional Tunisian salted meat products to select the most suitable for use as starters for fermenting meat.

Critical characteristics included acidification and enzymatic activities responsible for final sensory properties; also important were safety characteristics, including antagonistic activity against spoilage strains and antibiotic resistance. The researchers determined that all strains of *L. plantarum* had good acidifying activity; however they showed some differences in antimicrobial, proteolytic and enzymatic activities. Başyiğit Kılıç et al. (2011b) investigated the technological properties of twenty *L. plantarum* strains to evaluate their potential usage as starter cultures in the dairy industry. During two months in cold storage, there were no significant changes in the number of bacteria or the pH of the skim milk inoculated with *L. plantarum* strains. The authors suggested that *L. plantarum* AC3-10 and AB6-25 can be used in industrial yogurt manufacture, based on their technological properties such as proteolitic activity, acidifying ability, and production of flavour compounds.

Floros et al. (2012) tested 19 facultatively heterofermentative lactobacilli from Feta, Kasseri, and Graviera cheeses for potential probiotic strains. Data from this study revealed that isolates B1, G16, G22, E22, E35, and H30 from Feta; PB2.2 from Kasseri; and 631 from Graviera have promising probiotic properties *in vitro*. β-galactosidase, low proteolytic and coagulation activities, and antibacterial activities make them promising candidates as adjunct cultures for the food industry. In another study, yoghurt was produced using a mixture of potential probiotic *L. plantarum* AB6-25, AC18-82, AK4-11 and a commercial starter culture. The yoghurt was divided into four experimental batches to which were added 0.25%, 0.5%, 1%, and 1.5% β-glucan. The survivability of these potential probiotic strains and the physico-chemical properties of the yoghurts were analyzed during a 21-day storage period. The highest *L. plantarum* count was found in the yoghurt containing 0.25% β-glucan. The study found the best physico-chemical properties to be in the 0.25% and 0.5% β-glucan containing yoghurts. Therefore, the researchers suggested using 0.25% and 0.5% β-glucan in yoghurts produced using these potential probiotic bacteria and commercial starter culture (Başyiğit Kılıç, 2012).

Wang et al. (2010) identified and established the functional and technological characteristics of potential probiotic *Lactobacillus* strains isolated from two sources: the faeces of breast-fed infants and traditional Taiwanese pickled cabbages. The authors selected the strains *L. reuteri* F03, *L. paracasei* F08 and *L. plantarum* C06 for producing probiotic fermented milk, due to their acid and bile tolerance and ability to adhere to Caco-2 cells. The milks were fermented with these 3 strains separately, and rats were fed a daily dose of 10^8 CFU/day for 14 days. After the consumption of the *Lactobacillus*-fermented milk, the rats showed increased faecal lactobacilli counts, while the counts of coliform and *C. perfringens* were significantly decreased. On the other hand, Başyiğit Kılıç et al. (2010) investigated the effects of a probiotic culture mix (*L. fermentum*, *L. plantarum* and *E. faecium*) and alfa-tocopherol administration on the microbial flora in rat GIT and faeces during a 14-day feeding period. The results indicated that the probiotic culture and alfa-tocopherol administration had no significant effects on the microbial flora of the rat intestinal tract during the 14 days of intake. Minelli et al. (2004) reported that in rats administered milk fermented with L. casei, the faecal *E. coli* counts remained stable, but *Clostridia* counts decreased significantly. Yang et al. (2005) also reported decreased faecal coliform counts as one of advantages of *Lactobacillus* and *Bifidobacterium* proliferation in the rat gut. Such potentially probiotic bacteria colonizing the intestinal mucosa provide a barrier effect against pathogens by using a variety of mechanisms, such as occupation of niches, competition for nutrients, and production of antimicrobials (Ouwehand et al., 2001).

3.1. Methods to increase survival and viability of probiotics

Researchers have long been encouraged to find new, efficient methods of improving the viability of probiotics in food products (especially fermented types), since viability can be affected by the acidic-bile conditions of the gastrointestinal tract (Mortazavian et al., 2007). The latest developments focus on fermentation technologies for producing probiotic bacteria; new approaches for enhancing the performance of these fastidious organisms during fermentation, downstream processing, and utilization in commercial products; and improving functionality in the gut. Processes to optimize survival and functionality in the gut include sublethal stress applications during cell production and new fermentation technologies, such as immobilized cell biofilm-type fermentations, are promising in this respect (Lacroix & Yildirim, 2007).

3.1.1. Immobilized cell biofilm

Cell immobilization in fermentations is an attractive and rapidly expanding research area because of its technical and economic advantages, compared to a free cell system (Stewart & Russell, 1986). The immobilization method is cheap, simple and easy (Kourkoutas et al., 2006). The technology of cell immobilization allows an increase in cell stability and a decrease of the lethal effect on the microbial cells, providing protection from the conditions of the environment (Champagne et al., 1994; Grosso & Fávaro-Trindade, 2004). Thus immobilization techniques could provide protection to acid-sensitive LAB and increase their survival rate during the shelf life of the yoghurt and during their passage through the gastrointestinal tract (Cui et al., 2000; Fávaro-Trindade & Grosso, 2002). Kushal et al. (2006) determined that the process of co-immobilization of probiotic strains of L. acidophilus NCDC 13 and B. bifidum NCDC 255 resulted in better protection of the viability of the cultures during transit through the gastrointestinal tract. In another study conducted by Kourkoutas et al. (2006), L. casei cells were immobilized on apple pieces and the immobilized biocatalysts were used separately as adjuncts in producing probiotic fermented milk. The results showed that the immobilized biocatalyst was able to ferment after storage for 15, 98 and 129 days at 4 °C, while no infection was reported during storage periods. Denkova et al. (2007) determined that the immobilization of the cells of L. acidophilus A., L. helveticus H., L. casei subsp. casei C. and L. plantarum 226-15 in chitosan resulted in preparations with high concentration of viable cells. The immobilized LAB in the chitosan gel beads was resistant to the model conditions of digestion: low and neutral values of pH, enzyme presence, and high concentrations of bile salts.

3.1.2. Encapsulation

Encapsulation is the process of forming a continuous coating around an inner matrix that is wholly contained within the capsule wall as a core of encapsulated material (Kailasapathy,

2002). Encapsulation occurs naturally when bacterial cells grow and produce exo-polysaccharides. The microbial cells are entrapped within their own secretions that act as a protective structure or a capsule, reducing the permeability of material through the capsule, and making it less exposed to adverse environmental factors. Many LAB synthesise exo-polysaccharides, but they produce insufficient amounts to encapsulate themselves fully (Shah, 2002). Encapsulating probiotics in hydrocolloid beads has been investigated as a means of improving their viability and survival in food products and in the intestinal tract (Picot & Lacroix, 2004). Other benefits of encapsulation include reduction of cell injury, protection of probiotics from bacteriophages (Steenson et al., 1987), increased survival during freeze-drying and freezing (Kim & Yoon, 1995), and greater stability during storage (Kebary et al., 1998). Several methods of encapsulation have been used on probiotics in fermented milk products and biomass production: emulsion or two phase systems, the extrusion or droplet method, and spray drying and spray coating (Mortazavian et al., 2007). The common materials used for microencapsulation of probiotics are alginate and its derivatives, starch, mixtures of xanthan-gelan, carrageenan and its mixtures, gelatin, cellulose acetate phethalate, chitosan, and miscellaneous compounds such as whey proteins, soybean oil, gums, wax, and calcium chloride (Rao et al., 1989, Picot & Lacroix, 2004, Chandramouli et al., 2004).

Hou et al. (2003) demonstrated that encapsulation of *L. delbrueckii* spp. *bulgaricus* increased their bile tolerance, and viability was elevated by approximately four log units after encapsulation within artificial sesame oil emulsions. Encapsulation in spray dried whey protein microcapsules improved survival of *B. breve* R070 but not that of *B. longum* R023 during refrigerated storage in yoghurt (Picot & Lacroix, 2004). Ding & Shah (2007) stated that encapsulation improved the survival of probiotic bacteria including *L. rhamnosus, B. longum, L. salivarius, L. plantarum, L. acidophilus, L. paracasei, B. lactis* type Bl-O4, and *B. lactis* type Bi-07 when exposed to acidic conditions, bile salts, and mild heat treatment. Capela et al. (2006) found improved viability of probiotic organisms encapsulated in 3% v/w sodium alginate in freeze-dried yogurt after 6 months of storage at 4 and 21°C. Ozer et al. (2009) studied the viability of encapsulated bacteria in white-brined cheese; the researchers used *B. bifidum* BB-12 and *L. acidophilus* LA-5 that had been encapsulated in Na-alginate by either an extrusion or an emulsion technique. Both encapsulation techniques were found to be effective in keeping the numbers of probiotic bacteria higher than the level of the therapeutic minimum. While the counts of non-encapsulated probiotic bacteria decreased approximately by 3 logs, the decrease was more limited in the cheeses containing microencapsulated cells (approximately 1 log). Khater et al. (2010) tested the ability of twelve non-encapsulated and encapsulated lactic acid and bifidobacteria strains to assimilate cholesterol and to survive at a low pH and fairly high bile concentrations. The results obtained declared that encapsulation effectively protected the microorganisms from the hostile environment in the GIT, thus potentially preventing cell loss. The assimilative reductions of cholesterol by non-encapsulated and encapsulated strains were clearly different, varying from 32.6% to 89.3% and 27.9% to 85.1% respectively. Kim et al. (2008) stated that encapsulation reduces the ability of LAB to assimilate cholesterol.

4. Effects of probiotics on human health

Probiotics have the potential for contributing greatly to human and animal health via a wide range of applications. Historically, probiotics have been used in food for humans and animals without any side effects, while providing for the balance of intestinal flora (Holzapfel & Wood, 1998). The health-promoting effects of probiotics have been widely explored and include stabilization of the indigenous microbial population, boosting of the immune system, inhibition of the growth of pathogenic organisms, prevention of diarrhea from various causes, alleviation of lactose intolerance, increased nutritional value of foods, reduction of serum cholesterol levels, antimutagenicity and anticarcinogenicity, reduction of the risk of inflammatory bowel conditions, improvement of digestion of proteins and fats, synthesis of vitamins, and detoxification and protection from toxins (Klaenhammer, 1998; Perdigon et al., 2002; Gaudana et al., 2010).

Anderson & Gilliland (1999) conducted two controlled clinical studies to test effects of yoghurt on heart-related health. They reported an average reduction of serum cholesterol by 2.9% with regular consumption of yoghurt containing L. acidophilus and a 6-10% decrease in cardiac complications due to hypercholesterolemia. A study by Ouwehand et al. (2002) found that a multi-strain probiotic mixture composed of L. reuteri, L. rhamnosus and Propionibacterium freudenreichii proved effective in both increasing the number of bowel movements and decreasing mucin secretion in elderly subjects. The probiotic mixture was more effective than L. reuteri alone, although unfortunately it is difficult to draw conclusions about mixtures versus individual probiotics, since only one component of the mixture was tested and its dose was over 10 times lower than the total bacterial dose in the mixture. Agarwal & Bhasin (2002) have reported that the strain L. casei DN-114001 reduced diarrhoeal morbidity by 40% in children.

Isolauri et al. (1999) found significant improvement when a supplement of either L. rhamnosus or B. lactis was given to children from 4 to 6 years of age who had atopic eczema. Another study involving pregnant women and newborns suggested that consumption of probiotic L. rhamnosus GG reduced the rate of newborns having atopic dermatitis (Kalliomaki et al., 2001). In an Australian study, 178 newborns of women with allergies who received either L. acidophilus LAVRI-A1 or placebo daily for the first 6 months of life showed no difference in atopic dermatitiS. However, at 12 months, the rate of sensitization was significantly higher in the probiotic group. These results suggested that the probiotic treatment had increased the risk of subsequent cow's milk sensitization (Taylor et al., 2007).

Can (2003) used an experimental animal model to study the effects of a probiotic mixture and L. GG on immune responses in allergy. The OVA specific IgE levels of the study groups which were administred probiotics and reference strain were found lower than the skim milk fed groups. A double-blind, randomized, placebo controlled trial study was conducted by Abrahamsson et al. (2007) on 188 subjects with allergic disease, in which the mothers received L. reuteri ATCC 55730 daily from gestational week 36 until delivery, and their babies continued with the probiotic until 12 months. Probiotic supplemented babies showed less IgE-associated eczema during the second year. Several probiotic effects are mediated through immuneregulation, particularly through establishing and maintaining a balance between pro-and anti-inflammatory cytokines (Isolauri et al., 2001). TNF-a and IL-6 are pro-inflammatory cytokines, which are produced by the host in response to bacterial colonisation or invasion and hence are central to the host defense mechanism against pathogens (Solis-Pereyra et al., 1997). Though lipopolysaccharide of Gram-negative bacteria is known to stimulate their production, Miettinen et al. (1996) have reported an increase in IL-6 and TNF-a production in human PBMC exposed to lactobacilli and thereby suggested the use of probiotics as vaccine vectors and for the purpose of stimulating non-specific immunity. Kailasapathy & Chin (2000) proved that the synthesis of cytokines is increased as the probiotics adhere to the intestinal epithelium.

Ziarno et al. (2007) studied cholesterol assimilation by commercial starter cultures, reporting L. acidophilus monocultures to assimilate cholesterol by 49-55%. In another study involving hypercholesterolemic mice, the probiotic potential of L. plantarum PHO4 was established by Nguygen et al. (2007). The mice were fed with 10^7 CFU per day over two weeks. These mice had 7 to 10% lesser serum cholesterol and triglycerides than the control mice deprived of the probiotic feed.

Many probiotic species have been identified to be effective in children suffering from rotaviral diarrhea (Saavedra, 2000). Longdet et al. (2011) investigated the probiotic efficacy of L. casei isolated from human breast milk in the prevention of shigellosis in albino rats infected with clinical strains of Shigella dysenteriae. The results showed that the experimental rats infected with S. dysenteriae but not treated suffered from shigellosis, while the test groups infected and treated with the L. casei showed no sign of the disease as well as no clinical effect on the liver.

Senol et al. (2011a) investigated the protective effect of a probiotic mixture of 13 different bacteria and a-tocopherol on 98% ethanol-induced gastric mucosal injury. Levels of gastric mucosal pro-and anti-inflammatory cytokines, malondialdehyde, and secretory immunglobulin A were measured. Results showed that probiotic pretreatment significantly suppressed the ethanol-induced increase of gastric mucosal interleukin-4 levels. Pretreatment with either probiotic or a–tocopherol inhibited the ethanol-induced increase of mucosal malondialdehyde concentration. Probiotic pretreatment enhanced the gastric mucosal secretory immunoglobulin A concentration. The researchers indicated that the probitic mixture and a-tocopherol reduced ethanol-induced gastric mucosal lipid peroxidation, suggesting that these probiotics may be beneficial for helping heal gastric lesions induced by lower ethanol concentration. In another study, the role of a probiotic mixture, including 13 different bacteria, in the prevention of aspirin-induced gastric mucosal injury was investigated. Pretreatment with the probiotic mixture reduced aspirin-induced gastric damage and exerted a tendency toward downregulation of proinflammatory cytokines elicited by aspirin. Researchers also found that the probiotic mixture increased sIgA production approximately 7.5-fold in the stomach, and significantly reduced the malondialdehyde increase in the gastric mucosa elicited by aspirin. Additionally, pretreatment with the probiotic mixture alleviated aspirin-induced reduction of mast cell count in the gastric mucosa. Probiotic mixture pretreatment attenuates the aspirin-induced gastric lesions by reducing the lipid peroxidation, enhancing mucosal sIgA production, and stabilizing mucosal mast cell degranulation into the gastric mucosa (Senol et al., 2011b).

5. Final remarks

Significant data have been accumulated on probiotics and their beneficial health effects. Furthermore, more insights and key findings on the impact of processing and storage on probiotic viability and stability have been gained. A variety of microorganisms, typically food grade LAB, have been evaluated for their probiotic potential and are applied as adjunct cultures in various types of food products or in therapeutic preparations. In addition, further studies are needed to determine if preventive probiotic strategies are safe with regard to development of probiotic infections. Cooperation amongst food technologists, medical and nutrition scientists, and anticipation of future consumer demands are crucial for future success in probiotics.

Author details

Gülden Başyiğit Kiliç
Mehmet Akif Ersoy University, Department of Food Engineering,
Faculty of Engineering-Architecture, Burdur, Turkey

6. References

Abrahamsson, T.R., Jakobsson, T., Bottcher, M.F., Fredrikson, M., Jenmalm, M.C. & Bjorksten, B. (2007). Probiotics in prevention of IgE-associated eczema: a double-blind, randomized, placebo-controlled trial, *Journal of Allergy and Clinical Immunology* 119: 1174 1180.

Agarwal, K.N. & Bhasin, S.K. (2002). Feasibility studies to control acute diarrhoea in children by feeding fermented milk preparations, Actimel and Indian Dahi, *European Journal of Clinical Nutrition* 56: 56-59.

Ammor, M.S. & Mayo, B. (2007). Selection criteria for lactic acid bacteria to be used as functional starter cultures in dry sausage production: An update, *Meat Science* 76: 138–146.

Anderson, J.W. & Gilliland, S.E. (1999). Effect of fermented milk (yogurt) containing *Lactobacillus acidophilus* L1 on serum cholesterol in hypercholesterolemic humans, *Journal of the American College of Nutrition* 18(1): 43-50.

Başyiğit, G., Kuleaşan, H. & Karahan, A.G. (2006). Viability of human derived probiotic lactobacilli in ice-cream produced with sucrose and aspartame, *Journal of Industrial Microbiology and Biotechnology* 33: 796-800.

Başyiğit Kılıç, G., Kuleaşan, H., Eralp, I. & Karahan, A.G. (2009). Manufacture of Turkish Beyaz cheese added with probiotic strains, *LWT - Food Science and Technology* 42. 1003-1008.

Başyiğit Kılıç, G., Kılıç, B., Kuleaşan, H. & Karahan, A.G. (2010). Effect of Probiotics and α-tocopherol applications on microbial flora of rat gastrointestinal tract, *Journal of Animal and Veterinary Advances* 9(14): 1972-1977.

Başyiğit Kılıç, G. & Karahan, A.G. (2010). Identification of lactic acid bacteria isolated from the fecal samples of healthy humans and patients with dyspepsia and determination of their pH, bile and antibiotic tolerance properties, *Journal of Molecular Microbiology and Biotechnology* 18: 220-229.

Başyiğit Kılıç, G., Kuleaşan, H. & Çakmak, V.F. (2011a). Determination of probiotic properties of *L. plantarum* strains isolated from the human fecal samples, *Proceedings of Novel Approches in Food Industry, International Food Congress*, 26-29 May 2011, Çeşme-İzmir, Turkey, pp. 56.

Başyiğit Kılıç, G., Kuleaşan, H., Akpınar, D., Çakmak, V.F. (2011b). Characterization of technological properties of human originated probiotic *L. plantarum* strains, *Proceedings of International Scientific Conference on Probiotics and Prebiotics-IPC*, Slovakia, pp. 14.

Başyiğit Kılıç, G. (2012) (unpublished data). Determination of probiotic and technological properties of some *Lactobacillus plantarum* strains, TUBITAK 109 O 623, Ongoing Project.

Blanquet, S., Marol-Bonnin, S., Beyssac, E., Pompon, D., Renaud, M. & Alric, M. (2001). The "biodrug" concept: an innovative approach to therapy, *Trends in Biotechnology* 19(10): 393–400.

Bude-Ugarte, M., Guglielmotti, D., Giraffa, G., Reinheimer, J.A. & Hynes, E. (2006). Non starter lactobacilli isolated from soft and semi hard Argentinean cheeses: genetic characterization and resistance to biological barriers, *Journal of Food Protection* 69: 2983-2991.

Can, R. (2003). The effects of probiotics on allergy, *PhD Thesis*, Süleyman Demirel University, Medicinal Faculty, Departmant of Microbiology, Isparta, Turkey, p. 75 (unpublished).

Candela, M., Seibold, G., Vitali, B., Lachenmaier, S., Eikmanns, B.J. & Brigidi, P. (2005) Real-time PCR quantification of bacterial adhesion to Caco-2 cells: competition between bifidobacteria and enteropathogens, *Research in Microbiology* 156(8): 887-895.

Capela, P., Hay, T.K.C. & Shah, N.P. (2006). Effect of cryoproyectants, prebiotics and microencapsulation on survival of probiotic organisms in yoghurt and freeze-dried yoghurt, *Food Reseach International* 39: 203-211.

Carr, F.J., Chill, D. & Maida, N. (2002). The lactic acid bacteria: a literature survey, *Critical Reviews on Microbiology* 28: 281-370.

Champagne, C.P., Lacroix C. & Sodini-Gallot, I. (1994). Immobilized cell technology for the dairy industry, *Critical Reviews in Biotechnology* 14: 109-134.

Chandramouli, V., Kalasapathy, K., Peiris, P. & Jones, M. (2004). An improved method of microencapsulation and its evaluation to protect *Lactobacillus* spp. in simulated gastric conditions, *Journal of Microbiological Methods* 56: 27-35.

Charteris, W.P., Kelly, P.M., Morelli, L. & Collins, J.K. (1998). Development and application of an *in vitro* methodology to determinate the transit tolerance of potentially probiotic *Lactobacillus* and *Bifidobacterium* species in the upper human gastrointestinal tract, *Journal of Applied Microbiology* 84: 759-768.

Clementi, F. & Aquilanti, L. (2011). Recent investigations and updated criteria for the assessment of antibiotic resistance in food lactic acid bacteria, *Anaerobe* 17: 394-398.

Corzo, G. & Gilliland, S.E. (1999). Bile salt hydrolase activity of three strains of *Lactobacillus acidophilus*, *Journal of Dairy Science* 82: 472-480.

Cui, J.H., Goh, J.S., Kim, P.H., Choi, S.H. & Lee B.J. (2000). Survival and stability of bifidobacteria loaded in alginate poly-l-lysine microparticles, *International Journal of Pharmacy Research* 210: 51-59

Danielsen, M. & Wind, A. (2003). Susceptibility of *Lactobacillus* spp. to antimicrobial agents, *International Journal of Food Microbiology* 82: 1–11.

Denkova, Z., Slavchev, A., Blazheva, D. & Krastanov, A. (2007). The effect of the immobilization of probiotic lactobacilli in chitosan on their tolerance to a laboratory model of human gut, *Biotechnology & Biotechnology Equipment* 21 (4),442-450.

De Preter, V., Hamer, H.M., Windey, K. & Verbeke, C. (2011). The impact of pre-and/or probiotics on human colonic metabolism: Does it affect human health? *Molecular Nutrition & Food Research* 55: 46–57.

De Vries, M.C., Vaughan, E.E., Kleerebezem, M. & De Vos, W.M. (2006). *Lactobacillus plantarum* – survival, functional and potential probiotic properties in the human intestinal tract, *International Dairy Journal* 16: 1018–1028.

De Vuyst, L. (2000). Technology aspects related to the application of functional starter cultures, *Food Technology and Biotechnology* 38(2): 105–112.

Duary, R.K., Rajput, Y.S., Batish, V.K. & Grover, S. (2011). Assessing the adhesion of putative indigenous probiotic lactobacilli to human colonic epithelial cells, *The Indian Journal of Medical Research* 134: 664-671.

Essid, I., Medini, M. & Hassouna, M. (2009). Technological and safety properties of *Lactobacillus plantarum* strains isolated from a Tunisian traditional salted meat, *Meat Science* 81: 203–208.

Fávaro-Trindade, C.S. & Grosso, C.R.F. (2002). Microencapsulation of *L. acidophilus* (La-05) and *B. lactis* (Bb-12) and evaluation of their survival at the pH values of the stomach and in bile, *Journal of Microencapsulation* 19: 485-494.

FAO/WHO (2001). Evaluation of health and nutritional properties of powder milk and live lactic acid bacteria. Rome, Italy: Food and Agriculture Organization of the United Nations and World Health Organization Expert Consultation Report.

FAO/WHO (2002). Guidelines for the evaluation of probiotics in food, *Report of a Joint FAO/WHO Working Group on Drafting Guidelines for the Evaluation of Probiotics in Food*, London Ontario, Canada.

Fávaro-Trindade, C.S. & Grosso, C.R.F. (2002). Microencapsulation of *L. acidophilus* (La-05) and *B. lactis* (Bb-12) and evaluation of their survival at the pH values of the stomach and in bile, *Journal of Microencapsulation* 19: 485-494.

FB 1046, 2009. Probiotic Market- Advanced Technologies and Global Market (2009 - 2014), By: markets and markets.com. Publishing Date: September 2009. Report Code: FB 1046. [accessed on March 26, 2012]. URL: www.marketsandmarkets.com/Market-Reports/probiotic-market-advanced-technologies-and-global-market-69.html.

Floros, G., Hatzikamari, M., Litopoulou-Tzanetaki, E. & Tzanetakis, N. (2012). Probiotic and technological properties of facultatively heterofermentative lactobacilli from Greek traditional cheeses, *Food Biotechnology* 26: 85–105.

Forestier, C., De Champs, C., Vatoux, C. & Joly, B. (2001). Probiotic activities of *Lactobacillus casei rhamnosus*: in vitro adherence to intestinal cells and antimicrobial properties, *Research in Microbiology* 152(2): 167-173.

Ganguly, N.K., Bhattacharya, S.K., Sesikeran, B., Nair, G.B., Ramakrishna, B.S., Sachdev, H.P.S., Batish, V.K., Kanagasabapathy, A.S., Muthuswamy, V., Kathuria, S.C., Katoch, V.M. Satyanarayana, K., Toteja, G.S., Rahi, M., Rao, S., Bhan, M.K., Kapur, R. & Hemalatha, R. (2011). CMR-DBT guidelines for evaluation of probiotics in food, *The Indian Journal of Medical Research* 134(1): 22–25.

Gaudana, S.B., Dhanani, A.S., Bargchi, T. (2010). Probiotic attributes of *Lactobacillus* strains isolated from food and of human origin, *British Journal of Nutrition* 103: 1620–1628.

Georgieva, R., Iliev, I., Haertlé T., Chobert, J.M., Ivanova, I. & Danova, S. (2009). Technological properties of candidate probiotic *Lactobacillus plantarum* strains, *International Dairy Journal* 19: 696–702.

Grosso, C.R.F. & Fávaro-Trindade, C.S. (2004). Stability of free and immobilized *Lactobacillus acidophilus* and *Bifidobacterium Lactis* in acidified milk and of immobilized *B. lactis* in yoghurt, *Brazilian Journal of Microbiology* 35: 151-156.

Hobbs, C. (2000). Pro-life therapy with probiotics. Health World Online 2000. URL: healthy.net/asp/templates/article.asp?id=953.

Holzapfel, W. H., Geisen, R. & Schillinger, G. (1995). Biological-preservation of foods with reference to protective cultures, bacteriocins, and food grade enzymes, *International Journal of Food Microbiology* 24: 343-362.

Holzapfel, W.H. & Wood, B.J.B. (1998). *The genera of lactic acid bacteria*, Blackie Academic and Professional, London.

Holzapfel, W.H., Haberer, P., Geisen, R., Bjorkroth, J. & Schillinger, U. (2001). Taxonomy and important features of probiotic microorganisms in food and nutrition, *American Journal of Clinical Nutrition* 73: 365-373.

Hou, R.C., Lin M.Y., Wang, M.M. & Tzen, J.T. (2003). Increase of viability of entrapped cells of *Lactobacillus delbrueckii* spp. *bulgaricus* in artificial sesame oil emulsions, *Journal of Dairy Science* 86: 424-428.

Hutt, P., Shchepetova, J., loivukene, K., Kullisaar, T. & Mikelsaar, M. (2006). Antagonistic activity of probiotic lactobacilli and bifidobacteria against entero- and uropathogens, *Journal of Applied Microbiology* 100:1324-1332.

Huys, G., Vancanneyt, M., D'Haene, K., Vankerckhoven, V., Goossens, H. & Swings, J. (2006). Accuracy of species identity of commercial bacterial cultures intended for probiotic or nutritional use, *Research Microbiology* 157: 803–810.

Isolauri, E., Arvola, T., Sutas, Y. & Salminen S. (2001). Probiotics: effects on immunity, *American Journal of Clinical Nutrition* 73(2): 444-450.

Kailasapathy, K. & Chin, J. (2000). Survival and therapeutic potential of probiotic organisms with reference to *Lactobacillus acidophilus* and *Bifidobacterium* spp., *Immunology and Cell Biology* 78: 80–88.

Kailasapathy, K. (2002). Microencapsulation of probiotic bacteria: technology and potential applications, *Current Issues on Intestinal Microbiology* 3: 39-48.

Kalliomaki, M., Salminen, S., Arvilommi, H., Kero, P., Koskinen, P. & Isolauri, E. (2001). Probiotics in primary prevention of atopic disease: a randomised placebo-controlled trial, *Lancet* 357: 1076-1079.

Kebary, K.M.K., Hussein, S.A. & Badawi, R.M. (1998). Improving viability of *Bifidobacteria* and their effect on frozen ice milk, *Egyption Journal of Dairy Science* 26: 319-337.

Khater, K.A.A., Ali, M.A. & Ahmed, E.A.M. (2010). Effect of encapsulation on some probiotic criteria, *Journal of American Science* 6(10): 836-845.

Kim, K.I. & Yoon, Y.H. (1995). A study on the preparation of direct vat lactic acid bacterial starter, *Korean Journal of Dairy Science* 17: 129-134.

Kim, S.J., Cho, S.Y., Kim, S.H., Song, O.J., Shin, S., Chu, D.S. & Park, H.J. (2008). Effect of microencapsulation on viability and other characteristics in *Lactobacillus acidophilus* ATCC 43121, *LWT Food Science and Technology* 41: 493-500.

Kitazawa, H., Toba, T., Itoh, T., Kumano, N., Adachi, S. & Yamaguchi, T. (1991). Antitumoral activity of slime-forming encapsulated *Lactococcus lactis* subsp. *cremoris* isolated from Scandinavian ropy sour milk, "viili", *Animal Feed Science andTechnology* 62: 277-283.

Klaenhammer, T.R. (1998). Functional activities of *Lactobacillus* probiotics: genetic mandate, *International Dairy Journal* 8: 497–505.

Klein, G., Pack, A., Bonaparte, C. & Reuter, G. (1998). Taxonomy and physiology of probiotic lactic acid bacteria*International Journal of Food Microbiology*41: 103–125.

Kos, B., Šušković, J., Beganović, J., Gjuračić, K., Frece, J., Iannaccone, C. & Canganella, F. (2008). Characterization of the three selected probiotic strains for the application in food industry, *World Journal of Microbiology and Biotechnology* 24(5): 699–707.

Kourkoutas, Y., Kanellaki, M. & Koutinas, A.A. (2006). Apple pieces as immobilization support of various microorganisms, *LWT Food Science and Technology* 39: 980–986.

Kristo, E., Biliaderis, C.G. & Tzanetakis, N. (2003). Modelling of rheological, microbiological and acidification properties of a fermented milk product containing a probiotic strain of *Lactobacillus paracasei, International Dairy Journal*, 13: 517–528.

Kushal, R., Anand S.K. & Chander H. (2006). *In vivo* demonstration of enhanced probiotic effect of co-immobilized *Lactobacillus acidophilus* and *Bifidobacterium bifidum*, *International Journal of Dairy Technology* 59: 265-271.

Lacroix, C. & Yildirim, S. (2007). Fermentation technologies for the production of probiotics with high viability and functionality, *Current Opinion in Biotechnology* 18: 176–183.

Leroy, F. & deVuyst, L. (2004). Lactic acid bacteria as functional starter cultures for the food fermentation industry, *Trends in Food Science and Technology* 15: 67-78.

Lim, S.M. & Im, D.S. (2009). Screening and characterization of probiotic lactic acid bacteria isolated from Korean fermented foods, *Journal of Microbiology and Biotechnology* 19(2): 178-186.

Lo Curto, A., Pitino, I., Mandalari G., Dainty J.R., Faulks, R.M. & Wickham John, M.S. (2011). Survival of probiotic lactobacilli in the upper gastrointestinal tract using an *in vitro* gastric model of digestion, *Food Microbiology* 7: 1359-1366.

Longdet I.Y., Kutdhik, R.J. & Nwoyeocha, I.G. (2011). The probiotic efficacy of *Lactobacillus casei* from human breast milk against Shigellosis in Albino rats, *Advances in Biotechnology & Chemical Processes* 1: 12-16.

Lourens-Hattingh, A. & Viljoen, B.C. (2001). Review: yoghurt as probiotic carrier in food, *International Dairy Journal* 11: 1–17.

Marciani, L., Ramanathan, C., Tyler, D.J., Young, P., Manoj, P., Wickham, M.S.J., Fillery Travis, A., Spiller, R.C. & Gowland, P.A. (2001). Fat emulsification measured using NMR transverse relaxation, *Journal of Magnetic Resonance Imaging* 153: 1-6.

Marciani,L., Wickham, M.S.J., Wright, J., Bush, D., Faulks, R.M., Fillery Travis, A., Gowland, P.A. & Spiller, R.C. (2003). Magnetic resonance imaging (MRI) insights into how fat emulsion stability alters gastric emptying, *Gastroenterology* 124: A581-1.

Marciani, L., Bush, D., Wright, P., Wickham, M.S.J., Pick, B., Wright, J., Faulks, R.M., Fillery Travis, A.J., Spiller, R.C. & Gowland, P.A. (2005). Monitoring of gallbladder and gastric coordination by EPI. *Journal of Magnetic Resonance Imaging* 21: 82-85.

Marciani, L., Wickham, M.S.J., Bush, D., Faulks, R.M., Wright,J., Fillery-Travis, A., Spiller, R. & Gowland, P. (2006). Magnetic resonance imaging of the behaviour of oil-in water emulsions in the gastric lumen of man, *British Journal of Nutrition* 95: 331-339.

Mater, D.D.G., Bretigny, L., Firmesse, O., Flores, M.J., Mogenet, A., Bresson, J.L. & Corthier, G. (2005). *Streptococcus thermophilus* and *Lactobacillus delbrueckii* subsp. *bulgaricus* survive gastrointestinal transit of healthy volunteers consuming yogurt, *FEMS Microbiology Letters* 250: 185-187.

Mathur, S. & Singh, R. (2005). Antibiotic resistance in food lactic acid bacteria--a review, *International Journal of Food Microbiology* 105(3): 281-295.

Matijasic, B.B., Narat, M. & Zoric, M. (2003) Adhesion of two *Lactobacillus gasseri* probiotic strains on Caco-2 cells, *Food Technology and Biotechnology* 41(1): 83-88. Miettinen, M., Vuopio-Varkila, J. & Varkila, K. (1996). Production of human tumor necrosis factor a, interleukin-6, and interleukin-10 is induced by lactic acid bacteria, Infection and Immunity 64: 5403–5405.

Minekus, M., Marteau, P., Havenaar, R. & Huisin't Veld, J.H.J. (1995). A multi compartmental dynamic computer-controlled model simulating the stomach and smal lintestine, *Atla* 23: 197–209.

Minelli, E.B., Benini, A., Marzotto, M., Sbarbati, A., Ruzzenente, O. & Ferrario, R., (2004). Assessment of novel probiotic Lactobacillus casei strains for the production of functional dairy foods, *International Dairy Journal* 14: 723-736.

Molly, K.,Vande Woestyne, M. & Verstraete, W. (1993). Development of a5-step multi-chamber reactor as a simulation of the human intestinal microbial ecosystem, *Applied Microbiology and Biotechnology* 39: 254–258.

Mortazavian, A., Razavi, S.H., Ehsani, m.R. & Sohrabvandi, S. (2007). Principles and methods of microencapsulation of probiotic microorganisms, *Iranian Journal of Biotechnology* 5:1.

Moussavi, M. & Adams, M.C. (2009). An *in vitro* study on bacterial growth interactions and intestinal epithelial cell adhesion characteristics of probiotic combinations, *Current Microbiology* 60: 327-35.

Nawaz, M., Wang, J., Zhou, A., Ma, C., Wu, X. & Xu, J. (2011) Screening and characterization of new potentially probiotic lactobacilli from breast-fed healthy babies in Pakistan, *African Journal of Microbiological Researches* 5(12): 1428-1436.

Nguygen, T.D.T., Kang, J.H. & Lee, M.S. (2007). Characterization of *Lactobacillus plantarum* PH04, a potential probiotic bacterium with cholesterol lowering effects, *International Journal of Food Microbiology* 113(3): 358-361.

Ouwehand, A. C. (1998). Anti-microbial components from lactic acid bacteria, *in* Salminen, S. & Von wright, A. (ed). *Lactic Acid Bacteria: Microbiology and Functional Aspects*, 2nd Edition, Marcel Dekker Inc., NewYork, pp. 139-160.

Ouwehand, A.C., Salminen, S.J. (1998). The health effects of cultured milk products with viable and non-viable bacteria, *International Dairy Journal* 8: 749-758.

Ouwehand, A.C., Tuomola, E.M., Tölkkö, S. & Salminen, S. (2001) Assessment of adhesion properties of novel probiotics strains to human intestinal mucus, *International Journal of Food Microbiology* 64: 119-126.

Ouwehand, A.C., Lagstrom, H., Suomalainen, T. & Salminen, S. (2002). Effect of probiotics on constipation, fecal azoreductase activity and fecal mucin content in the elderly, *Annals of Nutrition and Metabolism* 46(3–4): 159–162.

Ozer, B., Kirmaci, H. A., Senel, E., Atamer, M. & Hayaloglu. A. (2009). Improving the viability of Bifidobacterium bifidum BB-12 and Lactobacillus acidophilus LA-5 in white-brined cheese by microencapsulation, *International Dairy Journal* 19: 22-29.

Park, Y.S., Lee, J.Y., Kim, Y.S. & Shin, D.H. (2002). Isolation and characterization of lactic acid bacteria from feces of newborn baby and from dongchimi, *Journal of Agricultural and Food Chemistry* 50: 2531–2536.

Perdigon, G., Maldonado Galdeano, C. & Valdez, J.C. & Medici, M. (2002). Interaction of lactic acid bacteria with the gut immune system, *European Journal of Clinical Nutrition* 56(4): 21–26.

Picot, A., & Lacroix, C. (2004). Encapsulation of bifidobacteria in whey protein-based microcapsules and survival in simulated gastrointestinal conditions and in yoghurt, *International Dairy Journal* 14 (6): 505-515.

Pitino, I., Randazzo, C.L., Mandalari, G., Lo Curto, A., Faulks, R.M., Le Marc, Y., Bisignano, C., Caggia, C. & Wickham, M.S.J. (2010). Survival of *Lactobacillus rhamnosus* strains in the upper gastrointestinal tract, *Food Microbiology* 27: 1121-1127.

Prasad, J., Gill, H., Smart, J. & Gopal, P.K. (1998). Selection and characterization of *Lactobacillus* and *Bifidobacterium* strains for use as probiotic, *International Dairy Journal* 8: 993–1002.

Rao, A.V., Shiwnarin, N. & Maharij, I. (1989). Survival of microencapsulated *Bifidobacterium pseudolongum* in simulated gastric and intestinal juices, *Canadian Institution of Food Science and Technology Journal* 22: 345-349.

Reid, G., Jass, J., Sebulsky, M.T. & McCormick, J.K. (2003). Potential uses of probiotics in clinical practice, *Clinical Microbiological Reviews* 16(4): 658–72.

Rodgers, S. (2008). Novel applications of live bacteria in food services: probiotics and protective cultures, *Trends in Food Science and Technology* 19: 188-197.

Rönkä E., Malinen E., Saarela M., Rinta-Koski M., Aarnikunnas J., Palva A. (2003). Probitic and milk technological properties of *Lactobacillus brevis*, *International Journal Food Microbiology* 83: 63–74.

Saarela, M., Mogensen, G., Fonden, R., Matto, J. & Mattila-Sandholm, T. (2000). Probiotic bacteria: safety, functional and technological properties, *Journal of Biotechnology* 84: 197–215.

Saavedra, J. (2000). Probiotics and infectious diarrhea, *American Journal of Gastroenterology* 95: 16-18.

Salminen, S., von Wright, A., Morelli L., Marteau P., Brassart D., de Vos, W.M., Fondén, R., Saxelin, M., Collins, K., Mogensen, G., Birkeland, S.E. & Mattila-Sandholm T. (1998). Demonstration of safety of probiotics – a review, *International Journal Food Microbiology* 44: 93–106.

Sandeep, B., Gaudana, Akhilesh, S., Dhanani & Tamishraha, B. (2010). Probiotic attributes of *Lactobacillus* strains isolated from food and of human origin, *British Journal of Nutrition* 103: 1620–1628.

Sanders, M.E. & Huis in't Veld, J.H.J. (1999). Bringing a probiotic containing functional food to the market: microbiological, product regulatory and labeling issues, *in* Konings, W.N., Kuipers, O.P. & Huis in't Veld, J.H.J. (ed.), *Proceedings of the 6th symposium on lactic acid bacteria: genetics, metabolism and applications*, Kluwer Academic Publishers, Antonie van Leeuwenhoek, Veldhoven, The Netherlands, pp. 293–316.

Sanders, ME. (2000). Considerations for use of probiotic bacteria to modulate human health, *Journal of Nutrition* 130: 384-390.

Sanders, M.E. (2008). Probiotics: Definition, sources, selection, and uses, *Clinical Infectious Diseases* 46: 58–61.

Schillinger, U., Yousif, N.M., Sesar, L. & Franz, C.M. (2003). Use of group-specific and RAPD-PCR analyses for rapid differentiation of Lactobacillus strains from probiotic yogurts, *Current Microbiology* 47: 453–456.

Shah, N.P. (2000) Probiotic bacteria: selective enumeration and survival in dairy foods, *Journal of Dairy Science* 83: 894-907.

Shah, N. (2002). The exopolysaccharides production by starter cultures and their influence on textural characteristics of fermented milks, *Proceeding of Symposium on New Developments in Technology of Fermented Milks*, International Dairy Federation, Comwell Scanticon, Kolding, Denmark, p. 5.

Sheu, S.J., Hwang, W.Z., Chen, H.C., Chiang, Y.C. & Tsen, H.Y. (2009.) Development and use of tuf gene-based primers for the multiplex PCR detection of *Lactobacillus acidophilus*, *Lactobacillus casei* group, *Lactobacillus delbrueckii*, and *Bifidobacterium longum* in commercial dairy products, *Journal of Food Protection* 72: 93–100.

Senol, A., Isler, M., Karahan, A.G., Kilic, G.B., Kuleasan, H., Kaya, S., Keskin, M., Goren, İ., Saritas, U., Aridogan, C.B. & Delibas, N. (2011a). Preventive effect of probiotics and alpha-tocopherol on the ethanol-induced gastric mucosal injury in rats, *Journal of Medicinal Food* 14(1-2): 173-179.

Senol, A., Isler, M., Karahan, A.G., Basyigit Kilic, G., Kuleasan, H., Goren, İ., Saritas, U., Kaya, S., Ciris, M., Akturk, O., Aridogan, C.B., Demirin, H. & Cakmakci, M.L. (2011b). Effect of the probiotics on aspirin-induced gastric mucosal lesions, *Journal of Turkish Gastroenterology* 22 (1):18-26.

Stewart, G.G. & Russell, I. (1986). One hundred years of yeast research and development in the brewing industry, *Journal of the Institute of Brewing* 92: 537–558.

Siripatrawan, U. & Harte, B.R. (2007). Solid phase microextraction/gas chromatography/mass spectrometry integrated with chemometrics for detection of *Salmonella typhimurium* contamination in a packaged fresh vegetable. *Analytica Chimica Acta* 581: 63–70.

Solis-Pereyra, B., Aattouri, N. & Lemonnier, D. (1997). Role of food in the stimulation of cytokine production, *American Journal of Clinical Nutrition* 66: 521–525.

Steenson, L.R., Klaenhammer, T.R. & Swaisgood, H.E. (1987). Calcium alginate-immobilized cultures of lactic streptococci are protected from attack by lytic bacteriophage, *Journal of Dairy Science* 70: 1121-1127.

Sumeri, I., Arike, L., Adamberg, K. & Paalme, T. (2008). Single bioreactor gastrointestinal tract simulator for study of survival of probiotic bacteria, *Applied Microbiology and Biotechnology* 80: 317-324.

Šušković, J., Kos, B., Goreta, J. & Matošić, S. (2001). Role of lactic acid bacteria and bifidobacteria in symbiotic effect, *Food Technology and Biotechnology* (39): 227–235.

Taylor, A.L., Dunstan, J.A. & Prescott, S.L. (2007). Probiotic supplementation for the first 6 months of life fails to reduce the risk of atopic dermatitis and increases the risk of allergen sensitization in high-risk children: a randomized controlled trial, *Journal of Allergy and Clinical Immunology* 119: 184-191.

Tharmaraj, N. & Shah, N.P. (2009). Antimicrobial effects of probiotics against selected pathogenic and spoilage bacteria in cheese-based dips, *International Food Research Journal* 16: 261-276.

Tilsala-Timisjärvi, A. & Tapanialtossava, T. (1998). Strain-specific identification of probiotic *Lactobacillus rhamnosus* with Randomly Amplified Polymorphic DNA-Derived PCR Primers, *Applied and Environmental Microbiolog* 64(12): 4816–4819.

Vaillancourt, J. (2006). Regulating pre- and probiotics: a U.S. FDA perspective, *Proceedings of ending the war metaphor: the changing agenda for unraveling the host-microbe relationship. in* Institute of Medicine of the National Academies. Washington, DC: National Academies Press, pp. 229–237.

Verdenelli, M.C., Ghelfi, F., Silvi, S., Orpianesi, C., Cecchini, C. & Cresci, A. (2009). Probiotic properties of *Lactobacillus rhamnosus* and *Lactobacillus paracasei* isolated from human faeces, *European Journal of Nutrition* 48: 355–363.

Wang, C., Lin, P., Ng, C. & Shyu, Y. (2010). Probiotic properties of *Lactobacillus* strains isolated from the feces of breast-fed infants and Taiwanese pickled cabbage, *Anaerobe* 16: 578-585.

Wei, H., Wolf, G. & Hammes, W.P. (2006). Indigenous microorganisms from iceberg lettuce with adherence and antagonistic potential for use as protective culture, *Innovative Food Science Emerging Technologies* 7: 294–301.

Wood, B.J.B. & Holzapfel W.H. (1995). *The Lactic Acid Bacteria, Vol. 2, The Genera of Lactic Acid Bacteria*, Blackie Academic & Professional, London, pp 55-124.

Yang, S.C., Chen, J.Y., Shang, H.F., Cheng, T.Y., Tsou, S.C. & Chen, J.R. (2005). Effects of symbiotics on intestinal microflora and digestive enzyme activies in rats, *World Journal of Gastroenterology* 11:7413-7417.

Zago, M., Fornasari, E., Carminati, D., Burns, P., Suàrez, V., Vinderola, G., Reinheimer, J. & Giraffa, G. (2011). Characterization and probiotic potential of *Lactobacillus plantarum* strains isolated from cheeses, *Food Microbiology* 28: 1033-1040.

Ziarno, M., Sekul, E. & LafrayaAguado, A. (2007). Cholesterol assimilation by commercial yoghurt starter cultures, *ACTA Scientiarum Polonorum-Food Science and Human Nutrition* 6(1): 83-94.

Chapter 9

Probiotics and Intestinal Microbiota: Implications in Colon Cancer Prevention

Katia Sivieri, Raquel Bedani, Daniela Cardoso Umbelino Cavallini and Elizeu A. Rossi

Additional information is available at the end of the chapter

1. Introduction

Colon cancer (CC) is one of the commonest causes of death among all types of cancers [1]. The development of cancer is a multifactorial process influenced by genetic, physiological, and environmental factors [2,3]. Regarding environmental factors, the lifestyle, particularly dietary intake, may affect the risk of CC developing [1,4]. Western diet, rich in animal fat and poor in fiber, is generally associated with an increased risk of colon cancer [5,6,7]. Thus, it has been hypothesized that the connection between the diet and CC, may be the influence that the diet has on the colon microbiota and bacterial metabolism, making both relevant factors in the etiology of the disease [8,9]. Additionally, it has been clearly demonstrated that the gut microbiota may be modulated by many factors including diet [10].

Several studies have indicated that the intestinal microbiota is an important determinant for general health of the human body [1]. Therefore, a beneficial modulation of the composition and metabolic activity of the gut microbiota might represent an interesting approach to improve health, reducing the risk of CC development. This modulation may be though about probiotic consumption.

Probiotics are defined as live microorganisms which when administered in adequate amounts confer a health benefit on the host [11]. Among the best known probiotic microorganisms are strains belonging to the *Lactobacillus* and *Bifidobacterium* genera. However, other microorganisms, such as *Enterococcus* spp., *Streptococcus* spp., *Escherichia coli* Nissle 1917, some bacilli, and *Saccharomyces cerevisiae* subsp. *boulardii* have also been considered for use as probiotics [12].

Even though the mechanisms by which probiotics may inhibit colon cancer are not fully elucidated, certain potential mechanisms have been disclosed, such as the alteration of the composition and the metabolic activities of the intestinal microbiota, the changing physicochemical conditions in the colon, the binding of dietary carcinogens, the production of short chain fatty acids (SCFA), the protection of the colonic mucosa and enhancement the immune system [1,3].

The anticarcinogenic effects of probiotic microorganisms *in vitro* and in animal studies are well documented [3]. In clinical trials, the probiotics are thought to play a protective role in the initial process of carcinogenesis. Nevertheless, it is important to determine whether the long-term administration of these microorganisms might result in changes in the incidence of CC in humans [13]. Additionally, there are several challenges for the development of probiotics, including the selection of the appropriate microorganisms, control of dietary intake, time and frequency of probiotic dosing and the use of accepted biomarkers for raised cancer risk that might be monitored during clinical trials [4,13]. Further experimental models are needed to understand the exact mechanisms involved in the influence of probiotics on colon cancer development.

Therefore, this chapter will discuss the effects of probiotics in colon cancer prevention and the possible mechanisms of action these microorganisms. Additionally, this chapter will also show the results of original work, carried out by our research group, about the effects of probiotic *Enterococcus faecium* CRL 183 (strain isolated from Tafí cheese, a homemade traditional highlands cheese the province of Tucumán, Argentina) on intestinal microbiota and colon cancer prevention.

2. Colon cancer

Social and economic transformations related to urbanization and industrialization in Brazil resulted in changes in the morbimortality profile of the population. While, in the first half of the 20th century infectious disease event were the most frequent, from the 1960 metabolic diseases and noncommunicable grievances occupied the first place, contributing to the process of demographic transition, which favours the spread of cardiovascular and respirator disease, cancer and diabetes, as does the nutritional transition, with a marked reduction of malnutrition and large growth in the number of overweight people [14].

Known for many centuries, cancer was widely regarded as a disease of developed countries with large financial resources. However, for approximately four decades, this situation has undergone transformation, and most of the global burden of cancer can be observed in developing countries, especially those with low to medium resources [15].

Cancer has become a global public health problem of course, since the World Health Organization (WHO) estimates that in the year 2030 there will be 27 mil new cases of cancer, 17 million deaths and 75 million people living with the disease [15].

Cancer of the colon and rectum is the third commonest type of cancer among men and the commonest in women. It is estimated that in 2011, in Brazil, 14,180 new cases of colon cancer and rectum, occurred in men and women. These values correspond to a perceived risk 15 new cases per 100 thousand men and 16 per 100 thousand women [15].

This neoplasia is considered to have a good prognosis when diagnosed in the early stages. Colon cancer like others forms of cancer develops as a result of interaction between endogenous and environmental factors. Among the factors that may affect the risk of developing this disease are age, eating habits, physical activity, alcohol consumption, smoking, nutritional status, presence of polyps, cancer history of self and family, cases of ulcerative enterocolitis and chronic constipation [15,16].

Most cases of CC occur sporadically, being the most common type of adenocarcinoma, which develops from glandular cells that cover the wall of the intestine [17]. Adenocarcinomas grow from normal epithelium through an accumulation of mutations that result in malignant transformation [19].

Genomic instability is fundamental to this process and is related to the rearrangement of genes, or loss of DNA fragments, aneuploidy and loss of heterozygosis [19]. In addition, inactivation of tumor suppressor genes, such as APC, DCC, DPC4 and p53, along with the activation of oncogenes, of which the family of *ras* genes are the best well described, play important parts in the appearance of malignancy [17].

Generally, the colon tumor is detected for the first time as a polyp (mass of cells growing out of the wall of the colon), although nowadays it is possible to detect small lesions affecting the crypts, called aberrant crypts foci (ACF) [18]. ACF are not only morphologically but also genetically distinct lesions and are precursors of adenoma and cancer. Tumors can appear anywhere in the colon, although most sporadic rectal colon cancers are located on the left side of the distal colon (including the rectum and sigmoid colon) [19].

Epidemiological studies have pointed to the high consumption of red meat, fat and low fiber intake, typical of the Western diet as risk factors in the etiology of this type of cancer [20].

One of the possible effects of a Western diet on colon cancer is related to increased excretion of bile acids [21]. In addition, the increased ammonia production in rats consuming a diet rich in protein has also been linked to an increased risk of cancer [22]. However, high consumption of fruits, cereals, fish and calcium may reduce the risk of developing colon cancer [23].

The effect of diet on carcinogenesis can be modulated by changes in metabolic activity and composition of the intestinal microbiota [23]. Several studies have trial to establish relationships between bacteria and colon cancer. We know that various bacterial metabolites are carcinogenic, examples being, the nitrosamines, phenol, indole, ammonia and amines [13].

There is multiple evidence that bacteria play a key role in the emergence of chronic inflammatory bowel diseases. Experimental studies demonstrate the impossibility of developing this inflammation in the absence of bacteria and researches have tried for many years trying to identify a possible causative agent of inflammatory bowel diseases. Studies suggest that chronic inflammatory intestinal activity seems, paradoxically to be triggered by bacteria belonging to the normal commensal which take on microbiota in situations as yet unknown, a pathological role that can activate the local immune apparatus [24].

There are many types of intestinal bacteria that produce a variety of metabolites that modulate the normal development and functioning of the host. On the other hand, the metabolic activity of intestinal microbiota can generate compounds that are harmful such as reactive oxygen intermediates. These molecules, which include superoxide, hydrogen peroxide, hypochlorous acid, singlet oxygen and hydroxyl radical, can cause oxidative damage to cellular DNA and increase the risk of colon cancer [25]. Studies have shown that *Enterococcus faecalis* can produce superoxide and hydrogen peroxide, causing damage to DNA in skin cells, in both *in vitro* and *in vivo* tests [26].

Given the role of intestinal microbiota in colon carcinogenesis, it is suggested that factors that modulate beneficially the composition and/or activity of the microbiota could inhibit the development of CC.

3. Evidences for relationship among intestinal microbiota, probiotics and colon cancer

3.1. The intestinal microbiota

The gastrontestinal (GI) microbiota undergoes changes in quantity and quality, depending on the location of colonization in the GI. Traditional culture-based characterization may take into account no more than 30% or so of the microorganisms that can be seen and enumerated by microscopic observation. The worldwide species diversity of commensal intestinal bacteria is immense. In that respect, the use of molecular tools has indicated that the majority of the dominant bacterial species observed in the faecal microbiota of an individual (approximately 80%) are specific to this individual [27]. Also, these species are not distributed homogeneously along the length of the GI, so the bacterial activities are considerably variable in different parts of the intestine [28].

The stomach and the small intestine contain few species, whereas the colon contains a complex and dynamic microbial ecosystem, with a great concentration of bacteria. Among these are the bifidobacteria and lactobacilli, considered non-pathogenic or beneficial bacteria [29]. The bacterial population in the large intestine is very large and reaches a maximum count of 10^{12} CFU.g^{-1}. In the small intestine, bacterial contents are considerably smaller from 10^4 to 10^7 CFU.g^{-1}, while in the stomach only 10^1 at 10^2 CFU.g^{-1} are found in function of low pH on this site. In total, the number of intestinal bacteria is approximately ten times the number of cells that make up the human body [30].

On the basis of rRNA sequencing 40,000 strains of intestinal bacteria can be indentified, including non-cultivable bacteria [31]. It was noted that 99% of intestinal bacteria consist of four phyla, Proteobacteria, Actinobacteria, and two main phyla Bacteroidetes and Firmicutes [32]. While the species in the phylum Bacteroidetes show a great variety between individuals, a large number of species in the phylum Firmicutes belong to clusters of clostridial butyrate producers [33].

With advances in molecular biology, it is known that the intestinal microbiome, contains 100 times more genes than the whole human genome [34]. Thus, a close relationship is evolving between the human gut microbiota. The human intestine exhibits to a symbiotic relationship that plays a key role in human homeostasis, including metabolism, growth and immunity [35].

One of the primary functions of the intestinal microbiota is the harnessing of energy from elements of the diet that could be lost through excretion [36]. The polysaccharides are not absorbed in the colon, but metabolized by resident microorganisms to short chain fatty acids (SCFA), such as propionate and butyrate, which are absorbed by passive diffusion [37]. SCFA production is dependent on the available fermentation of substrate, such as, starch or other polysaccharides, results butyrate, acetate and propionate [37]. SCFA concentrations are higher on the right side of the colon than on the left and this is probably due to the greater availability of carbohydrates [29]. The SCFA have an important role in the maintenance of the epithelial layer. Studies show that epithelial cells acquire about 70% of their butyrate oxidation [29]. The butyrate also acts as a trophic factor for cells in intact tissues [38]. In addition, it has been proposed that butyrate lowers the risk of colon cancer by its ability to inhibit the genotoxic activity of nitrosamines and hydrogen peroxide, as well as to induce various levels of apoptosis, differentiation and the cell cycle stop colon cancer in animal models [39].

Other researchers also cite the effect of butyrate on mediators of inflammation, it has been proved that this SCFA is able to inhibit the expression of some cytokines (TNF, IL-6, IL-1) and to inhibit the activation of nuclear factor κB (NF- κB) [40]. Other functions of the gastrointestinal microbiota include digestion of poorly digested nutrients, modification of bile acids, and nutritional supplementation by auxotrophic of mutants additional compounds that cannot be acquired by food consumption, such as folic acid and biotin [41].

The non-pathogenic commensal microbiota has a profound impact on the normal physiology of the GI tract. It ensures the efficiency of bowel motility, intestinal growth and immunity, as well as digestion, nutrient absorption and fortification of the mucus barrier [42].

Researchers have made advances in the characterization of GI microbiota defining the responses that may contribute to the development of inflammatory bowel diseases, such as, colon cancer [43]. Given the importance of a better understanding of intestinal microbiota, the TGI has been often studied. In recent decades, various intestinal simulators have been and are being developed, to facilitate the study of the intestinal microbial ecosystem and its interactions [44, 45].

3.2. Methods for *in vitro* evaluation of effects of probiotics on intestinal microbiota

The FAO/WHO refers to probiotics as live microorganisms that administered in adequate doses, benefit the health of the host [11]. The beneficial effects of ingesting probiotics enhanced relief of the symptoms of lactose intolerance, treatment for diarrhea, reduction of serum cholesterol, enhanced immune response and anticarcinogenic effects [46].

The rising consumption of probiotic products by Europeans is mainly is in the form of dairy products containing generally *Lactobacillus* spp. and *Bifidobacterium* spp. However there are products in which the microorganisms used are strains of *Enterococcus* spp. or yeasts such as *Saccharomyces boulardii* [47]. Foods for human consumption containing lactic acid bacteria (LAB) include fermented milk, fruit juices, wine and sausages. Simple cultures or mixed microorganisms are used in probiotic preparations [48].

Several experimental observations have pointed to the potential protective effect of LAB against the development of tumors in the colon [49]. Within the intestinal microbiota, the LAB complex constitutes part of those bacteria able to promote a beneficial effect. They have an important role in retarding colon carcinogenesis by possibly of influencing metabolic, protective and immunological functions in the intestine [39]. The effect of intake of probiotics on intestinal native microbiota can be assessed through *in vivo* or *in vitro* models. *In vivo* models may involve healthy human volunteers, hospitalized patients or an animal model, but these models have some limitations such as high cost, delay in obtaining results and the type of food or drugs administered [50], whereas, *in vitro* models enable you to simplify the system and study separately the metabolism of native and added microbiota, in the presence of specific substrates [50].

In vitro fermentation models range from a simple batch system to more complex systems of continuous flow and multi-stage. *In vitro* gut fermentation models enable the stable cultivation of a complete intestinal microbiota for a defined and model-specific period of time. Selection of the appropriate model requires careful evaluation of the study objectives given the advantages and limitations exhibited by each type of system. Some existing systems are included in the batch, continuous culture, multi-stage continuous culture, continuous artificial digestive system and stationary systems [51].

Batch fermentation is the growth of a pure or mixed bacterial suspension in a carefully selected medium without the further addition of nutrients. These models are generally closed systems is sealed bottles or reactors containing suspensions of fecal material which are maintained under anaerobic conditions. Several studies have already been carried out, using this type of model in research on the prebiotic potential of fructans. This template is particularly useful to investigate metabolic profiles of SCFAs arising from active metabolism of dietary compounds by intestinal microbiota [50].

Continuous culture fermentation models exist as either single- or multistage systems and are necessary to perform long-term studies, as substrate replenishment and toxic product removal are facilitated. Single-stage continuous fermentation models are often used to elucidate proximal colon function and metabolic activity as the mixing of digest from both the caecum and ascending colon is well simulated in these models [52].

These models have several advantages, such as: the ease of use of the system, the possibility of using radioactive substances and the low operation cost [28].

A major advance for *in vitro* fermentation systems was the development of continuous multi-stage models, which allow the simulation of horizontal processes. This type of system makes it easy to study the nutritional and physicochemical properties of intestinal microbiota, through the combination of three reactors connected in series, simulating the proximal, distal and transversal colon (see Figure 1). Later, Molly et al. [44] developed the human microbial ecosystem simulator (SHIME ®), which consists of a succession of five connected reactors, which represent the different parts of the human gastrointestinal tract with their respective values of pH, residence time and volumetric capacity (Figure 1). The five reactors are continually agitated and kept at a temperature of 37 ºC by means of a thermostat. The medium is kept in the anaerobic state, by daily injection of N_2. The appropriate pH for each portion of the GI tract is controlled automatically by adding 1N NaOH or concentrated HCl [44, 45].

Figure 1. Computer controlled simulation of human microbial ecosystem (SHIME ®) housed in the Probiotics Research Laboratory of FCF/UNESP-Brazil. Sivieri et al.[53]

The adaptation, survival and proliferation of a human intestinal microbiota in continuous fermentation *in vitro* models are depended on environmental parameters such as pH, retention time, temperature, flow rate and oxygen depletion. The rigorous control of these factors allows steady established state in conditions the microbial composition and metabolic activity, creating a reproducible system.

The continuous cultivation model has been used in research on the metabolism and ecology of intestinal microbiota, with an emphasis on the use of probiotics [51, 54], prebiotics [55, 56] and the formation of fermentation products [57]. The *in vitro* modeling of host digestive functions in vitro coupled with multistage continuous fermentation, represents the most advanced attempt thus far at simulating interdependent physiological functions within the human gut, stomach lumen and small intestine. Human digestive functions that are reproduced in the TIM-1 small intestine model include bile secretion, motility, pH and absorption capacity of the upper intestine. Proximal colon simulator models such as TIM-2 include other host functions such as peristaltic mixing and water and metabolite absorption. The combination of TIM-1 and TIM-2 models led to the creation of an artificial digestive system which has been used to investigate pharmaceutical drug delivery and advanced nutritional studies [58, 59].

The use of a multidisciplinary biological systems approach, in combination with '-omics' platforms as outlined will facilitate the most advanced system for unraveling the complex microbial and host factors governing human gut microbiota functionality [60].

In vitro fermentation models are an innovative technological platform where the greatest advantages are exhibited by the virtually limitless experimental capacity as experimentation is not restricted by ethical concerns. Host intestinal function is only partially simulated in some model designs (e.g. TIM-1 and TIM-2) and together with microbial population balancing remains a major challenge of in vitro gut fermentation modeling.

3.3. Inhibition on colon cancer by probiotics and the possible action mechanisms of these microorganisms

The evidence pointing to the beneficial effects of probiotics on colon cancer comes from *in vitro* tests, experiments with animals and clinical trials. Additionally, these has been much discussed on which step in the process of carcinogenesis might the effect by probiotics. It is likely that different probiotic strains act on different stages of carcinogenesis [20].

In general, the probiotics do not colonize the human gut, but some strains are can permanently colonize the indigenous microbiota [61].

The mechanisms by which probiotics may inhibit colon cancer are not yet fully characterized. However, several explanations have been suggested including: alteration of the metabolic activities of the intestinal microbiota; quantitative and qualitative changes in the intestinal microbial compositin; alteration of physicochemical conditions in the colon; binding and/or degradation of potential carcinogens; SCFA production; production of anti-tumorigenic or anti-mutagenic compounds; modulation of hosts's immune response, and/or physiology [3,62, 63].

Probiotics may modulate the metabolic activities of the intestinal microbiota by three possible mechanisms: competing with and displacing other components of the microbiota; producing antibacterial substances, including bateriocins, to control the growth of other members of the microbiota; producing lactic and other organic acids, which might lower the luminal pH and thus modulate enzyme activity [20,64].

Several investigations have shown that probiotics can influence bacterial enzymes activity related to the production of carcinogenic compounds, such as beta-glucuronidase, nitroreductase and azoreductase [65, 66, 67].

Bacterial glucuronidase appears to have an important role in the initiation of colon cancer, due to its ability to hydrolyze several glucuronides and carcinogenic aglycones in the intestinal lumen [65,68]. The nitroreductase and azoreductase take past in to the formation of aromatic amines harmful to the body [69].

Both harmful and beneficial bacteria are commonly found in the intestines and differ in their enzymatic activity [70]. In general, bacteria from the genera *Bifidobacterium* and *Lactobacillus* produced a very little activity of enzymes that convert pro-carcinogens into carcinogens, compared with bacteria from the genera *Bacteroides* and *Clostridium* [71]. Therefore, the activities of these enzymes in the lumen might be correlated with the number of lactic acid bacteria (LAB) in the intestine [72]. This suggests that increasing the proportion of LAB in the gut could diminish the levels of xenobiotic metabolizing enzymes [71]. Thus, the effect of probiotic microorganisms on fecal enzyme activities might be explained by this mechanism.

In a preliminary study, on feces of small animal, the animal supplementation of a high cholesterol diet with a mixture of probiotic strains of *L. johnsonii* and *L. reuteri* for 5 weeks significantly decreased the activity of fecal-glucuronidase and azoreductase [67].

Gorbach and Goldin [65] studied, in humans, the effect of ingestion of *L. acidophilus* NCFM strains about the activity of-glucuronidase, nitroreductase and azoreductase. Both strains had a similar effect and caused a significant decline in the activity of these three enzymes. A reverse effect was found 10 to 30 days after the end of the intake of these bacteria, suggesting that continuous consumption of *L. acidophilus* is necessary for maintaining.

Benno and Mitsuoka [73] and Spanhaak et al. [66] also found in humans, a significant reduction in the activity glucuronidase after intake of *Bifidobacterium longum* and *L casei* Shirota, respectively. On the other hand, Marteau et al. [74] verified in healthy volunteers that the regular consumption of a fermented dairy product (100 g three times per day) containing *L. acidophilus*, *B. bifidum*, *Streptococcus thermophilus* and *S. cremoris* for 3 weeks decreased the feces nitroreductase activity from baseline but not that of β-glucuronidase or azoreductase.

Feces metabolites are also indicators of bacterial activity. Changes in enzyme activities and the concentration of ammonia, phenol and cresol have been detected in volunteers who consumed Lactobacilli [65]. Other metabolites with possible adverse effects are N-nitroso compounds, diacylglycerol and secondary bile acids [49].

A wide variety of microrganisms can produce ammonia, for example, enterobacteria, bacteroides and clostridia. Ammonia is considered a potential promoter of tumor in the colon and it can increase the rate of neoplastic transformation in the intestine. According to Benno and Mitsuoka [73], reducing the proportion of clostridia and bacteroides could explain the decrease in the concentration of ammonia in individuals who consumed fecal *B. longum*.

Epidemiological studies indicate an association between the risk of developing colon cancer and the consumption of high fat diets [7,75, 76]. For the digestion of fats, bile acids conjugated to glycine or taurine molecules are released into the small intestine and reabsorbed in the same location. It is believed that the deoxycholic acids may be cytotoxic to the epithelial cells, which could lead to the development of colon cancer [71]. Probiotic modulation of the intestinal microbiota may affect the activity of one of the enzymes (7a-dehydroxylase) forming these toxic products, but probiotics may also reduce the toxicity of bile salts that bind to them [77]. Lidbeck et al. [68] found that administering L. acidophilus to colon cancer patients for 6 weeks resulted in reduction in the concentration of soluble bile acids in the stool.

The consumption of fermented milk containing L. acidophilus may reduce the population of harmful bacteria, such as coliforms, and increased levels of lactobacilli in the intestine [78], suggesting that supplementation with this microorganism can have a beneficial effect since it inhibits the growth of bacteria that harmful are possibly involved in the production of tumor promoters and pro-carcinogens. Savard et al. [79] assessed the impact of four week's consumption of commercial yoghurt with Bifidobacterium animalis subsp. lactis (BB-12) and Lactobacillus acidophilus (LA-5) on fecal bacterial counts in healthy adults. The yoghurt had a positive effect on the bacterial population in that a the increase in beneficial bacteria and the reduction of potentially pathogenic bacteria was observed.

Not all studies show a correlation between the administration of probiotics and the activity of intestinal microbiota. Bartram et al. [80] argued that the fecal microbiota is relatively stable and generally unaffected by the administration of probiotics. In an intervention study, 12 individuals consumed yogurt (500 mL) enriched with B. longum. No significant difference was found in fecal weight, pH, concentration of fecal short chain fatty acids, bile acids and neutral sterols after 3 weeks of intervention. Despite the rise in the fecal concentration of B. longum, the results suggested litlle or no modulation of resident microbiota.

Some researchers have suggested that a high intestinal pH may be related to increased risk of colon cancer, whereas acidification of the colon could prevent the formation of carcinogens. Benno and Mitsuoka [73] found a significant reduction of faecal pH in health men who ingested B. longum for 5 weeks.

Evidence indicates that a high concentration of short chain fatty acids (acetate, propionate and butyrate) can assist in maintaining an appropriate pH in the lumen of the colon for the expression of many bacterial enzymes that probably metabolize carcinogens in the gut [81]. The activity of some dietary carcinogens, such as nitrosamines (resulting from commensal bacterial metabolic activity in individuals who consume a diet rich in proteins) can be neutralized by butyric acid produced by some probiotics [82]. Furthermore, production of ammonia, nitrosamines and secondary bile acids in the intestinal environment can be reduced by lowering the pH [83].

Butyrate, particularly, has received much attention as a potential chemopreventive agent [1,84]. While acting as an energy source for untransformed cells, butyrate possibly reduces survival of tumor cells by inducing apoptosis and differentiation, as well as by inhibiting proliferation. These mechanisms may play an important role in the reduction and/or inhibition of promotion and progression of cancer [1, 85].

Studies show that the LAB may be involved in the detoxification of various carcinogens such as polycyclic aromatic hydrocarbons and heterocyclic aromatic amines [86]. The mechanisms of action of these bacteria are poorly known, but it is possible that the LAB bind directly to the carcinogen and catalyze detoxification reactions [62]. It is worth noting that the protective effects conferred by LAB only appear when these are at a high density and when there is a regular intake [87].

Evidence is accumulating that heterocyclic aromatic amines (HCAs), which are derived from amino acids in meat during cooking, might be involved in the etiology of human cancer [88]. Zsivkovits et al. [89] showed that *L. bulgaricus* 291, *S. thermophilus* F4, *S. thermophilus* V3 and *B. longum* BB536 are highly protective against the genotoxic effects of HCAs in rats. Additionally, the inhibition of HCAs induced DNA damage was dose dependent and significant when 1×10^7 cells/animal were administered. Other authors showed that *L. casei* DN 114001 may metabolize or adsorb HCAs and reduce their genotoxicity *in vitro* [89].

In vivo evidence that probiotics bond the carcinogens are still not conclusive. Hayatsu Hayatsu (1993) demonstrated the marked suppressive effect of orally administered *L. casei Shirota* (LcS) on the urinary mutagenicity arising from ingestion of fried ground beef by humans. In another clinical trial, the consumption of *L. acidophilus* decreased the urinary and fecal excretion of mutagens [68]. In view of the *in vitro* results, it is possible that the LAB supplements are influencing excretion of mutagens by simply binding them in the intestine [62]. Even though the binding of carcinogens is a possible mechanism for the inhibition of genotoxicity and mutagenicity by LAB in vitro, some researchers have reported that it does not appear to have any influence *in vivo* [90]. Additionally, the extent of the binding depends on the mutagen and bacterial strain used [71].

Several studies have also reported the effect of probiotics on the promotion phase of carcinogenesis. Rowland et al. [91] found that administration of *B. longum* (6×10^9 CFU/day) inhibited the formation of aberrant crypt foci (ACF) in rats that received an induced of carcinogenesis (azomethane). As the probiotic treatment began 1 week after exposure to the carcinogen, these results indicate an effect on the early promotional phase of carcinogenesis [71].

Goldin et al. [92] observed a lower incidence of colonic tumors in rats who consumed *Lactobacillus* GG before, during and after chemical induction with dimethylhydrazine (DMH) than in animals that were fed the probiotic after receiving carcinogen. The researchers concluded that probiotics acted by inhibiting the initiation stage of carcinogenesis.

Kumar et al. [93] tested the efficacy of *L. plantarum* AS1 in the suppression of colorectal cancer induced by DMH in rats and formed that AS1 was capable of diminishing colon tumor through its antioxidant activity. However, long-term administration of this strain was necessary to achieve the maximum inhibitory effect.

On the other hand, not all studies have shown significant effects of probiotic on carcinogen-induced ACF. Gallanger et al. [94] using na ACF promotion protocol together with *B. longum* and *L. acidophilus*, obtained inconsistent results, which they attributed to differences in the ages of rats when DMH was administered.

Several studies have correlated the effect of probiotic on colon cancer with the modulation of the immune system. There is evidences that probiotics may contribute to the development of the mucosal immune system by influencing the innate inflammatory response and reducing mucosal inflammation. Additionally, probiotics also act on dendritic and epithelial cells and native T cells in the lamina propria of the gut and can thus influence adaptive immunity [13, 95].

Probiotics may influence the immune system by the action of products, such as metabolites, cell-wall components and DNA. Thus, immune modulatory effects might even be achieved by dead probiotic microorganisms or just probiotic derived components such as peptidoglycan fragments or DNA. Probiotic products are recognized by host cells sensitive to them these because they are equipped with recognition receptors adhesion. The main target cells in this context are therefore gut epithelial and gut-associated immune cells. The adhesion of probiotics to epithelial cells might itself might already trigger a signaling cascade leading to immune modulation [96].

Recent advances in the understanding of the immunomodulatory activity of probiotics have resulted from the discovery of Toll-like pattern recognition receptors (TLRs). These are transmembrane proteins present on the surface of cells such as macrophages, monocytes, dendritic cells and epithelial cells [97].

The innate immune system recognizes a large number of molecular structures from bacteria, such as, lipopolysaccharides and lipoteichoic acid, and is able to distinguish whether a particular microorganism is part of its microbiota or not. Different structures can activate different TLRs [98]. For example, TLR-2 recognizes the peptidoglycan, lipoteichoic acid, which is a component of the wall of Gram-positive bacteria such as lactobacilli and bifidobacteria [99], whereas TLR-4 is the most important receptor for lipopolysaccharide, the main component of the wall of Gram-negative bacteria [100].

Rachmilewitz et al. [101] using a probiotic mixture of 8 strains of freeze-dried lactic acid bacteria (*Bifidobacterium longum, B. infantis, B. breve, Lactobacillus acidophilus, L. casei, L. delbrueckii subsp. bulgaricus, L. plantarum, Streptococcus salivaris subsp. thermophilus*), reported that the chromosomal DNA of this mixture was responsible, via TLR-9 receptors for an anti-inflammatory effect observed in mice with colitis.

The connection of components of microorganisms to these receptors can lead to a cascade of inflammatory reactions via the activation of nuclear factor-kB (NF-kB), with subsequent release of cytokines, epitope chemokines and lipid mediators of reactive oxygen and nitrogen species [102]. Studies have shown that probiotics can activate elements responsible for the formation of cytokines and epitope chemokine's, although that response was weaker for *L. rhamnosus* if than for a Gram-positive pathogen (*Streptococcus pyogenes*) [103]. Some authors have suggested that a possible mechanism of action of probiotics would be the inhibition of NF-kB activation by reducing intestinal inflammation [104]. However, the possible mechanisms of probiotics against carcinogenesis, regarding the modulation of the immune system, are complex and still need to be better further elucidated.

An inflammatory immune response produces monocytes and macrophages, activated by cytokines that release cytotoxic molecules capable of the lyzing tumor cells *in vitro* [105]. The cytokines IL-1 and inflammatory TNF (tumor necrosis factor) exert cytotoxic and cytostatic effects on neoplastic cells *in-vitro* [106]. Natural-killer cells (NK) are effective against tumor cells and low activity of this cell type has been linked to a risk of cancer [107]. Matsuzaki and Chin [108] found that in mice, NK cell activity and inflammatory responses increased with the administration of probiotic strains.

Several studies in humans have shown an increase of NK cells in response to the consumption of probiotics [109, 110], and the same has been in animal models. When Takagi et al. [111] administered the strain *L. casei* Shirota, in order to inhibit tumor development induced by methylcholantracene in mice, there were high levels of NK cells in the group treated with the probiotic, which slowed the early development of the tumor, compared to the control group.

On the other hand, Berman et al. [112] did not observed any increasing in NK cells in healthy subjects who consumed during 8 weeks a formulation containing 4 species of probiotics (*L. rhamnosus, L. plantarum, L. salivarus* and *B. bifidum*). However, the researchers did note an increase in phagocytosis by neutrophils and monocytes.

Evidence has shown that the probiotic *Lactobacillus casei* Shirota has anti-tumor effects and antineoplastic action in rodents (biologically or chemically induced). Intrapleural administration of the strain in mice with tumor induced the production of various cytokines, such as interferon IL-1 and TNF in the thoracic cavity, which resulted in tumor inhibition and increased survival [113]. A study on *B. longum* and *B. animalis* showed that these bacteria induce the production of inflammatory cytokines (IL-6 and TNF-) [114].

In a clinical trial, the effect of *L. casei* Shirota on NK cell activity in humans was investigated. The activity of NK was increased as a likely consequence of *L. casei* Shirota-induced IL-12 production which was detected in *in vitro* assays [115].

According to the results of the various studies mention here, the probiotic microorganisms are capable of modulating the immune system in a strain-specific manner [116]. Therefore, different strains may induce different immune responses that might lead to the inhibition of carcinogenesis.

4. Effects of *Enterococcus faecium* CRL 183 on intestinal microbiota and colon cancer

Enterococcus spp. are Gram-positive, non-sporulating, catalase and oxidase negative facultative anaerobes [72]. Species of this genus are natural constituents of the intestinal microbiota of humans and comprise the third-largest genus of lactic acid bacteria (LAB), after *Lactobacillus* spp. and *Streptococcus* spp. [117].

It is hard to determine the exact number of enterococci species, but from a microbiological and functional point of view, *Enterococcus faecalis* and *Enterococcus faecium* are considered the most important [117, 118].

Some strains of *Enterococcus* spp. exhibit antibiotic resistance, possess virulence factors (adhesions, invasins, pili and haemolysin) and may cause bacteremia, endocarditis and other infections [117]. However, commercial pharmaceutical preparations of enterococci include *Enterococcus faecium* SF68® (NCIMB 10415, produced by Cerbios-Pharma SA, Barbengo, Switzerland) and *Enterococcus faecalis* Symbioflor 1 (SymbioPharm, Herborn, Germany), are on the market without reported health problems. Since 2008, *Enterococcus faecium* has been authorized for use in food and recognized as a probiotic microorganism in Brazil [119].

Currently, several strains of *Enterococcus faecium* are considered safe for human consumption, being used as starter cultures in cheese making and other fermented products and recognized as probiotic microorganisms [120]. The use of *Enterococcus faecium* as a starter culture in various fermented foods can be explained by its resistance to high concentrations of NaCl and low pH, and its ability to produce different aromas.

The strain of *Enterococcus faecium* CRL 183 was isolated by researchers from at the Reference Center for Lactobacillus (Cerela-Argentina), from cheese samples of Tafí – a traditional homemade cheese from the highland province of Tucuman, Argentina [121]. *In vitro* and *in vivo* studies showed that *Enterococcus faecium* CRL 183 is able to adhere to the intestinal cells, resists the gastrointestinal environment and colonizes the large intestine of rats, thus satisfying the requirement for a probiotic microorganism [122, 123]. Furthermore, this strain has no antibiotic resistance and no virulence factors, ensuring its safe use as a starter culture [121].

Enterococcus faecium CRL 183 has been investigated by our research group for about 20 years, with the objective of defining its functional properties in the free form or associated with food products [122,123,124,125,126, 127, 128, 129, 130].

The best functional effects of *Enterococcus faecium* CRL 183 were obtained when this microorganism was used as a starter culture of a yogurt-like fermented soy product (soy yogurt) [129]. This product has sensorial and technological properties similar to fermented-milk yogurt drinks and has exhibited functional properties in animal tests and clinical trials. Among the beneficial effects of the soy product fermented with *Enterococcus faecium* CRL 183, the following deserve special attention: improved of lipid profile, modulation of the immune system, positive changes in the intestinal microbiota, and reduction of colon cancer development [123, 125,128,130] .

Sivieri et al [123] studied the effect of daily ingestion of *Enterococcus faecium* CRL 183 (8 log CFU/mL) on the incidence of colorectal tumors induced by 1,2 dimethylhydrazine (DMH) in rats (20 mg/kg body weight, in a weekly dose, for 14 weeks). The experiment was conducted over 42 weeks and the rats were allocated to three groups: G1 - Control (not induced); G2 – Induced with DMH; G3 – Induced with DMH + *E. faecium* CRL 183. Thioglycollate-elicited peritoneal exudate cells (PECs) were harvested from animals in PBS and the adherent cells were obtained after incubation with LPS or RPMI-1640 (CO_2 - 95:5, v/v). The cytokine levels (IL-4, IFN- γ and TNF- α) were determined in the supernatant of of the cell culture by ELISA. After euthanasia, colons were removed for histological analysis. The animals with induced colorectal cancer and that received the suspension of *Enterococcus faecium* CRL 183 (G3) showed a 50% reduction in average number of tumors compared to G2 ($P < 0.001$) (Figures 2 and 3). The total number of aberrant crypt foci (ACF), the total ACF/mm^2, the number of crypts per ACF and the adenocarcinoma were also reduced in G3. In addition, G3 exhibited increased production of IL-4, IFN- γ and TNF- α by PECs compared to G2.

Anti-tumor activities of probiotic acid lactic bacteria have been attributed to an enhanced immune response [132]. The induction of TNF- α by probiotic bacteria would be necessary to initiate cross-talk between the immune cells associated with the lamina propria and the intestinal epithelial cells. IFN- γ is involved in the maturation of immune cells (dendritic cells), controls their cellular proliferation at the intestinal level and induce other cytokines, especially IL-4, IL-5 and IL-10. Because of its role in mediating macrophage and NK cell activation, IFN- γ is important in the host defense against intracellular pathogens, viruses and tumors [133]. According to Perdigón et al. [134] IL-4 exerts control over the inflammatory response induced by the carcinogen. In that study, the antitumor activity of *Enterococcus faecium* CRL 183 was attributed to its ability to modulate the immune response.

It has been suggested that increasing the consumption of red meat and animal fat lead to an increased risk of developing cancer colon, in comparison with a vegetarian diet [23]. Several studies have demonstrated that the microbiota of the colon is involved in the etiology of the colon cancer and that the some strains of probiotic microorganism can have beneficial effects on the composition of the intestinal microbiota, stimulates the production of short chain fatty acids (SCFA) and inhibit the activity of enzymes that convert pro-carcinogens into carcinogens [39,49, 135, 136].

Based in these evidence, a study was carried out to determine if consumption of a soy product fermented with *E. faecium* CRL 183 was able to modify the fecal microbiota of rats fed a diet containing red meat [122]. The experiment was conducted during over days and the animals were randomly divided into six groups: GI - standard casein-based rodent feed; GII to GVI - beef-based feed. From the 30th day, G III–VI also ingested the following products: G III, *E. faecium*-fermented soy product; G IV, pure suspension of *E. faecium*; G V, sterilized fermented soy product; and G VI, unfermented soy product (3 mL kg^{-1} BW day^{-1}). The feces of each animal were collected at the start (T0) and on the 30th (T30) and 60th (T60) days of the experiment, to determine the viable cell counts of total aerobic and anaerobic bacteria, *Enterococcus* spp., Enterobacteria, *Lactobacillus* spp., *Clostridium* spp., *Bacteroides* spp. and *Bifidobacterium* spp.

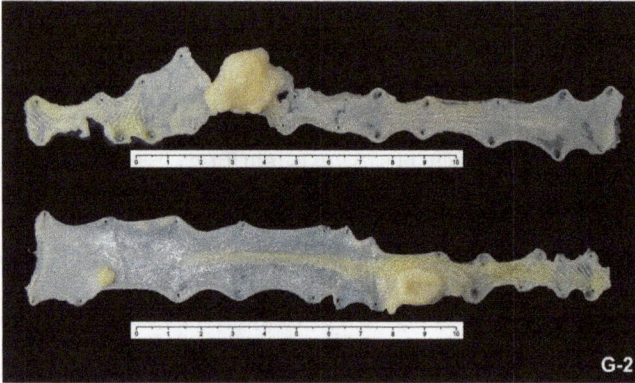

Figure 2. Topographic view of macroscopic growths by G2 – Induced with DMH. Sivieri et al [123]

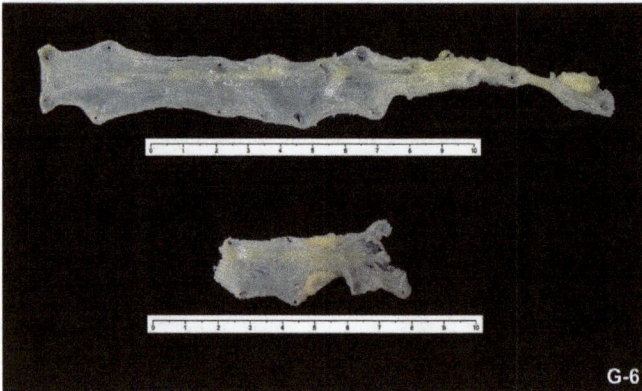

Figure 3. Topographic view of macroscopic growths by G3 – Induced with DMH + *E. faecium* CRL 183. Sivieri et al [123]

By day T30 days of experiment, rats on a red meat-based diet exhibited an increase in the population of total anaerobes, enterobacteria and enterococci and a decrease in the numbers of lactobacilli and bifidobacteria. From T30 to T60, the obtained results showed that fermented soy product and pure *Enterococcus faecium* CRL 183 suspension promoted an increase in the numbers of lactobacilli (0.45 log CFU g^{-1} and 1.83 log CFU g^{-1}, respectively). During the same period, only the animals treated with pure *Enterococcus faecium* CRL 183 suspension showed a rise in the fecal bifidobacterium population. The fermented soy product promoted a slight fall in the *Bacteroides* spp. population (2.80 ± 0.20 to 2.34 ± 0.07 log CFU g-1), but the counts of *clostridia*. and enterobacteria were unchanged.

Another study, using New Zealand rabbits with induced hypercholesterolemia as an animal model, was conducted to investigate the possible correlation between fecal microbiota, serum lipid parameters and atherosclerotic lesion development. It was show that, after 60 days of the experiment, intake of the probiotic soy product (with or without isoflavones) was correlated with significant increases ($P<0.05$) in *Lactobacillus* spp., *Bifidobacterium* spp. and *Enterococcus* spp. and a decrease in the enterobacteria population (Cavallini et al., 2011).

The studies conducted by Bedani et al (2010) and Cavallini et al. (2011) suggest that daily ingestion of the soy product fermented with *Enterococcus faecium* CRL 183, or the pure culture of this probiotic microorganism, may contribute to a beneficial balance of the fecal microbiota.

Currently, other studies, using animal models and an *in vitro* simulator of human intestinal microbial ecosystem (SHIME), are being conducted by our research group in order to elucidate the possible mechanisms involved in the protective effect of *Enterococcus faecium* CRL 183 against colon cancer and the importance of the modulation of fecal microbiota and stimulation of the immune system in the disease pathogenesis [53].

5. Conclusions

From the above discussion, it is evident that probiotics have the capacity to modulate the intestinal microbiota and the immune system, to the benefit of the host organism, reducing the risk of many chronic degenerative diseases, among them colon cancer. It appears also that the actions performed by probiotics are species-strain-dependent, so that several effects or actions can occur with the same bacterial genus. However, the results of several experiments reported in the literature, highlight a degree of controversy concerning the effects observed, especially regarding the various types of cancers and it is difficult to compare these studies. Such controversies are due mainly to large variations in the time of the experiment - usually prevailing those of short duration the experimental models, bacterial strains and the doses and frequencies of administration of probiotics. In this sense, it is important that further studies be done to define and standardize these variables mentioned, and especially to elucidate the mechanisms involved in each of the observed effects.

It showed also be mentioned that, according there is also in the literature, that probiotics studied are taken almost exclusively in milk as can be observed in the products available on the market. This condition often makes them inappropriate for certain lactose intolerant population groups on those and allergic to milk proteins. Thus, alternative vehicles for probiotics, free of lactose and of β-lactoglobulin, such as the aqueous extract of soybeans, for example, deserve special attention from researchers seeking to develop products with a good nutritional profile and suitable to transport the probiotic specified for the purpose desire. It is expected that in the near future, as results of the interaction of various fields of study such as food science and technology, nutrition, microbiology, genetic engineering and molecular biology the market can offer consumers products that are more accessible and effective, reducing the risk of certain diseases, particularly certain types of cancer, and acting as adjuvants in specific treatments for existing diseases.

Finally, from the results obtained by our research group in the studies of probiotics in relation to colon cancer, and even other diseases, it appears that these was always variability between individuals, either in clinical trials or in studies with animal models, suggesting a possible specificity of these individuals in relation to consumption of given probiotics. This leads us to wonder, if today nutrigenomics is already a reality, is it not the moment to propose studies on something like "probiogenomics" or even about self-probiotics? Certainly the future will provide an answer to that question

Author details

Katia Sivieri,
Department of Food & Nutrition, Faculty of Pharmaceutical Sciences, São Paulo State University, Araraquara, SP, Brazil

Raquel Bedani
University of São Paulo, Faculty of Pharmaceutical Sciences, São Paulo, SP, Brazil.

Daniela Cardoso Umbelino Cavallini
Department of Food & Nutrition, Faculty of Pharmaceutical Sciences, São Paulo State University, Araraquara, SP, Brazil

Elizeu A. Rossi
Department of Food & Nutrition, Faculty of Pharmaceutical Sciences, São Paulo State University, Araraquara, SP, Brazil

6. References

[1] Stein K, Borowicki A, Scharlau D, Schettler A, Scheu K, Obst U, Glei M (2011). Effects of synbiotic fermentation products on primary chemoprevention in human colon cells. J Nutr Biochem. (in press).

[2] Turpin W, Humblot C, Thomas M, Guyot JP (2010).Lactobacilli as multifaceted probiotics with poorly disclosed molecular mechanisms. Int J Food Microbiol. 143:87–102.

[3] Lyra A, Lahtinen S, Ouwehand AC. Gastrointestinal benefits of probiotics – clinical evidence. In Salminen, S.; von Wright, A.; Lahtinen, S.; Ouwehand, A., eds. Lactic acid bacteria: microbiological and functional aspects. 4th ed. Boca Raton: CRC Press, 2012. p. 509-523.

[4] Pearson JR, Gill CIR. Rowland, IR (2009). Diet, fecal water, and colon cancer – development of biomarker. Nutr Rev. 67:509-526.

[5] Reddy BS (2000). Novel approaches to the prevention of colon cancer by nutritional manipulation and chemoprevention. Cancer Epidem Biomar. 9:239–247.

[6] Norat T, Lukanova A, Ferrari P, Riboli E (2002). Meat consumption and colorectal cancer risk: dose-response meta-analysis of epidemiological studies. Int J Cancer. 98:241–256.

[7] O'Keefe SJD, Ou J, Aufreiter S, O'Connor D, Sharma S, Sepulveda J, Fukuwatari T, Shibata K, Mawhinney T (2009). Products of the colonic microbiota mediate the effects of diet on colon cancer risk. Journal of Nutrition, v. 139, p. 2044–2048, 2009.

[8] McGarr SE, Ridlon JM, Hylemon PB (2005). Diet, anaerobic bacterial metabolism, and colon cancer. J Clin Gastroenterol. 39:98–109.

[9] Hatakka K, Holma R, El-Nezami H, Suomalainen T, Kuisma M, Saxelin M, Poussa T, Mykkänen H, Korpela R (2008). The influence of Lactobacillus rhamnosus LC705 together with Propionibacterium freudenreichii ssp. shermanii JS on potentially carcinogenic bacterial activity in human colon. Int J Food Microbiol. 128:406–410.

[10] Gong J, Yang C (2012). Advances in the methods for studing gut microbiota and their relevance to the research of dietary fiber functions. Food Res Int. (in press).

[11] FAO/WHO. Working Group Report on Drafting Guidelines for the Evaluation of Probiotics in Food. London, Ontario, Canada, april 30 and may 1, 2002.

[12] Rauch M, Lynch SV (2011). The potential for probiotic manipulation of the gastrointestinal microbiome. Curr Opin Biotech.23:191-201.

[13] Zhu Y, Luo TM, Jobin C, Young, H.A (2011). Gut microbiota and probiotics in colon tumorigenesis. Cancer Letters. 309: 119–127.

[14] CVE. CENTRO DE VIGILÂNCIA EPIDEMIOLÓGICA. Doenças crônicas não transmissíveis. São Paulo, 2012. Disponível em: www.cve.saude.sp.gov.br/htm/cve_dcnt.html Acesso em: 18 jan. 2012.

[15] INCA – INSTITUTO NACIONAL DE CÂNCER. Ações de prevenção primária e secundária no controle do câncer. Rio de Janeiro: Inca, 2008. 628p. Disponível em: www1. inca.gov.br/enfermagem/docs/cap5.pdf Acesso em: 08 jan. 2012.

[16] Neves FJ, Mattos IE, Koifman RJ (2005). Mortalidade por câncer de cólon e reto nas capitais brasileiras no período 1980-1997. Arquivos de Gastroenterologia, 42:63 – 70.

[17] Hope ME, Hold GL, Kain R (2005). Sporadic colorectal cancer – role of the commensal microbiota. FEMS Microbiol. Lett. 244:1-7.

[18] Grady WM (2004). Genomic instability and colon cancer. Genomic Metastasis Reviews, 23:11 – 27.

[19] Rabeneck L, Davila JA, El-Serag HB (2003). Is there a true "shift" to the right colon the incidence of colorectal cancer? Am. J. Gastroenterol. 98:1400-1409.

[20] Commane D, Hughes R, Shortt C (2005). The potential mechanisms involved in the anti-carcinogenic action of probiotics. Mutat. Res.591:276-289.

[21] Bartram HP, Scheppach W, Schmid H (1993). Proliferation of human colonic mucosa as an intermediate biomarker of carcinogenesis: effects of butyrate, deoxycholate, calcium, ammonia, and pH. Cancer Res. 53:3283-3288.

[22] Macbain AJ, Macfarlane GT (2007). Ecological and physiological studies on large intestinal bacteria in relation to production of hydrolytic and reductive enzymes involved in formation of genotoxic metabolites. J. Med. Microbiol. 47: 407-416.

[23] Guarner F, Malagelada JR (2003). Gut flora in health and disease. Lancet 361:512–519.

[24] Pinho MA (2008). biologia molecular das doenças inflamatórias intestinais. Rev bras colo-proctol. 28:119 – 123.

[25] Owen RW, Spiegelhalder B, Bartsch H (2000). Generation of reactive oxygen species by the faecal matrix. Gut. 46:225-232.

[26] Huycke MM, Abrams V, Moore DR (2002). Enterococcus faecalis produces extracellular superoxide and hydrogen peroxide that damages colonic epithelial cell DNA. Carcinogenesis. 23:529-536.

[27] Montalto M, D'Onofrio F, Gallo A, Cazzato A, Gasbarrini G. Intestinal microbiota and its functions (2009). Digestive and Liver Disease Supplements, 3:30-34.

[28] Gibson GR, Fuller R (2000). Aspects of in vitro and in vivo research approaches directed toward identifying probiotics and prebiotics for human. J. Nutr.130:S391-S395.

[29] Cummings JH, Macfarlane GT (1997). Colonic microflora: nutrition and health. Nutrition.13:476-478.

[30] Whitman W.B., Coleman D.C., Wiebe WJ (1998). Prokaryotes: the unseen majority. Proc. Natl. Acad. Sci. USA. 95:6578–6583

[31] Frank DN, St Amand AL, Feldman RA, Boedeker EC, Harpaz N, Pace NR (2007). Molecular-phylogenetic characterization of microbial commu- nity imbalances in human inflammatory bowel diseases. Proc. Natl. Acad. Sci. USA. 104: 13780 –13785.

[32] Mason KL, Huffnagle GB, Noverr MC, Kao JY (2008). Overview of gut immunology. In: GI Microbiota and Regulation of the Immune System, edited by GB Huffnagle and MC Noverr. Austin, TX: Landes Bioscience and Springer Science∞Business Media, 2008, p. 1 - 14.

[33] Eckburg PB (2005). Diversity of the human intestinal microbial flora. Science. 308:1635–1638

[34] Tsai F, Coyle WJ. (2009) The microbiome and obesity: is obesity linked to our gut flora? Curr Gastroenterol Rep 11:307–313.

[35] Metges CC (2000). Contribution of microbial amino acids to amino acid homeostasis of the host. J. Nutr. 130:1857S–1864S

[36] Cummings JH., Macfarlane GT, Englyst HN (2001). Prebiotic digestion and fermentation. Am. J. Clin. Nutr. 73:415S–20S.

[37] Pryde SE, Duncan SH, Hold G.L (2002). The microbiology of butyrate formation in the human colon. FEMS Microbiol. Lett. 17:133-139.

[38] Csordas A (1996). Butyrate, aspirin and colorectal cancer. Eur. J. Cancer. Prev. 5:221-231.

[39] Wollowski I, Rechkemmer G, Pool-Zobel B(2001). Protective role of probiotics and prebiotics in colon cancer. Am. J. Clin. Nutr. 73:451S–455S

[40] Segain JP, Raingeard de la Bletiere D, Bourreille A (2000). Butyrate inhibits inflammatory responses through NFkappaB inhibition: implications for Crohn's disease. Gut: 47:397-403.

[41] Jones BV, Begley M, Hill C, Gahan CG, Marchesi JR (2008). Functional and comparative metagenomic analysis of bile salt hydrolase activity in the human gut microbiome. Proc Natl Acad Sci USA 105: 13580–13585.

[42] Round JL, Mazmanian SK (2009). The gut microbiota shapes intestinal immune responses during health and disease. Nat Rev Immunol 9: 313–323.

[43] Azcárate-Peril MA, Sikes M, Bruno-Bárcena JM (2011). The intestinal microbiota, gastrointestinal environment and colorectal cancer: a putative role for probiotics in

prevention of colorectal cancer? Am. J. Physiol. Gastrointest. Liver Physiol. 301: G401–G424.

[44] Molly K, Woestyne MV, Verstraete (1993). Development of a 5-step multichamber reactor as a Simulation of the Human Intestinal Microbial Ecosystem. Appl. Microbiol. Biotechnol, 39:254–2583.

[45] Possemiers S, Verthé K, Uyttendaele S, Verstraete W (2004). PCR-DGGE-based quantification of stability of the microbial community in a simulator of the human intestinal microbial ecosystem. Fems Microbiol Ecol. 49: 495–507.

[46] Saavedra L, Taranto MP, Sesma F, Valdez GF (2003). Homemade traditional cheeses for the isolation of probiotic Enterococcus faecium strains. Int J Food Microbiol. 88:241-245.

[47] Mercenier A, Pavan S, Pot B (2002). Probiotic as biotherapeutic agents: present knowledge and future prospects. Curr. Pharm. Design.8:99-110.

[48] Berg RD (1998). Probiotic, probiotics or conbiotics? Trends Microbiol. 6:89-92.

[49] Salminen S, Bouley C, Boutron-Ruault MC, Cummings JH, Franck A, Gibson GR (1998). Functional food science of gastrointestinal physiology and function. Br. J. Nutr. 80:147S–171S

[50] MacFarlane GT, MacFarlane S (2007). Models for intestinal fermentation: association between food components, delivery systems, biovailability and functional interactions in the gut. Curr Opin Biotechnol. 18:156-162.

[51] Payne S, Gibson G, Wynne A, Hudspith B, Brostoff J, Tuohy K (2003). In vitro studies on colonization resistance of the human gut microbiota to Candida albicans and the effects of tetracycline and Lactobacillus plantarum LPK. Curr. Issues Intest. Microbiol. 4:1-8.

[52] Gibson GR, Roberfroid MB (1995). Dietary modulation of the human colonic microbiota: introducing the concept of prebiotics. J Nutr. 125:.1401 - 1412.

[53] Sivieri K, Bianchi F, Rossi EA (2011). Fermentation by gut microbiota cultured in a simulator of the human intestinal microbial ecosystem is improved probiotic Enterococcus faecium CRL 183. Functional foods in health and disease, 10:389-402.

[54] Pereira DIA, MCCartney, AL, Gibson G (2000). An in vitro study of the probiotic potential of a bile salt hydrolyzing Lactobacillus fermentum strain, and determination of its cholesterol-lowering properties. Appl Environ Microbiol. 69:4743-4752

[55] Rycroft CE, Jones MR, Gibson GR, Rastall RA (2001). Acomparative "in vitro" evaluation of the fermentation properties of prebiotic oligosaccharides. J. Applied Microbial. 91:878-887.

[56] Tzortzis G, Goulas AK, Gee GM, Gibson GR (2005). A Novel Galactooligosaccharide Mixture Increases the Bifidobacterial Population Numbers in a Continuous In Vitro Fermentation System and in the Proximal Colonic Contents of Pigs In Vivo. J. Nutr. 135:1726-1731.

[57] McFarlane S, Furrie E, Cummings JH, McFarlane, GT (2004). Chemotaxonomic analysis of bacterial populations colonizing the rectal mucosa in patients with ulcerative colitis. Clin Infect Dis. 38:1690-1699.

[58] Minekus, M. (1995). A multi compartmental dynamic computer-controlled model simulating the stomach and small intestine. Altern. Lab. Anim. (ALTA) 23:197–209

[59] Souliman, S. (2007) Investigation of the biopharmaceutical behavior of theophylline hydrophilic matrix tablets using USP methods and an artificial digestive system. Drug Dev. Ind. Pharm. 33, 475–483.

[60] Payne A N , Zihler A, Chassard C, Lacroix C (2012). Advances and perspectives in in vitro human gut fermentation modeling. Trends Biotechnol. 30:17–25

[61] Guiemonde M, Salminen S (2006). New methods for selecting and evaluating probiotics. Dig. Liver. Dis. 38: 242 S-247 S.

[62] Rafter J (2003). Probiotics and colon cancer. Best. Pract. Res. Clin. Gastroenterol. 17:849-859.

[63] Fotiadis C, Stoidis CN, Spyropoulos BG, Zografos ED (2008). Role of Probiotics, Prebiotics and Synbiotics in Chemoprevention for Colorectal Cancer. World J. Gastroenterol. 14:6453-6457.

[64] Oelschlaeger T (2010). Mechanisms of probiotic actions. International Journal of Medical Microbiology 300:57-62.

[65] Goldin BR, Gorbach SL (1984). Alterations of the intestinal microflora by diet, oral antibiotics and Lactobacillus: decreased production of free amines from aromatic nitro compounds, azo dyes and glucuronides. J. Natl. Cancer. Inst. 73:689-695.

[66] Spanhaak S, Havenaar R, Schaafsma G (1998). The effect of consumption of milk fermented by Lactobacillus casei strain Shirota on the intestinal microflora and immune parameters in humans. Eur. J. Clin. Nutr. 52:899-907.

[67] Haberer P, Toit M, Dicks LMT (2003). Effect of potentially probiotic lactobacilli on faecal enzyme activity in minipigs on high-fat, high-cholesterol diet – a preliminary in vivo trial. Int. J. Food Microbiol. 87:287-291.

[68] Lidbeck A, Nord CE, Gustafsson JA, Rafter J (1992). Lactobacilli, anticarcinogenic activities and human intestinal microflora. Eur. J. Cancer. Prev. 1:341-353.

[69] Ling WH, Hänninen O, Mykkänen H, Heikure M (1992). Colonization and fecal enzyme activities after oral Lactobacillus GG administration in eldery nursing home residents. Ann. Nutr. Metab. 36:162-166.

[70] Mital BK, Garg SK (1995). Anticarcinogenic, hypocholesterolemic, and antagonistic activities of Lactobacillus acidophilus. Crit. Rev. Microbiol. 21:175-214.

[71] Burns AJ, Rowland I (2000). Anti-Carcinogenicity of Probiotics and Prebiotics. Curr. Issues Intest. Microbiol. 1: 13-24.

[72] Foulquié Moreno MR, Sarantinopoulos P, Tsakalidou E, De Vuyst L (2006). The role and application of enterococci in food and health. Intern J Food Microbiol. 106:1-24.

[73] Benno Y, Mitsuoka T (1992). Impact of Bifidobacterium on human fecal microflora. Microbiol. Immunol. 36:683-694.

[74] Marteau P, Pochart P, Flourié B, Pellier P, Santos L, Desjeux JF, Rambaud JC (1990). Effect of Chronic Ingestion of a Fermented Dairy Product Containing Lactobacillus acidophilus and Bifidobacterium bifidum on Metabolic Activities of the Colonic Flora in Humans. Am J. Clin. Nutr. 52:685-688.

[75] Reddy BS (2000). Novel approaches to the prevention of colon cancer by nutritional manipulation and chemoprevention. Cancer Epidemiol Biomarkers Prev. 9:239–247.

[76] Norat T, Lukanova A, Ferrari P, Riboli E (2002). Meat consumption and colorectal cancer risk: dose-response meta-analysis of epidemiological studies. Int. J. Cancer. 98:241–256.

[77] Iannitti T, Palmieri B (2010). Therapeutical Use of Probiotic Formulations in Clinical Practice. Clin. nutr. 29:701-725.

[78] Shahani KM, Ayebo AD (1980). Role of dietary lactobacilli in gastrointestinal microecology.Am J Clin Nutr. 33:2448-57.

[79] Savard P, Lamarche B, Paradis M.E, Thiboutot H, Laurin E, Roy D (2011). Impact of Bifidobacterium animalis subsp. lactis BB-12 and, Lactobacillus acidophilus LA-5-containing yoghurt, on fecal bacterial counts of healthy adults. Int. j. food microbiol. 149:50-57.

[80] Bartram HP, Scheppach W, Gerlach S. (1994). Does yoghurt enriched with Bifidobacterium longum affect colonic microbiology and fecal metabolites. Am. J. Clin. Nutr. 59:428-432.

[81] Kopp-Hoolihan L (2001). Prophylactic and therapeutic uses of probiotics: a review. J. Am. Diet. Assoc.101:229-241.

[82] Wollowski I, Rechkemmer G, Pool-Zobel B(2002). Protective role of probiotics and prebiotics in colon cancer. Am. J. Clin. Nutr. 73:451S–455S

[83] Zampa A, Silvi S, Fabiani R, (2004). Effects of different digestible carbohydrates on bile acid metabolism and SCFA production by human gut micro-flora grown in an in vitro semi-continuous culture. Anaerobe. 10:19-26.

[84] Scharlau D, Borowicki A, Habermann N, Hofmann T, Klenow S, Miene C (2009). Mechanisms of primary cancer prevention by butyrate and other products formed during gut flora-mediated fermentation of dietary fibre. Mutat. Res. 682:39-53.

[85] Sengupta S, Muir JG, Gibson PR (2006). Does butyrate protect from colorectal cancer? J Gastroenterol Hepatol; 21:209–18.

[86] Knasmuller S, Steinkellner H, Hirschl AM. Impact of bacteria in dairy products and of the intestinal microflora on the genotoxic and carcinogenic effects of heterocyclic aromatic amines. Mutat Res 2001;480:129-138.

[87] Tannock GW, Munro K, Harmsen HJM (2000). Analysis of the fecal microflora of human subjects consuming a probiotic product containing Lactobacillus rhamnosus DR20. Appl Environ Microbiol. 66:2578-2588.

[88] Nowak A, Libudzisz Z (2009). Ability of Probiotic Lactobacillus casei DN 114001 to Bind or/and Metabolise Heterocyclic Aromatic Amines In Vitro. Eur. J. nutr. 48:419–427.

[89] Zsivkovits M, Fekadu K, Sontag G, Nabinger U, Huber W.W, Kundi M, Chakraborty A, Foissy H, Kasmuller S (2003). Prevention of heterocyclic amine-induced DNA damage in colon and liver of rats by different Lactobacillus strains. Carcinogenesis. 24:1913-1918.

[90] Bolognani F, Rumney C.J, Rowland I.R (1997). Influence of carcinogen binding by lactic acid producing bacteria on tissue distribution and in vivo mutagenicity of dietary carcinogens. Food Chem. Toxicol. 35: 535-545.

[91] Rowland I, Rumney C, Counts J, (1998). Effects of Bifidobacterium longum and inulin on gut bacterial metabolism and carcinogen induced aberrant crypt foci in rats. Carcinogenesis.19:281-285.

[92] Goldin B, Gualtieri L, Moore R (1996). The effect of Lactobacillus GG on the initiation and promotion of DMH induced intestinal tumours in the rat. Nut. Cancer 25:197-204.

[93] Kumar RS, Kanmani P, Yuvaraj N, Paari KA, Pattukumar V, Thirunavukkarasu C, Arul V (2012). Lactobacillus plantarum AS1 Isolated from South Indian Fermented Food Kallappam Suppress 1,2-Dimethyl Hydrazine (DMH)-Induced Colorectal Cancer in Male Wistar Rats. Appl. biochem. biotechnol 166:620–631.

[94] Gallaher DD, Stallings WH, Blessing LL, Busta FF, Brady LJ (1996). Probiotics cecal microflora, and aberrant crypts in the rat colon. J. Nutr. 126: 1362-1371.

[95] Saad SMI, Bedani R, Mamizuka EM. Benefícios à saúde dos probióticos e prebióticos. In: Saad, S.M.I.; Cruz, A.G.; Faria, J.A.F. eds. Probióticos e prebióticos em alimentos: fundamentos e aplicações tecnológicas. São Paulo: Livraria Varela, 2011.p.51-84.

[96] Oelschlaeger T (2010) Mechanisms of probiotic actions. International Journal of Medical Microbiology 300:57-62.

[97] Corthésy B, Gaskins HR, Mercenier A (2007). Cross-talk between probiotic bacteria and the host immune system. J Nutr. 137:781S-790S.

[98] Britti MS, Roselli M, Finamore A (2006). Regulation of immune response at intestinal and peripheral sites by probiotics. Biologia 61:735-740.

[99] Neuhaus FC, Baddiley J (2003). A continuum of anionic charge: structures and functions of d-alanyl-teichoic acids in gram-positive bacteria. Microbiol. Mol. Biol. Rev. 67:686-723.

[100] Chow JC, Young DW, Golenbock DT (1999). Toll-like Receptor-4 Mediates Lipopolysaccharide-induced Signal Transduction. J. Biol. Chem. 274:10689-10692.

[101] Rachmilwitz D, Katakura K, Karmeli F, Hayashi T, Reinus C, Rudensky B, Akira S, Takeda K, Lee J, Takabayashi K, Raz E (2004). Toll like receptor 9 signaling mediates the antiinflammatory effects probiotics in murine experimental colitis. Gastroenterology, 126:520-528.

[102] Watson JL, Mckay DM. The immunophysiological impact of bacterial CpG DNA on the gut Clin Chim Acta 2006;364:1-11.

[103] Veckman, V, Miettinen, M, Siren, J, (2004). Streptococcus pyogenes and Lactobacillus rhamnosus differentially induce maturation and production of Th1-type cytokins and chemokines in human monocyte-derived dendritic cells. J. Leukoc. Biol.75:764-771.

[104] Petroff EO, Kojima K, Ropeleski MJ (2004). Probiotics inhibit nuclear factor-κB and induce heat shock proteins in colonic epithelial cells through proteasome inhibition.Gatroenterology.127:1474-1487.

[105] Philip R, Epstein I(1986). Tumour necrosis factor as immunomodulator and mediator of monocyte citotoxicity induced by itself, · -interferon and interleukin-1. Nature. 323:86-89.

[106] Raitano A, Kore M (1993). Growth inhibition of a human colorectal carcinoma cell line by interleukin-1 associated with enhanced expression of · -interferon receptors. Cancer Res. 53:636-640.

[107] Takeuchi H, Maehara Y, Tokunaga E (2001). Prognostic significance of natural killer cell activity in patients with gastric carcinoma: a multivariate analysis. Am. J. Gastroenterol. 96:574-578.

[108] Matsuazaki T, Chin J (2000). Modulating immune responses with probiotic bacteria. Immunol. Cell Biol. 78:67-73.

[109] Nagao F, Nakayama M, Muto T (2000). Effects of a fermented milk drink containing Lactobacillus casei strain Shirota on the immune system in healthy human subjects. Biosci. Biotechnol. Biochem. 64:2706-2708.

[110] Gill HS, Rutherford KJ, Cross ML (2001). Enhancement of immunity in the elderly by dietary supplementation with the probiotic Bifidobacterium lactis HN019. Am J Clin Nutr 74:833-839.

[111] Takagi A, Matsuzaki T, Sato M (2001). Enhancement of natural killer cell cytotoxicity delayed murine caecinogenesis by a probiotic microorganism. Carcinogenesis;22:599-605.

[112] Berman SH, Eichelsdoerfer P, Yim D (2006). Daily ingestion of a nutritional probiotic supplement enhances innate immune function in healthy adults. Nutr. Res.26:454-459.

[113] Matsuzaki, T (1998). Immunomodulation by treatment with Lactobacillus casei strain Shirota. Int. J. Food Microbiol. 16:133–140.

[114] Sekine K, Kawashima T, Hashimoto Y (1994). Comparison of the TNF-α levels induced by human-derived Bifidobacterium longum and rat-derived Bifidobacterium animalis in mouse peritoneal cells. Microflora;39:79-89.

[115] Takeda K, Suzuki T, Shimada SI, Shida K, Nanno M, Okumura K (2006). Interleukin-12 is involved in the enhancement of human natural killer cell activity by Lactobacillus casei Shirota. Clin. exp.immunol. 146:109–115.

[116] Oelschlaeger T (2010) Mechanisms of probiotic actions. International Journal of Medical Microbiology 300:57-62.

[117] Franz CMAP, Huch M, Abriouel H, Holzapfel W, Gálvez A (2011). Enterococci as probiotics and their implications in food safety. Intern. J. Food. Microbiol. 151:125-140.

[118] Foulquié-Moreno MR, Sarantinopoulos P, Tsakalidou E, De Vuyst L (2006). The role and application of enterococci in food and health. Int. J. Food Microbiol.106:1–24.

[119] AGÊNCIA NACIONAL DE VIGILÂNCIA SANITÁRIA. Alimentos com alegações de propriedades funcionais e ou de saúde, novos alimentos/ingredientes, substâncias bioativas e probióticos. Disponível em: http://www.anvisa.gov. br/alimentos/ comissoes/tecno_lista_alega.htm. Acesso em: 01 abr. 2012.

[120] Gardiner GE, Ross RP, Wallace JM, Scanlan FD, Jägers PPJM, Fitzgerald GF, Collins K, Stanton C (1999). Influence of a Probiotic Adjunct Culture of Enterococcus faecium on the Quality of Cheddar Cheese. J. Agric. Food Chem. 47:4907-4916.

[121] Saavedra JM (2001). Clinical application of probiotic agents. Am. J Clin. Nutr. 73:1147S-1151S.

[122] Bedani R, Pauly-Silveira, ND, Roselino, MN, Valdez GF, Rossi EA. Effect of fermented soy product on the fecal microbiota of rats fed on a beef-based animal diet (2010). J Sci Food Agric. 90:233–238.

[123] Sivieri K, SpinardI-Barbisan ALT, Barbisan LF, Bedani R, Pauly ND, Carlos IZ, Benzatti F, Vendramini RC, Rossi EA (2008). Probiotic Enterococcus faecium CRL 183 inhibit chemically induced colon cancer in male Wistar rats. Eur Food Res Technol. 228:231-237.

[124] Cavallini DCU, Bedani R, Pauly ND, Bondespacho LQ, Vendramini RC, Rossi EA (2009). Effects of probiotic bacteria, isoflavones and simvastatin on lipid profile and atherosclerosis in cholesterol-fed rabbits. Lipids Health Dis.8:1-10.

[125] Manzoni MSJ (2008). Fermented soy product supplemented with isofl avones affects adipose tissue in a regional-speccifi c manner and improves HDL cholesterol in rats fed on a cholesterol-enriched diet. Eur. Food Res. Technol. 227:1591-1597.

[126] Rossi EA (1994). In vitro effect of Enterococcus faecium and Lactobacillus acidophilus on cholesterol. Microbiol. Alim. Nutr. 12: 267-270.

[127] Rossi EA, Cavallini DCU, Carlos IZ, Vendramini RC, Dâmaso AR, Valdez GF (2008). Intake of isoflavone-supplemented soy yogurt fermented with Enterococcus faecium lowers serum total cholesterol and non-HDL cholesterol of hypercholesterolemic rats. European Food Research and Technology. 228:275-282.

[128] Rossi EA, Cavallini DCU, Carlos IZ, Oliveira MG, Valdez GF (2003). Efeito de um novo produto fermentado de soja sobre lípides séricos de homens adultos normocolesterolêmicos. Arch. Lat. Nutr. 53:47-51.

[129] Rossi EA, Vendramini RC, Carlos IZ, Pei, YC, Valdez GF (1999). Development of a novel fermented soymilk product with potential probiotic properties. Eur Food Res Technol , 209:305-307.

[130] Rossi EA, Vendramini RC, Carlos IZ, Ueiji IS, Squinzari MM, Silva Júnior SI, Valdez GF (2000). Effects of a Novel Fermented Soy Producton the Serum Lipids of Hypercholesterolemic Rabbits. Arq Bras Cardiol. 74:213-216.

[131] Cavallini DCU, Suzuki JY, Abdalla DSP, Vendramini RC, Pauly-Silveira, ND, Pinto R, Rossi, EA (2011). Influence of a probiotic soy product on fecal microbiota and its association with cardiovascular risk factors in an animal model. Lipids Health Dis.10:1-9.

[132] Vinderola G, Matar C, Perdigon G (2007). Milk fermentation products of L. helveticus R389 activate calcineurin as a signal to promote gut mucosal immunity. BMC Immunol. 8:19.

[133] Galdeano CM, LeBlanc AM, Vinderola G, Bibas Bonet ME, Perdigón G (2007). Proposed model: mechanisms of immunomodulation induced by probiotic bacteria. Clin Vaccine Immunol. 14:485–492.

[134] Perdigón G, Locascio M, Medici M, Pesce de Ruiz Holgado A, Oliver G (2003). Interaction of bifidobacteria with the gut and their influence in the immune function. Biocell, 27:1–9.

[135] Augenlicht LH, Mariadason JM, Wilson A, Arango D, Yang W, Heerdt BG (2002). Short chain fatty acids and colon cancer. J. Nutr. 132:3804S–3808S.

[136] Macgarr SE, Ridlon JM, Hylemon PB (2005). Diet, anaerobic bacterial metabolism, and colon cancer: a review of the literature. J. Clin. Gastroenterol. 39:98–109.

Chapter 10

Lactic Fermentation and Bioactive Peptides

Anne Pihlanto

Additional information is available at the end of the chapter

1. Introduction

Fermented milk products have naturally high nutritional value, and as an extra benefit many health-promoting effects, such as improvement of lactose metabolism, reduction of serum cholesterol and reduction of cancer risk [1]. The beneficial health effects associated with some fermented dairy products may, in part, be attributed to the release of bioactive peptide sequences during the fermentation process. Numerous peptides and peptide fractions, having bioactive properties have been isolated from fermented dairy products. These activities include immunomodulatory, cytomodulatory, hypocholesterolemic, antioxidative, antimicrobial, mineral binding, opioid and bone formation activities. Many recent articles and book chapters have reviewed the release of various bioactive peptides from milk proteins through microbial proteolysis [2-5].

Many industrially utilized dairy starter cultures are highly proteolytic. The use of bioactive peptides producers microbial cultures (starter and non-starter) may allow the development new fermented dairy products. The proteolytic system of lactic acid bacteria e.g. *Lactococcus (L.) lactis, Lactobacillus (Lb.) helveticus* and *Lb. delbrueckii* ssp. *bulgaricus*, is already well characterized. This system consists of a cell wall-bound proteinase and a number of distinct intracellular peptidases, including endopeptidases, aminopeptidases, tripeptidases and dipeptidases [6]. *Lb. helveticus* are known to have high proteolytic activities [7], causing the release of oligopeptides from digestion of milk proteins [8]. These oligopeptides can be a direct source of bioactive peptides following hydrolysis by gastrointestinal enzymes. Rapid progress has been made in recent years to elucidate the biochemical and genetic characterization of these enzymes. The fact that the activities of peptidases are affected by growth conditions makes it possible to manipulate the formation of peptides to a certain extent [9].

Cardiovascular disease (CVD) is the single leading cause of death for both males and females in technologically advanced countries in the world. In lesser-developed countries it generally ranks among the top five causes of death. The World Health Organization estimates that by 2020, heart disease and stroke will have surpassed infectious diseases to become the leading cause of death and disability worldwide [10]. Consequently, there has been an increased focus on improving diet and lifestyle as a strategy for CVD risk reduction.

Elevated blood pressure is one of the major independent risk factors for CVD [11]. Angiotensin I-converting enzyme (ACE) plays a crucial role in the regulation of blood pressure as it promotes the conversion of angiotensin I to the potent vasoconstrictor angiotensin II as well as inactivates the vasodilator bradykinin. By inhibiting these processes, synthetic ACE inhibitors (ACEI) have long been used as antihypertensive agents. In recent years, some food proteins have been identified as sources of ACEI peptides and are currently the best-known class of bioactive peptides [12, 13]. These nutritional peptides have received considerable attention for their effectiveness in both the prevention and the treatment of hypertension.

Oxidant stress, the increased production of reactive oxygen species (ROS) in combination with outstripping endogenous antioxidant defense mechanisms, is another significant causative factor for the initiation or progression of several vascular diseases. ROS can cause extensive damage to biological macromolecules like DNA, proteins and lipids. Specifically, the oxidative modification of LDL results in the increased atherogenicity of oxidized LDL. Therefore, prolonged production of ROS is thought to contribute to the development of severe tissue injury [14]. Some peptides derived from hydrolyzed food proteins exert antioxidant activities against enzymatic (lipoxygenase-mediated) and nonenzymatic peroxidation of lipids and essential fatty acids [15]. The antioxidant properties of these peptides have been suggested to be due to metal ion chelation, free radical scavenging and singlet oxygen quenching.

This review centers on liberation during fermentation, of bioactive peptides with properties relevant to cardiovascular health including the effects on blood pressure and oxidative stress. The focus is mainly to those peptides with in vivo blood pressure lowering effects. Moreover, bioavailability of peptides and aspects of necessary further information is given.

2. Release and identification of peptides

2.1. Peptides in cheese

Proteolysis in cheese has been linked to its importance for texture, taste and flavour development during ripening. Changes of the cheese texture occur due to breakdown of the protein network. It contributes directly to taste and flavour by the formation of peptides and free amino acids as well as by liberation of substrates for further catabolic changes and thereby formation of volatile flavour compounds. Besides sensory quality aspects of proteolysis, formation of bioactive peptides as a result of proteolysis during cheese ripening has been reported. Cheese contains phosphopeptides as natural constituents [16, 17], and secondary proteolysis during cheese ripening leads to the formation of other bioactive peptides, such as those with ACEI activity. The findings by Meisel et al. [18] showed that inhibitory activity increased as proteolysis developed, however, the bioactivity decreased when proteolysis during ripening exceeded a certain level. Another link between potential antihypertensive peptides and proteolysis was found in Parmesan cheese [19]. A bioactive peptide derived from α_{s1}-casein was isolated from 6-month old cheese, but it was degraded further during maturation and was not detectable after 15 month of ripening. ACEI peptide fractions having different potencies have been isolated from various Italian cheeses, e.g. Crescenza (37% inhibition), mozzarella (59% inhibition), Gorgonzola (80% inhibition) and Italico (82% inhibition) [20]. ACEI peptides have also been found in enzyme-modified cheeses [21], in a low-fat cheese made in Finland [22] and Manchego cheeses manufactured with different starter cultures [23]. Mexican Fresco cheese manufactured with *Enterococcus faecium* or a *L. lactis* ssp. *lactis-Enterococcus faecium* mixture showed the largest number of fractions with ACEI activity among tested lactic acid strains [24]. Pripp et al. [25] investigated the relationship between proteolysis and ACE inhibition in Gamalost, Castello, Brie, Pultost, Norvegia, Port Salut and Kesam. The traditional Norwegian cheese Gamalost had per unit cheese weight higher ACE inhibition potential than Brie, Roquefort and Gouda-type cheese. However, ACE inhibition expressed as IC_{50} per unit peptide concentration from ethanol soluble fraction assessed by the OPA-assay was highest for Kesam, a Quark-type cheese with a low degree of proteolysis.

When β-casomorphins were looked from commercial cheese products, no peptides were found or their concentration in the cheese extract was below 2 µg/ml [26]. They further noted that the enzymatic degradation of β-casomorphins was influenced by a combination of pH and salt concentration at the cheese ripening temperature. Therefore, if formed in cheese, β-casomorphins may be degraded under conditions similar to Cheddar cheese ripening. Precursors of β-casomorphins, on the other hand, have been identified in Parmesan cheese [19]. β-Casomorphins were found at a higher level in the mould cheeses (166–648 mg/100 g), whereas the opioid peptides with antagonistic activity (casoxin-6) were identified at a higher level in the semi-hard cheeses (136–276 mg/100 g) and a low quantity of casomorphins (4–100 mg/100 g) [27]. Immunomodulating properties in water-soluble extracts from traditional French Alps cheeses, Abondance and Tomme de Savoie have been observed [28]. However, no correlation between peptide composition and *in vitro* immunomodulation of T-lymphocyte cells could be established.

A limited number of bioactive peptides have been isolated and identified in Gouda, Manchego, Festivo and Crescenza cheeses (Table 1). Several ACEI peptides have been identified from N-terminal of α_{s1}-casein of Gouda, Festivo, Cheddar and Fresco cheeses [22, 24, 29, 30]. In addition, peptides from β–casein, Tyr-Pro-Phe-Pro-Gly-Pro-Ile-Pro-Asn (β-cn, f(60–68)); and Met-Pro-Phe-Pro-Lys-Tyr-Pro-Val-Gln-Pro-Phe (β-cn, f(109–119)) from Gouda [29] and Tyr-Gln-Glu-Val-Leu-Gly-Pro-Val-Arg-Gly-Pro-Phe-Pro-Ile-Ile-Val (β-cn, f(193-209)) from Cheddar [30] have been identified. Antihypertensive peptides Val-Pro-Pro (VPP) (β-cn, f(84–86)) and Ile-Pro-Pro (IPP) (β-cn, f(74–76) and κ-cn, f(108–110)), have also been identified and quantified in different cheese varieties [31-33]. In some varieties physiologically relevant amounts was observed, however, a large variation exists between samples of the same cheese variety, as well as between different varieties. The concentrations of VPP and IPP were in the range of 0-224 mg/kg and 0-95 mg/kg, respectively, indicating that some cheese varieties contain similar concentrations of VPP and IPP to fermented milk products. Milk pretreatment, cultures, scalding conditions, and ripening time were identified as the key factors influencing the concentration of these two naturally occurring bioactive peptides in cheese. Thus, it is necessary to develop a reproducible cheese-making process with selected cultures to produce higher concentrations of these peptides that could be used for clinical trials.

Cheese variety	Milk protein fragment	Peptide sequence	ACE-inhibition IC$_{50}$ µM	Ref
Gouda	α_{s1}-cn f(1-9)	RPKHPIKHQ	13.4	29
	α_{s1}-cn f(1-13)	RPKHPIKHQGLPQ	ND	
	β-cn f(68-66)	YPFPGPIPN	14.8	
	β-cn f(109–119)	MPFPKYPVQPF	ND	
Manchego	ovine α_{s1}-cn f(102-109)	KKYNVPQL	77.2	23
	ovine α_{s1}-cn f(205-208)	VRYL	24.1	
Cheddar (with probiotics)	α_{s1}-cn f(1-9)	RPKHPIKHQ	ND	30
	α_{s1}-cn f(1-7)	RPKHPIK		
	α_{s1}-cn f(1-6)	RPKHPI		
	α_{s1}-cn f(24-32)	FVAPFPEVFGK		
	β-cn f(193-209)	YQEPVLGPVRGPFPIIV		
Swiss cheese varieties	β-cn, f(84–86)	VPP	9	31-
	β-cn, f(74–76) and κ-cn, f(108–110)	IPP	5	34
Fresco cheese	α_{s1}-cn f(1-15)	RPKHPIKHQGLPQEV	ND	24
	α_{s1}-cn f(1-22)	RPKHPIKHQGLPQEVLNEN		
	α_{s1}-cn f(14-23)	LLR		
	α_{s1}-cn f(24-34)	EVLNENLLRF		
	β-cn f(193-205)	FVAPFPEVFGK		
	β-cn f(193-207)	YQEPVLGPVRGPF		
	β-cn f(193-209)	YQEPVLGPVRGPFPI YQEPVLGPVRGPFPIIV		

ND: Not described
IC50: Peptide concentration that shows 50% inhibition of ACE activity
One letter amino acid codes used

Table 1. Examples of identified bioactive peptides in different cheese varieties

2.2. Fermented milk

During fermentation process, lactic acid bacteria hydrolyze milk proteins, mainly caseins, into peptides and amino acids which are used as nitrogen sources necessary for their growth. Hence, bioactive peptides can be generated by starter and non-starter bacteria used in the manufacture of fermented dairy products (Table 2). Proteolytic system of *Lb. helveticus*, *Lb. delbrueckii* ssp *bulgaricus*, *L. lactis* ssp. *diacetylactis*, *L. lactis* ssp. *cremoris*, and *Streptococcus (Str.) salivarius* ssp. *thermophilus* strains have demonstrated to hydrolyze milk proteins and release ACEI peptides. Among lactic acid bacteria, *Lb. helveticus* has high extracellular proteinase activity and the ability to release large amount of peptides in fermented milk. As a result, among various kinds of fermented milk, antihypertensive effect related to ACEI peptides were found in milk produced by *Lb. helveticus*. Two ACEI peptides have been purified from sour milk and identified as VPP and IPP [34].

Organisms	ACE-inhibition IC_{50} mg/ml	Identified peptides Sequence	IC_{50} μM	Dose	Response (Δ SBP mmHg)	Ref.
Lb. helveticus and Str. thermophilus	ND	VPP IPP	9 5	5 ml/kg	-21.8 ±4.2 after 6 h	34
Lb. helveticus		VPP IPP	9 5	27 ml/day	-21 after 4 weeks	67
Lb.helveticus CPN4	ND	YP	720	10 ml/kg	32.1 ±7.4 after 6 h	42
Lb. helveticus CHCC637	0.16			10ml/kg	-12 after 4-8 h	37
Lb. helveticus CHCC641	0.26				-11 after 4-8 h	
Lact. delbrueckii ssp. *bulgaricus* *Str. salivarius* ssp *thermophilus* and *L.lactis* biovar *diacetylactis*		SKVYPFPGPI SKVYP	1.7 mg/ml 1.5 mg/ml		ND	43
Lb. jensenii	0.52	LVYPFPGPIHNSLP QN LVYPFPGPIH	71 89	0.2 kg/kg	approx -12 after 2 h	38
Enterococcus faecalis CECT 5727	0.053	LHLPLP LVYPFPGPIPNSLP QNIPP	5.5 5.2	2 mg/kg 6 mg/kg	-21.87 ±4.51 after 4h[1)] approx -15 after 4 h	44
Lb. delbrueckii subsp. *bulgaricus* SS1 *L. lactis* subsp. *cremoris* FT4	ND	NIPPLTQTPV LNVPGEIVE DKIHPF	173.3 300.1 256.8		ND	36
Mixed lactic acid bacteria (*Lb. casei, acidophilus, bulcaricus, Str. themophilus, Bifidobacterium*) and protease	0.24	GTW GVW	464.4 240.0	5 mg/ml	SBP -22 after 8 weeks	76

One letter amino acid codes used
ND Not described
1) Pure synthetic peptides were used in the study

Table 2. ACE-inhibitory and antihypertensive activity in spontaneously hypertensive rats of peptides produced by fermentation of milk

Pihlanto-Leppälä et al. [35] studied the potential formation of ACEI peptides from cheese whey and caseins during fermentation with various commercial dairy starters used in the manufacture of yogurt, ropy milk and sour milk. No ACEI activity was observed in these hydrolysates. Further digestion of the above samples with pepsin and trypsin resulted in the release of several strong ACEI peptides derived primarily from α_{s1}-casein and β-casein. The formation of ACEI peptides was demonstrated in two dairy strains, Lb. delbrueckii ssp. bulgaricus and L. lactis ssp. cremoris, after fermentation of milk separately with each strain for 72 hours [36]. The most inhibitory fractions of the fermented milk mainly contained β-casein-derived peptides with inhibitory concentration (IC_{50}) values ranging from 8.0 to 11.2 µg/ml. Fuglsang et al. [37] tested a total of 26 strains of wild-type lactic acid bacteria, mainly belonging to L. lactis and Lb. helveticus, for their ability to produce a milk fermentate with ACEI activity. All tested strains produced ACEI substances in varying amounts, and two of the strains exhibited high ACE inhibition and a high OPA index, which correlates well with peptide formation. In another study 25 lactic acid strains of Lactobacillus, Lactococcus and Leuconsotoc were used [38]. The strains were tested alone or in combination and the highest activities were observed in Lb. jensenii, Lb. acidophilus and Leuc. mesenteroides strains and all strains showed correlation between ACE inhibition and degree of proteolysis. In a recent study, milk was fermented to defined pH values with 13 strains of lactic acid bacteria. The highest ACEI activity was obtained with two highly proteolytic strains of Lb. helveticus and with the Lactococcus strains. Fermentation from pH 4.6 to 4.3 with these strains slightly increased the ACEI activity, whilst fermentation to pH 3.5 with Lb. helveticus reduced the ACEI activity [39]. Moreover, four different Enterococcus faecalis strains, isolated from raw milk, produced fermented milk with potent ACEI activity [40]. In a recent research it was found that L. lactis strains isolated from artisanal dairy starters or commercial starter cultures are potential for the production of fermented dairy products with ACEI properties. Especially, a strain isolated from artisanal cheese presented the lowest IC_{50} (13µg/ml) [41].

Bioactive peptides isolated from skim milk and whey fermented using a range of organisms are summarized in Table 2. The majority of identified peptides are casein-derived ACEI peptides having IC_{50} values ranging from 5 to 500 µM. The best characterized ACEI and antihypertensive peptides liberated with Lb. helveticus alone or in combination with Saccharomyces cerevisiae are the tripeptides IPP, and VPP. Yamamoto et al. [42] identified an ACEI dipeptide (Tyr-Pro) from a yogurt-like product fermented with Lb. helveticus CPN4 strain. This peptide sequence is present in all major casein fractions, and its concentration was found to increase during fermentation, reaching a maximum concentration of 8.1 µg/ml in the product. Ashar and Chand [43] identified an ACEI peptide from milk fermented with Lb. delbrueckii ssp. bulgaricus. The peptide showed the sequence Ser-Lys-Val-Tyr-Pro-Phe-Pro-Gly-Pro-Ile from β–casein with an IC_{50} value of 1.7 mg/ml. In combination with Str. salivarius ssp. thermophilus and L. lactis biovar. diacetylactis, a peptide structure with a sequence of Ser-Lys-Val-Tyr-Pro was obtained from β-casein with an IC_{50} value of 1.4 mg/ml. Both peptides were markedly stable to digestive enzymes, acidic and alkaline pH, as well as during storage at 5 and 10 ºC for four days. Two β-casein-derived peptides were identified from water soluble fraction of milk fermented with Lb. jensenii. The identified peptides were Leu-Val-Try-Pro-Phe-Pro-Gly-Pro-Ile-His-Asn-Ser-Leu-Pro-Gln-Asn, and

Leu-Val-Try-Pro-Phe-Pro-Gly-Pro-Ile-His [38]. Quirós et al. [44] identified two peptides in fermented milk with *Enterococcus faecalis* that corresponded to β -casein fragments Lys-His-Leu-Pro-Leu-Pro and Lys-Val-Tyr-Pro-Phe-Pro-Gly-Pro-Ile-Pro-ASn-Ser-Leu-Pro-Gln-Asn-Ile-Pro-Pro, with potent ACEI activity.

Many kinds of proteolytic enzymes have been reported from lactic acid bacteria, and have been reviewed extensively [6, 45]. The components of the proteolytic systems of lactic acid bacteria are divided into three groups, including the extracellular proteinase that catalyzes casein breakdown to peptides, peptidases that hydrolyze peptides to amino acids and a peptide transport system. The extracellular proteinase activity was almost correlated with ACEI activity in the fermented milk, suggesting that the proteolysis of casein by the extracellular proteinase is the most important parameter in the processing of active components [46]. The importance of the proteinase was also supported by the fact that a proteinase negative mutant was not able to generate antihypertensive peptides in the fermented milk, whereas the wild-type strain had the ability to release strong antihypertensive peptides in the fermented milk [47]. The enzymatic process generating the antihypertensive peptides VPP and IPP in *Lb. helveticus* has been elucidated. By the proteolytic action of the extracellular proteinase long peptide with amino acid residue including VPP and IPP sequences were generated. Next the long peptide would be hydrolyzed to shorter peptides by intracellular peptidases. A key enzyme that can catalyze C-terminal processing of Val-Pro-Pro-Phe-Leu and Ile-Pro-Pro-Leu-Thr to VPP and IPP has been purified from *Lb. helveticus* CM4. The endopeptidase has sequence homology in amino terminal sequence to a previously reported pepO-gene product [48]. Kilpi et al. [49] found out higher ACE inhibition in milk fermentation using peptidase-deletion mutants compared to the wild-type of *Lb. helveticus* strain. Unlike with the wild type strain, ACEI remained constant during the course of fermentation with the proline-specific peptidase mutant. The mutant strains had also different peptide profiles than the wild-type strain.

2.3. Other

Various types of fermented soybean foods are consumed in Asian countries such as Korea, China, Japan, Indonesia and Vietnam. Soybeans are traditionally fermented primarily by *Bacilli* species during the early stage of fermentation followed by *Aspergillus* species, which predominate during the remaining fermentation period [50]. ACEI peptides have been found in many traditional Asian fermented soy foods, such as soybean paste, soy sauce, natto and tempeh. ACEI peptide His-His-Leu was isolated from Korean fermented soybean paste [51]. Rye gluten sourdoughs fermented with *Lb. reuteri* and added protease were found to contain the lactoripeptides VPP, IPP [52]. Moreover, our recent studies showed that fermentation of rapeseed or flaxseed meals with *Bacillus subtils* or *Lb. helveticus* strains produced ACEI activity [53].

2.4. Other activities

It is reasonable to expect that lactic acid bacteria produce scavengers for hydroxyl radical, which can be metabolic compounds produced by bacteria or degradation products of milk proteins. The results have demonstrated that the antioxidant production is commonly higher within the group of obligately homofermentative lactobacilli, than within the facultatively or obligately heterofermentative strain groups. Also heterofermentative *Lactobacillus* sp. have been reported to exhibit antioxidative activity. *Lb. acidophilus, Lb. bulgaricus, Str. thermophilus* and *Bifidobacterium longum* exhibited antioxidative activity by various mechanisms, like metal ion chelating capacity, scavenging of reactive oxygen species (ROS), reducing activity and superoxide dismutase activity [54, 55]. Peptides liberated during fermentation can be partially responsible for the reported antioxidative properties. An antioxidative peptide derived from κ-casein was detected in milk after fermentation with *Lb. delbrueckii* subs. *bulgaricus* [56]. Moreover, Hernández-Ledesma et al. [57] found a moderate ABTS radical scavenging capacity in commercial fermented milk from Europe. Further studies of this radical scavenging activity in different HPLC fractions showed low TEAC values. Virtanen et al. [58] found that fermentation with *Leuc. mesenteroides* ssp. *cremoris, Lb. jensenii* and *Lb. acidophilus* strains produced compounds that showed both radical scavenging activity and inhibition of lipid peroxidation.

Inflammation plays a key role in the development of cardiovascular disease. It often begins with inflammatory changes in the endothelium, which begins to express the adhesion molecule VCAM-1. VCAM-1 attracts monocytes, which then migrate through the endothelial layer under the influence of various proinflammatory chemoattractants [59]. Accordingly, fermentation by lactic acid may be able to release components that possess immunomodulatory properties. Most of the studies have been done with synthetic peptides derived from enzymatic treatment of milk proteins using different *in vitro* models. Leblanc et al. [60] investigated the effect of peptides released during the fermentation of milk by *Lb. helveticus* on the humoral immune system and on the growth of fibrosacromas. The study showed that bioactive components were released during fermentation that contributed to the immunoenhancing and antitumor properties. Antimutagenic compounds were produced during fermentation by *Lb. helveticus*, and release of peptides is one possible explanation [61]. The permeate fraction obtained from milk fermented by *Lb. helveticus* was able to modulate the *in vitro* proliferation of lymphocytes by acting on the production of cytokines [62]. Tompa et al. [63] found that peptide fractions form *Lb. helveticus* BGRA43 fermented milk have anti-inflammatory potential. Matar et al. [64] fed milk fermented with a *Lb. helveticus* strain to mice for three days and detected significantly higher numbers of IgA secreting cells in their intestinal mucosa, compared with control mice fed with similar milk incubated with a non-proteolytic variant of the same strain. The immunostimulatory effect of fermented milk was attributed to peptides released from the casein fraction.

3. Antihypertensive effects in vivo

The search for *in vitro* ACEI is the most common strategy followed in the selection of potential antihypertensive peptides derived from food proteins. *In vitro* ACEI activity is generally measured by monitoring the conversion of an appropriate substrate by ACE in the presence and absence of inhibitors. The antihypertensive effects have been assessed by *in vivo* experiments using spontaneously hypertensive rats (SHR) as an animal model to study human essential hypertension [7]. Following a positive response in animal studies human studies may be carried out to ascertain the ACEI potential

3.1. Animal studies

A great number of studies have addressed the effects of both short-term and long-term administration of potential antihypertensive peptides using this animal model. Fermented milks with different IC_{50}-values ranging from from 0.08 to 1.88 mg/ml have been shown to decrease blood pressure in SHR from 10 to 32 mmHg (Table 2).

The first antihypertensive effect of milk casein-derived peptides was first demonstrated by casein hydrolysate formed by purified proteinase from *Lb. helveticus* CP790 and milk fermented with the same bacteria [65]. The authors concluded that peptides deliberated from casein by extracellular proteinases were responsible for the antihypertensive effect. The active substances were liberated during fermentation of milk with *Lb. helveticus* and *Saccharomyces cerevisiae* and were identified to be IPP and VPP. Oral administration of fermented milk or pure tripeptides were shown to produce strong antihypertensive effect in SHR after single-dose [34, 66]. Thereafter, several animal studies have been conducted to characterize the long-term effects of lactotripeptides or fermented milk containing them. These studies were mainly conducted with SHR but also Goto-Kakizaki (GK) rats and double transgenic rats (dTGR) with malignant hypertension have been used. The development of hypertension was attenuated significantly in rats receiving fermented milk product containing lactotripeptides, attenuation in systolic blood pressure was 12-21 mmHg in SHR, 10 mmHg in high salt-fed GK rats and 19 mmHg in dTGR in comparison to control group [67-69]. Pure tripeptides did not produce as strong antihypertensive effect as the milk products containing them. In addition, minerals alone did not attenuate the development of blood pressure as much as the fermented milk products [68]. These studies indicate that the bioavailability of peptides may be better from milk in comparison of water or is improved by other milk components.

After the blood pressure monitoring has been completed the effect of long-term intake of lactotripeptides on vascular function has been assessed [68,70,71]. Jauhiainen et al. [70], showed improved endothelium-dependent relaxation in mesenteric arteries and aortas of rats that had received minerals and lactotripeptide. Endothelial function of mesenteric arteries was strongly impaired in all groups of salt-loaded GK rats, and significantly improved endothelium-dependent relaxations were observed after treatment with different fermented milk products [68]. Protection of endothelial function after incubation with tripeptides IPP and VPP for 24 h was found in a study with isolated SHR mesenteric arteries [71].

Evidence from ACE inhibition was gained by Masuda et al. [72], who found that after receiving a single-dose of Calpis™ sour milk, ACE activity was decreased in SHR aorta. The lactotripeptides were detected in solubilized fraction from the abdominal aorta of SHR but not from WKY given the sour milk. Moreover, in SHR, plasma rennin activity increased after long-term treatment of fermented milk product containing the lactotripeptides [67]. In addition, treatment with fermented milk containing lactotripeptides and plant sterols decreased serum ACE activity [73]. In salt-loaded GK rats, fermented milk with lactotripeptides decreased serum ACE and aldosterone levels [68].

Besides the most extensively studied lactotripeptides, also other fermented milk products and peptides have been found. Different strains of lactic acid bacteria, such as *Lb. helveticus* CPN4, *Lb. bulgaricus, Lb. jensenii* and *Str. thermophilus*, have been also shown to provoke liberation of peptides with antihypertensive activity in SHR [36, 37, 41]. Two peptides, corresponding to β -casein fragments Leu-Val-Tyr-Pro-Phe-Pro-Gly-Pro-Ile-Pro-Asn-Ser-Leu-Pro-Gln-Asn-Ile-Pro-Pro and Leu-His-Leu-Pro-Leu-Pro, have been isolated in fermented milk with *Enterococcus faecalis* and their antihypertensive effect in SHR, after acute and long-term administration has been proved. The administration of 2 mg/kg of peptide Leu-His-Leu-Pro-Leu-Pro resulted in a significant decrease of the SBP in SHR 4 h post-administration [74,75]. Fermentation of milk with one or more lactic acid bacteria strains followed by hydrolysis using food-grade enzymes liberated tripeptides (Gly-Thr-Trp and Gly-Val-Trp). Oral administration of this fermented whey lowered significantly SBP in SHR from 9 to 15 weeks of age. Bioactive substances, tripeptides and γ-aminobutyric acid (GABA), contributed to lowering blood pressure of SHR [76].

Some of ACE-inhibitory peptide fractions from cheese have shown *in vivo* activities. A water-soluble peptide preparation isolated from Gouda ripened for 8 months was found to have the most potent antihypertensive activity (maximum decrease in SBP = 24.7 (± 0.3) mmHg (P ≤ 0.01) after 6 h) when administered to SHR by gastric intubation at doses between 6.1 and 7.5 mg/kg body weight. Three peptide fractions were isolated from water-soluble extract by hydrophobic chromatography using different concentrations of acetonitrile. The fractions eluting between 15% and 30%, 30–45% and 60–75% acetonitrile decreased SBP in SHR by 15.0, 29.3 and 18.8 mmHg (P ≤ 0.01), respectively, 6 h after gastric intubation. The peptide fraction eluting between 30% and 45% acetonitrile was shown to contain the sequences (αs1-cn f(1–9)) Arg-Pro-Lys-His-Pro-Ile-Lys-His-Gln and (β-cn f(60–68)) Tyr-Pro-Phe-Gly-Pro-Ile-Pro-Asn (Table 1), which, respectively, decreased SBP in SHR by 9.3 (± 4.8) and 7.0 (± 3.8) mmHg 6 h after gastric intubation [29].

Several sequences have been proposed as responsible for the antihypertensive activity of soy protein hydrolysates and fermented products, but only the peptide His-His-Leu derived from fermented soy paste was assayed in pure form in SHR, where a decrease of 32 mm Hg of SBP was reached at a dose of 100 mg/kg. Moreover, the synthetic tripeptide His-His-Leu resulted in a significant decrease of ACE activity in the aorta [77]. Soybean-derived products contain isoflavones, which are thought to possess a favourable effect in reducing cardiovascular risk factors as well as vascular function [78]. However, on the basis of *in vitro* results and literature review, Wu and Muir [79] have indicated that the contribution of isoflavones to a blood-pressure-lowering effect in soybean ACEI peptides may be negligible. Similarly, it has been reported that the reduction of hypertension of a fermented product from soy milk was contributed mainly by peptides of 800–900 Da but it could be also attributable to GABA [80]. Moreover, fermented soy product, miso, with added tripeptides

(VPP and IPP) from casein was reported to act as antihypertensive agents in SHR [81]. Recently, Nakahara et al. [82] used the Dahl salt-sensitive rats as a model of salt-sensitive hypertension to evaluate the antihypertensive effect of a peptide-enriched soy sauce-like seasoning. The results of these tests have highlighted an important lack of correlation between the *in vitro* ACEI activity and the *in vivo* action. This fact has provided doubts on the use of the *in vitro* ACEI activity as the exclusive criteria for potential antihypertensive substances, since physiological transformations may occur *in vivo*, and because other mechanisms of action than ACE inhibition might be responsible for the antihypertensive effect.

3.2. Effects in clinical studies

Evidence of the beneficial effects of bioactive peptides has to be based on clinical data. Most research has been focused in lactotripeptides, VPP and IPP, and their antihypertensive properties. About twenty human studies have been published linking the consumption of products containing lactotripeptides with significant reductions in both SBP and DBP. Oral administration of these tri-peptides included in different formulas, fermented milk, dried product, fruit juice, etc., products. However, recent studies have provided some conflicting results. Most clinical trials have assessed BP-lowering effects at multiple points over time. Most of the BP studies with lactotripeptides have been done in Japanese subjects, and several studies have been done in Finnish subjects [83-88]. Generally, maximum duration of treatment was 8 weeks at doses between 3 and 52 mg/day (Table 3). From these data, it becomes apparent that the largest part of the total BP reduction takes place in the first 1–2 weeks of treatment. Thereafter, a further gradual lowering is seen, but to a lesser extent than in the first period [84-86]. The first significant effects of lactotripeptides on BP in hypertensive subjects were observed after 1–2 weeks of treatment with dosages as low as 3.8 mg/d. Maximum BP-lowering effects of lactotripeptides approximate 13 mmHg SBP and 8 mmHg DBP active treatment v. placebo, and are likely reached after 8–12 weeks of treatment. Lactotripeptides exert a gradual effect on BP lowering after start of intake and return of BP after end of treatment as well [85, 86, 89]. The highest effective dosage of lactotripeptides was evaluated in a safety study, and consisted of 52.5 mg/d [88]. After 10 weeks of active treatment, mean SBP in subjects with hypertension decreased by 4.1 mmHg and DBP by 1.8 mmHg. The next highest dose of lactotripeptides that was tested amounted to 13.0 mg/d [89]. After 4 weeks of active treatment, SBP in subjects with mild hypertension decreased by 11.2 mmHg compared to placebo, and DBP tended to decrease by 6.5 mmHg. In none of the trials with normotensives were statistically significant BP changes found [90-92]. Even at the highest dosage of lactotripeptides used in normotensives, which included a total of 29.2 mg/d during a period of 7 d, no BP lowering effects by lactotripeptides were observed [93]. Thus lactotripeptides only seem to be active at elevated BP values. Evidence indicates that effectiveness is positively associated with BP level, which is in line with existing data for BP-lowering medication [94].

Design	Duration (weeks)	Study population	IPP mg/d	VPP mg/d	Source of peptides	Formula	SBP	DBP	Ref.
R, p-c, s-bld, parallel	8	30 eldery hypertensive patients	1.1	1.5	*Lb. helv + Str. cer*	1 x 95 ml milk drink	-14.1	-6.9	83
R, p-c, d-bld, parallel	8	64 subjects with SBP 140-159 and DBP 90-99 mmHg	1.58	2.24	*Lb. helv + Str. cer.*	2 x 150 g milk drink	-13	-8.4	84
R, p-c, d-bld, parallel	8	32 subjects with SBP 140 - 180 and DBP 90-105 mmHg	1·60	2·66	*Lb. helv + Str. cer.*	1 x 120 g milk drink	-12.1	-5.8	85
R, p-c, d-bld, parallel	8	18 hypertensive and 26 normotensive subjects	1.1	1.5	*Lb. helv + Str. cer.*	2 x 100 g milk drink	-7.6	-2	91
R, p-c, d-bld, parallel	8	30 subjects with SBP 140-180 and DBP 90-105 mmHg	1.52	2.53	*Lb. helv + Str. cer.*	2 x 160 g milk drink	-13.2	-7.8	92
R, p-c, d-bld, parallel [1]	21	39 subjects with SBP 133-176 and DBP 86-108 mmHg	2.25	3.0	*Lb. helv* LBK-16H	2 x 150 ml milk drink	-6.7	-3.6	86
R, p-c, d-bld, parallel Cross-over [2]	10 7	60 Finnish subjects with SBP 140-180 and DBP 90-110 mmHg	2.4-2.7	2.4-2.7	*Lb. helv* LBK-16H	1 x 150 ml milk drink	-2.3 -12.3	-0.5 -3.7	87
R, p-c, d-bld, parallel	10	94 hypertensive patients	30	22.5	*Lb. helv* LBK-16H	2 x 150 ml milk drink	4.1	1.8	88
R, p-c, d-bld, parallel	1	20 healthy volunteers normal blood pressure (<130 mmHg SBP and <85 mmHg DPB).	11.5	17.7	*Lb. helv* CM4	1 x 14 tablets	2.6	2	93
R, p-c, d-bld, parallel	8	135 Dutch subjects with untreated high-normal BP or mild hypertension	4.2	5.8	Fermentation	1 x 200 ml yoghurt drink	-0.5	-1.2	97
R, p-c, d-bld, crossover	4	70 Caucasian subjects with prehypertension or stage 1 hypertension	15	-	Hydrolysis by endopeptidase	2 x 7.5 mg capsules	-3.8	-2.3	102

1) Results reported as changes in SBP and DBP after each month of treatment for all subjects (intention-to-treat analysis), and as mean changes over the total intervention period among subjects who had BP measurements for each month (per protocol analysis); 2) First part of the study was carried out in parallel design and second part of the study was carried out in crossover design.

Table 3. Hypotensive effects of fermented milks with bioactive peptides in humans

The results have been included in two meta-analysis [95, 96], which described decreases around 5 mmHg for SBP and 2.3 mmHg for DBP. In general, the effects described in Japanese studies on lactotripeptides are larger than those reported in Finnish studies. However, it is unlikely that genetic differences can account for these differential effects. Moreover, clinical trials in Dutch and Danish subjects have described controversial results since no effect on blood pressure was found [97, 98]. In a recent meta-analysis with a total of 18 trials, it was found a reduction of 3.73 mm Hg for SBP and 1.97 mm Hg for DBP but it was highlighted that the effect was more evident in Asian subjects that in Caucasian ones [99]. The relevance of these findings in genetics or dietary patterns should be further investigated. Comparative studies on antihypertensive medication in different races/ethnic groups have demonstrated that pharmacokinetic parameters and haemodynamic effects are essentially the same in Chinese and Japanese subjects compared with Caucasian subjects [100].

Hypertension is a complex multifactor disorder that is thought to result from an interaction between environmental factors and genetic background. Subject characteristics such as age and race/ethnicity can affect BP, including the BP response to specific antihypertensive medication. For certain antihypertensive drugs, it has been reported that a polymorphism found in humans can affect the clinical effectiveness, and similarly, these differences could be also affecting clinical trials of functional ingredients [101]. Although ACE inhibition has been postulated as the underlying mechanism of these lactotripeptides, results about the inhibition of this enzyme are not conclusive in humans. Several studies have shown that rennin or ACE activity was not affected by the oral administration of the tripeptides [95, 102]. Therefore, other mechanisms could be implicated in the observed blood pressure reduction. It has been found that the intake of fermented milk containing these peptides may decrease sympathetic activity, leading to a diminished heart rate variability, heart rate and total peripheral resistance, although differences did not reach statistical significance [98].

4. Bioavailability

Bioavilability of bioactive peptides is an important target to establish the relationship between *in vitro* and *in vivo* activities. The likelihood of any bioactive peptide released during fermentation mediating a physiological response is dependent on the ability of that peptide to reach an appropriate target site. Therefore, peptides may need to be resistant to further degradation by proteolytic and peptidolytic enzymes in digestive tract. Thereafter peptides should be absorbed and enter systemic circulation. Resistance to hydrolysis is one of the main factors influencing the bioavailability of bioactive peptides. The effects of digestive enzymes on bioactive peptides, in particular ACEI peptides derived from different food matrices, have been evaluated *in vitro* gastrointestinal simulated systems. The common purpose of these experiments was to assess the effects of the peptidases of the stomach and the pancreas on the preservation of the ACEI activity of different hydrolysates. Studies have shown that the ACEI is low after fermentation but increases during hydrolysis that simulates gastrointestinal digestion [35,103]. The ACEI peptides in rapeseed hydrolysate exhibited good stability in an *in vitro* digestion model using human gastric and duodenal fluids [104]. The digestion of some peptides have been reported. For example, Ile-Val-Tyr was hydrolysed by pepsin, trypsin and chymotrypsin alone or in combination and IC_{50}-value did not change significantly during digestion [105]. Proline- and hydroxyproline-containing peptides are usually resistant to degradation by digestive enzymes. Tripeptides containing C-terminal proline-proline are generally resistant to proline-specific peptidases [106]. In some cases, pancreatic digestion is needed to produce active peptide. For instance, the active form of peptide Lys-Val-Leu-Pro-Val-Pro-Glu is generated by hydrolysis of the glutamine residue at the C-terminal during pancreatic digestion [107]. The results are not completely predictive of the resistance of the bioactive peptides because they do not mimic all the physiological factors affecting food digestion, as pH variations, the relative amounts of the enzymes, the interactions with other molecules, and the ratio peptidase/tested compound. These variations may affect the rate of enzymatic degradation of the bioactive peptides under study, therefore affecting the estimated bioavailability of these bioactive peptides. Moreover, commercial enzymes appear to digest whey proteins more efficiently compared with human digestive juices when used at similar enzyme activities [108]. This could lead to conflicting results when comparing human *in vivo* protein digestion with digestion using purified enzymes of non-human species.

Peptides have been reported to have poor permeation across biological barriers (e.g. intestinal mucosa) [109]. Peptides can be transported by active transcellular transport or by passive processes. Although substantial amino acid absorption occurs in the form of di- and tripeptides at the apical side of enterocytes, efflux of intact peptides via the basolateral membrane into the general circulation seems to be negligible [110]. The intestinal absorption of peptides have been performed using *in vitro* tests with monolayer of intestinal cell lines, simulating intestinal epithelium, as well as analysis of peptides and derivatives in blood samples after animal and clinical studies. Foltz et al. [111] investigated the transport of IPP and VPP by using three different absorption models and demonstrated that these tripeptides are transported in small amounts intact across the barrier of the intestinal epithelium. The major transport mechanisms of IPP and VPP were demonstrated to be paracellular transport and passive diffusion [112]. Another ACEI peptide, Leu-His-Leu-Pro-Leu-Pro resisted gastrointestinal simulation but was degraded to His-Leu-Pro-Leu-Pro by cellular peptidases before crossing Caco-2 cell monolayer. The pentapeptide was rapidly transported through Caco-2 cell monolayers through paracellular route [113].

Vascular endothelial tissue peptidases and soluble plasma peptidases further contribute to peptide hydrolysis. As a consequence, for most peptides, the plasma half-life is limited to minutes as shown for endogenous peptides such as angiotensin II and glucagon-like peptide 1 [114]. In order to exert antihypertensive effect ACEI peptides need to resist different peptidases such as ACE. In this regard ACEI peptides can be classified into three groups: the inhibitor type, of which the IC_{50}-value is not affected by preincubation with ACE; the substrate type, peptides that are hydrolysed by ACE to give peptides with a weaker activity; the pro-drug type inhibitor, peptides that are converted to true inhibitors by ACE or other proteases/peptidases. Only peptides belonging to pro-drug or inhibitor type exert antihypertensive properties after oral administration. There are some examples showing that peptides are absorbed and can exert *in vivo* activities. As regard to casein-derived IPP,

Jauhiainen et al. [115] used radiolabelled tripeptide and showed that it absorbed partly intact from the gastrointestinal tract after a single oral dose to rats. Considerable amounts of radioactivity were found from several tissues, e.g., liver, kidney and aorta. The excretion of IPP was slow; even after 48 hours the radiolabelled peptide had not been completely excreted. IPP did not bind to albumin or other plasma proteins *in vitro*. Considering this and the long-lasting retention of the radioactivity in the tissues, accumulation of IPP may occur in sufficient concentrations to cause blood pressure lowering effects e.g., by ACE-inhibition in the vascular wall. In another study the absolute bioavailability of the tripeptides in pigs was below 0.1%, with an extremely short elimination half-life ranging from 5 to 20 min [116]. In humans, maximal plasma concentration did not exceed picomolar concentration [117].

The improvement of limited absorption and stability of peptides has been a goal when evaluating their effectiveness. For example, some carriers interact with the peptide molecule to create an insoluble entity at low pH which later dissolves and facilitates intestinal uptake, by enhancing peptide transport over the non-polar biological membrane [118]. Bioavailability of bioactive tripeptides (VPP, IPP, LPP) was improved by administering them with a meal containing fiber, as compared to a meal containing no fiber. High methylated citrus pectin was used as a fiber [119]. Ko et al. [120] applied emulsification, microencapsulation and lipophilization to enhance the antihypertensive activity of a hydrolysate of tuna cooking juice. Among these treatments, lipophilization was the most effective, followed by microencapsulation and lecithin emulsification, getting for each of them a stronger effect than the obtained with the double untreated dosage. Antihypertensive effect of ovokinin (Phe-Arg-Ala-Asp-Pro-Phe-Leu) increased four-times compared to the untreated dosage after administration with egg yolk [121]. In this case, phospholipids were identified as responsible for enhancing the antihypertensive effect, particularly phosphatidylcholine, that could improve intestinal absorption or by protecting ovokinin of peptidases. Among drug delivery systems, emulsions have been used to enhance oral bioavailability or promoting absorption through mucosal surfaces of peptides and proteins [118]. Individually, various components of emulsions have been considered as candidates for improving bioavailability of peptides.

5. General conclusions

The interest on foods possessing health-promoting or disease-preventing properties has been increasing. An increasing number of foods sold in developed countries bears nutrition and health claims. Fermented milk with putative antihypertensive effect in humans could be an easy applicable lifestyle intervention against hypertension. In fact, much work has been done with dietary antihypertensive peptides and evidence of their effect in animal and clinical studies. Moreover, there are numerous available patents of products containing antihypertensive bioactive peptides. However, certain aspects, such as identification of the active form in the organism and the different mechanisms of action that contribute in the antihypertensive effect still need to be further investigated. Recent advances on specific analytical techniques able to follow small amounts of the peptides or derivatives from them in complex matrices and biological fluids will allow performing these kinetic studies in model animals and humans. Similarly, advances in new disciplines such as nutrigenomic and nutrigenetic will open new ways to follow bioactivity in the organism by identifying novel and more complex biomarkers of exposure and/or of activity. There is still poor knowledge on the resistance of peptides to gastric degradation, and low bioavailability of peptides has been observed. This reinforces the need of various strategies to improve the oral bioavailability of peptides.

More emphasis has been put on the legal regulation of the health claims attached to the products. Authorities around the world have developed systematic approaches for review and assessment of scientific data. Evidence on the beneficial effects of a functional food product should be enough detailed, extensive and conclusive for the use of a health claim in the product labeling and marketing. Besides being based on generally accepted scientific evidence, the claims should be well understood by the average consumer. First, it is necessary to identify and quantify the active sequences. Antihypertensive peptides are only minor constituents in highly complex food matrices and, therefore, a monitoring of the large-scale production by hydrolytic or fermentative industrial process is mandatory. Second, extensive investigations to prove the antihypertensive effect in humans as well as the minimal dose to show this effect are necessary to fulfill the requirements of the legislation concerning functional foods. Japan was the pioneer with the Foods for Special Health Use (FOSHU) legislation in 1991. Europe adopted a joint Regulation on Nutrition and Health Claims made on Foods in 2006 being the European Food Safety Authority (EFSA). At present, EFSA have concludes that the evidence is insufficient to establish a cause and effect relationship between the consumption of the tripeptides VPP and IPP and the maintenance of normal blood pressure. Bearing in mind that 'essential hypertension' consists of disparate mechanisms that ultimately lead to elevations in systemic BP, it is most probably that that products containing lactotripeptides offer a valuable option as a non-pharmacological, nutritional treatment of elevated blood pressure for some groups of people.

Author details

Anne Pihlanto
MTT Agrifood Finland, Biotechnology and Food Research, Jokioinen, Finland

6. References

[1] Shah N (2007) Functional cultures and health benefits. Int. dairy j. 17:1262-1277
[2] Takano T (2002) Anti-hypertensive activity of fermented dairy products containing biogenic peptides. Anton. leeuw. 82: 333-340.
[3] Korhonen HJT, Pihlanto-Leppälä A (2004) Milk-derived bioactive peptides : formation and prospects for health promotion. In: Edited by Colette Shortt and John O'Brien.

Handbook of functional dairy products. Functional foods and nutraceuticals series 6.0: p. 109-124.

[4] FitzGerald RJ, Murray BA (2006) Bioactive peptides and lactic fermentations. Int. j. dairy technology 59: 118-125.

[5] Jäkälä P, Vapaatalo H (2010) Antihypertensive peptides from milk proteins. Pharmaceuticals 3: 251-272.

[6] Christensen JE, Dudley EG, Pederson JA, Steele JL (1999) Peptidases and amino acid catabolism in lactic acid bacteria. Anton. leeuw. 76: 217-246.

[7] Luoma S, Peltoniemi K, Joutsjoki V, Rantanen T, Tamminen M, Heikkinen I, Palva A (2001) Expression of six peptidases from Lactobacillus helveticus in Lactococcus lactis. Appl. environ. microb. 67: 1232–1238.

[8] Foucaud C, Juillard V (2000) Accumulation of casein-derived peptides during growth of proteinase-positive strains of Lactococcus lactis in milk: their contribution to subsequent bacterial growth is impaired by their internal transport. J dairy res. 67: 233-240.

[9] Williams AG, Noble J, Tammam J, Lloyd D, Banks JM (2002) Factors affecting the activity of enzymes involved in peptide and amino acid catabolism in non starter lactic acid bacteria isolated from Cheddar cheese. Int. dairy j. 12: 841–852.

[10] Lopez AD, Murray CC (1998) The global burden of disease, 1990–2020. Nat. med. 4:1241–1243.

[11] Harris T, Cook EF, Kannel W, Schatzkin A, Goldman L (1985) Blood pressure experience and risk of cardiovascular disease in the elderly. Hypertension 7:118–24.

[12] Pihlanto A, Korhonen H (2003) Bioactive peptides and proteins. Adv. food res. 47: 175–276.

[13] Hernández-Ledesma B, del Mar Contreras M, Recio I (2011) Antihypertensive peptides: production, bioavailability and incorporation into foods. Adv. colloid interface sci. 165:23-35

[14] Van Gaal LF, Mertens IL, De Block CE (2006) Mechanisms linking obesity with cardiovascular disease. Nature 444: 876-880.

[15] Pihlanto A (2006) Antioxidative peptides derived from milk proteins. Int. dairy j. 16: 1306–1314.

[16] Roudot-Algaron F, Bars DL, Kerhoas L, Einhorn J, Gripon JC (1994) Phosptiopeptides from Comté Cheese: Nature and origin. J. food sci. 59: 544–547.

[17] Singh TK, Fox PF, Healy A (1997) Isolation and identification of further peptides from diafiltration retentate of the water-soluble fraction of Cheddar cheese. J. dairy res. 64:433-443.

[18] Meisel H, Goepfert A, Günter S (1997) ACE-inhibitory activities in milk products. Milchwissenschaft 52: 307–311.

[19] Addeo F, Chianes L, Salzano A, Sacchi R, Cappuccio U, Ferranti P, Malorni A (1992) Characterization of the 12% tricholoroacetic acid-insoluble oligopeptides of Parmigiano–Reggiano cheese. J. dairy res. 59: 401–411.

[20] Smacchi E, Gobbetti M (1998) Peptides from several Italian cheeses inhibitory to proteolytic enzymes of lactic acid bacteria, Pseudomonas fluorescens ATCC 948 and to the angiotensin I-converting enzyme. Enzyme microb.tech. 22: 687–694.

[21] Haileselassie SS, Lee B H, Bibbs BF (1999) Purification and identification of potentially bioactive peptides from enzyme modified cheese. J. dairy sci. 82: 1612–1617.

[22] Ryhänen E-L, Pihlanto-Leppälä A, Pahkala E (2001) A new type of ripened low-fat cheese with bioactive properties. Int. dairy j. 11: 441–447.

[23] Gomez JA, Ramos M, Recio I (2002) Angiotensin-converting enzyme-inhibitory peptides in Manchego cheeses manufactured with different starter cultures. Int. dairy j. 12: 697–706.

[24] Torres-Llanez MJ, González-Córdova AF, Hernandez-Mendoza A, Garcia HS, Vallejo-Cordoba B (2011) Angiotensin-converting enzyme inhibitory activity in Mexican Fresco cheese. J. dairy sci. 94:3794–3800.

[25] Pripp AH, Sørensen R, Stepaniak L and Sørhaug T (2006) Relationship between proteolysis and angiotensin I-converting enzyme inhibition in different cheeses LWT 39: 677–683.

[26] Muehlenkamp MR, Warthesen JJ (1996) β-casomorphins: Analysis in cheese and susceptibility to proteolytic enzymes from Lactococcus lactis ssp. cremoris. J. dairy sci. 79: 20-26.

[27] Sienkiewicz- Szlapka E, Jarmolowska B, Krawczuk S, Kostyra E, Kostyra H, Iwana M (2009) Contents of agonistic and antagonistic opioid peptides in different cheese varieties. Int. dairy j. 19: 258-263.

[28] Durrieu C, Degraeve P, Chappaz S, Martial-Gros A (2006) Immunomodulating effects of water-soluble extracts of traditional French Alps cheeses on a human T-lymphocyte cell line. Int. dairy j. 16: 1505-1514.

[29] Saito T, Nakamura T, Kitazawa H, Kawai Y, Itoh T (2000) Isolation and structural analysis of antihypertensive peptides that exist naturally in Gouda cheese J. dairy sci. 83: 1434–1440

[30] Ong L, Shah NP (2008) Release and identification of angiotensin-converting enzyme-inhibitory peptides as influenced by ripening temperatures and probiotic adjuncts in Cheddar cheeses. LWT - Food sci. technol. 41: 1555-1566.

[31] Bütikofer U, Meyer J, Sieber, R, Wechsler D (2007) Quantification of the angiotensin-converting enzyme-inhibiting tripeptides Val-Pro-Pro and Ile-Pro-Pro in hard, semi-hard and soft cheeses. Int. dairy j. 17: 968-975.

[32] Bütikofer U, Meyer J, Sieber R, Walther B, Wechsler D (2008) Occurrence of the angiotensin-converting enzyme–inhibiting tripeptides Val-Pro-Pro and Ile-Pro-Pro in different cheese varieties of Swiss origin. J. dairy sci. 91:29–38.

[33] Meyer J, Bütikofer U, Walther B, Wechsler D, Sieber R (2009) Hot topic: Changes in angiotensin-converting enzyme inhibition and concentrations of the tripeptides Val-Pro-Pro and Ile-Pro-Pro during ripening of different Swiss cheese varieties. J. dairy sci. 92: 826-836.

[34] Nakamura Y, Yamamoto N, Sakai K, Okubo A, Yamazaki S, Takano T (1995) Purification and characterization of angiotensin I-converting enzyme inhibitors from sour milk. J. dairy sci. 78: 777-783.

[35] Pihlanto-Leppälä A, Rokka T, Korhonen H (1998) Angiotensin I converting enzyme inhibitory peptides derived from bovine milk proteins. Int. dairy j. 8:325-331.

[36] Gobbetti M, Ferranti P, Smacchi E, Goffredi F, Addeo F (2000) Production of angiotensin-I-converting-enzyme-inhibitory peptides in fermented milks started by *Lactobacillus delbrueckii* subsp. *bulgaricus* SS1 and *Lactococcus lactis* subsp. *cremoris* FT4. Appl. environ. microb. 66: 3898–3904.

[37] Fuglsang A, Rattray FP, Nilsson D, Nyborg CB (2003) Lactic acid bacteria: inhibition of angiotensin-converting enzyme in vitro and in vivo. Anton. leeuw. 83: 27–34.

[38] Pihlanto A, Virtanen T, Korhonen H (2010) Angiotensin I converting enzyme (ACE) inhibitory activity and antihypertensive effect of fermented milk. Int. dairy j. 20: 3-10.

[39] Nielsen MS, Martinussen T, Flambard B, Sørensen KI, Otte J (2009) Peptide profiles and angiotensin-I-converting enzyme inhibitory activity of fermented milk products: Effect of bacterial strain, fermentation pH, and storage time Int. dairy j. 19: 155-165.

[40] Muguerza B, Ramos M, Sánchez E, Manso MA, Miguel M, Aleixander, A, Lopez-Fandino R (2006) Antihypertensive activity of milk fermented by *Enterococcus faecalis* strains isolated from raw milk. Int. dairy j. 16: 61–69.

[41] Rodríguez-Figueroa JC, Reyes-Díaz R, González-Córdova AF, Troncoso-Rojas R, Vargas-Arispuro I, Vallejo-Cordoba B (2010) Angiotensin-converting enzyme inhibitory activity of milk fermented by wild and industrial *Lactococcus lactis* strains. J. dairy sci. 93:5032–5038.

[42] Yamamoto N, Maeno M, Takano T (1999) Purification and characterization of an antihypertensive peptide from a yogurt-like product fermented by *Lactobacillus helveticus* CPN4. J. dairy sci. 82: 1388-1393.

[43] Ashar MN, Chand R (2004) Antihypertensive peptides purified form milks fermented with *Lactobacillus belbrueckii* ssp. *bulgaricus*. Milchwissenschaft 59:14-17.

[44] Quiros A, Ramos M, Muguerza B, Delgado MA, Miguel M, Aleixandre A, Recio I (2007) Identification of novel antihypertensive peptides in milk fermented with *Enterococcus faecalis*. Int. dairy j. 17: 33–41.

[45] Kunji ER, Mierau I, Hagting A, Poolman B, Koning, WN (1996) The proteolytic systems of lactic acid bacteria. Anton. leeuw. 70: 187-221.

[46] Yamamoto N, Ishida Y, Kawakami N, Yada H (1991) *Lactobacillus helveticus* bacterium having high capability of producing tripeptide, fermented milk product, and process for preparing the same. EU Patent, 1016709A1.

[47] Yamamoto N, Akino A, Takano T (1993) Purification and specificity of a cell-wall associated proteinase from *Lactobacillus helveticus* CP790. J. biochem, 114: 740-745.

[48] Ueno K, Mizuno S, Yamamoto N (2004) Purification and characterization of an endopeptidase has an important role in the carboxyl terminal processing of antihypertensive peptides in *Lactobacillus helveticus* CM4. Lett. appl. microbiol. 39: 313-318

[49] Kilpi E, Kahala M, Steele JM, Pihlanto A, Joutsjoki V (2007) Angiotensin I-converting enzyme inhibitory activity in milk fermented by wild-type and peptidase-deletion derivatives of *Lactobacillus helveticus* CNRZ32. Int. dairy j. 17: 976-984.

[50] Kwon DY, Daily JW, Kim HJ, Park S (2010) Antidiabetic effects of fermented soybean products on type 2 diabetes. Nutr. res. 30:1–13.

[51] Shin ZI, Yu R, Park SA, Chung DK, Ahn CW, Nam HS, Kim KS, Lee HJ (2001) His-His-Leu, an angiotensin I converting enzyme inhibitory peptide derived from Korean soybean paste, exerts antihypertensive activity in vivo. J. agric. food chem. 49: 3004-3009.

[52] Hu Y, Stromeck A, Loponen J, Lopes-Lutz D, Schieber A, Gänzle MG (2011) LC-MS/MS quantification of bioactive angiotensin I-converting enzyme inhibitory peptides in rye malt sourdoughs. J. agric. food chem. 59: 11983-11989.

[53] Pihlanto A, Johansson T, Mäkinen S (2012) Inhibition of angiotensin I-converting enzyme and lipid peroxidation by fermented rapeseed and flaxseed meal. Eng. life sci. 12 DOI: 10.1002/elsc.201100137

[54] Lin MY, Yen CL (1999) Antioxidative ability of lactic acid bacteria J. agric. food chem. 47: 1460-1466.

[55] Ou CC, Lu TM, Tsai JJ, Yen, JH, Chen, HW, Lin MY (2009) Antioxidative effect of lactic acid bacteria: Intact cells vs. intracellular extracts. J. food drug anal. 17: 209-216.

[56] Kudoh Y, Matsuda S, Igoshi K, Oki T (2001) Antioxidative peptide from milk fermented with *Lactobacillus delbrueckii* subsp. *bulgaricus* IFO13953. Nippon Shokuhin Kagaku Kaishi 48: 44-55.

[57] Hernández-Ledesma B, Miralles B, Amigo L, Ramos M, Recio, I (2005) Identification of antioxidant and ACE-inhibitory peptides in fermented milk. J. sci. food agric. 85: 1041-1048.

[58] Virtanen T, Pihlanto A, Akkanen S, Korhonen H (2007) Development of antioxidant activity in milk whey during fermentation with lactic acid bacteria. J. appl. microbial. 102: 106–115.

[59] Libby P (2006) Inflammation and cardiovascular disease mechanisms. Am. j. clin. nutr. 83: 456S– 460S.

[60] LeBlanc AM, Matar C, Valdéz JC, LeBlanc N, Perdigón G (2002) Immunomodulatory effects of peptidic fractions issued from milk fermented with *Lactobacillus helveticus*. J. dairy sci. 85: 2733–2742.

[61] Matar C, Nadathur SS, Bakalinsky AT, Goulet J (1997) Antimutagenic effects of milk fermented by Lactobacillus helveticus L89 and a protease-deficient derivative. J. dairy sci. 80: 1965-1970.

[62] Laffineur E, Genetet N, Leonil J (1996) Immunomodulatory activity of β-casein permeate medium fermented by lactic acid bacteria. J. dairy sci. 79: 2112–2120.

[63] Tompa G, Laine A, Pihlanto A, Korhonen H, Rogel, I, Marnila P (2011) Chemiluminescence of non-differentiated THP-1 promonocytes: developing an assay for screening anti-inflammatory milk proteins and peptides. Luminescence 26: 251-258,

[64] Matar C, Valdez JC, Medina M, Rachid M, Perdigon G (2001) Immunomodulating effects of milks fermented by *Lactobacillus helveticus* and its non-proteolytic variant. J. dairy res. 68: 601-609.

[65] Yamamoto N, Akino A, Takano T (1994) Antihypertensive effect of the peptides derived from casein by an extracellular proteinase from *Lactobacillus helveticus* CP790. J. dairy sci. 77: 917-922.

[66] Nakamura Y, Yamamoto N, Sakai K, Takano T (1995) Antihypertensive effect of sour milk and peptides isolated from it that are inhibitors to angiotensin I-converting enzyme. J. dairy sci. 78: 1253-1257.

[67] Sipola M, Finckenberg P, Korpela R, Vapaatalo H, Nurminen ML (2002) Effect of long-term intake of milk products on blood pressure in hypertensive rats. J. dairy res. 69: 103–111.

[68] Jäkälä P, Hakala A, Turpeinen A, Korpela R, Vapaatalo H (2009) Casein-derived bioactive tripeptides Ile-Pro-Pro and Val-Pro-Pro attenuate the development of hypertension and improve endothelial function in salt-loaded Goto-Kakizaki rats. J. funct. foods 1: 366–374.

[69] Jauhiainen T, Pilvi T, Cheng ZJ, Kautiainen H, Müller DN, Vapaatalo H, Korpela R, Mervaala E (2010) Milk products containing bioactive tripeptides have an antihypertensive effect in double transgenic rats (dTGR) harbouring human renin and human angiotensinogen genes. J. nutr. metab. doi:10.1155/2010/287030.

[70] Jauhiainen T, Collin M, Narva M, Cheng ZJ, Poussa T, Vapaatalo H, Korpela R (2005) Effect of long-term intake of milk peptides and minerals on blood pressure and arterial function in spontaneously hypertensive rats. Milchwissenschaft 60: 358–362.

[71] Jäkälä P, Jauhiainen T, Korpela R, Vapaatalo H (2009) Milk protein-derived bioactive tripeptides Ile-Pro-Pro and Val-Pro-Pro protect endothelial function in vitro in hypertensive rats. J. funct. foods 1: 266–273.

[72] Masuda O, Nakamura Y, Takano T (1996) Antihypertensive peptides are present in aorta after oral administration of sour milk containing these peptides to spontaneously hypertensive rats. J. nutr. 126: 3063–3068.

[73] Jäkälä P, Pere E, Lehtinen R, Turpeinen A, Korpela R, Vapaatalo H (2009) Cardiovascular activity of milk casein-derived tripeptides and plant sterols in spontaneously hypertensive rats. J. physiol. pharmacol. 60: 11–20.

[74] Miguel M, Recio I, Ramos M, Delgado MA, Aleixandre A (2006) Antihypertensive effect of peptides obtained from *Enterococcus faecalis*-fermented milk in rats J. dairy sci. 89: 3352–3359.

[75] Quiros A, Ramos M, Muguerza B, Delgado MA, Migue M, Aleixandre A, Recio I (2007) Identification of novel antihypertensive peptides in milk fermented with Enterococcus faecalis Int. dairy j. 17: 33–41.

[76] Chen GW, Tsai JS, Pan BS (2007) Purification of angiotensin I-converting enzyme inhibitory peptides and antihypertensive effect of milk produced by protease-facilitated lactic fermentation. Int. dairy j. 17: 641–647.

[77] Shin ZI, Yu R, Park SA, Chung DK, Ahn CW, Nam HS, Kim KS, Lee HJ (2001) His-His-Leu, an angiotensin I converting enzyme inhibitory peptide derived from Korean soybean paste, exerts antihypertensive activity in vivo. J. agric. food chem. 49: 3004-3009.

[78] Sacks FS, Lichtenstein A, Van Horn L, Harris W, Kris-Etherton P, Winston M (2006) Soy protein, isoflavones, and cardiovascular health. Circulation 113: 1034-1044.

[79] Wu J, Muir AD (2008) Hypotensive and physiological effect of angiotensin converting enzyme inhibitory peptides derived from soy protein on spontaneously hypertensive rats J. agric. food chem. 56: 9899–9904.

[80] Tsai JS, Lin YS, Pan BS, Chen TJ (2006) Antihypertensive peptides and γ-aminobutyric acid from prozyme 6 facilitated lactic acid bacteria fermentation of soymilk. Process biochem. 41: 1282–1288.

[81] Inoue K, Gotou T, Kitajima H, Mizuno S, Nakazawa T, Yamamoto N (2009) Release of antihypertensive peptides in miso paste during its fermentation, by the addition of casein. J. biosci. bioeng. 108: 111–115.

[82] Nakahara T, Sano A, Yamaguchi H, Sugimoto K, Chikata H, Kinoshita E, Uchida R (2010) Antihypertensive effect of peptide-enriched soy sauce-like seasoning and identification of its angiotensin I-converting enzyme inhibitory substances, J. agric. food chem.58: 821-827.

[83] Hata Y, Yamamoto M, Ohni M, Nakajima K, Nakamura Y, Takano T (1996) A placebo-controlled study of the effect of sour milk on blood pressure in hypertensive subjects. Am. j. clin. nutr. 64: 767-771.

[84] Kajimoto O, Kurosaki T, Mizutani J, Ikeda, N, Kaneko K, Yabune M, Nakamura Y (2002) Antihypertensive effects of liquid yogurts containing 'lactotripeptides (VPP, IPP)' in mild hypertensive subjects. J. nutr. food 5: 55–66.

[85] Hirata H, Nakamura Y, Yada H, Moriguchi S, Kajimoto O, Takahashi T (2002) Clinical effects of new sour milk drink on mild or moderate hypertensive subjects. J. new. rem. clin 51: 61–69.

[86] Seppo L, Jauhiainen T, Poussa T, Korpela R. (2003) A fermented milk high in bioactive peptides has a blood pressure-lowering effect in hypertensive subjects. Am. j. clin. nutr. 77: 326-330.

[87] Tuomilehto J, Lindstrom J, Hyyrynen J, Korpela R, Karhunen ML, Mikkola L, Jauhiainen T, Seppo L, Nissinen A (2004) Effect of ingesting sour milk fermented using Lactobacillus helveticus bacteria producing tripeptides on blood pressure in subjects with mild hypertension. J. hum. hypertens. 18: 795–802.

[88] Jauhiainen T, Vapaatalo H, Poussa T, Kyrönpalo S, Rasmussen M, Korpela R (2005) Lactobacillus helveticus fermented milk lowers blood pressure in hypertensive subjects in 24-h ambulatory blood pressure measurement. Am. j. hypertens. 18: 1600–1605.

[89] Arihara K, Kajimoto O, Hirata H, Takahashi R, Nakamura Y (2005) Effect of powdered fermented milk with Lactobacillus helveticus on subjects with high-normal blood pressure or mild hypertension. J. am. coll. nutr. 24: 257–265.

[90] Kajimoto O, Aihara K, Hirata H, Takahashi R, Nakamura Y (2001) Hypotensive effects of the tablets containing lactotripeptides (VPP, IPP). J. nutr. food 4: 51–61.

[91] Itakura H, Ikemoto S, Terada S, Kondo K (2001) The effect of sour milk on blood pressure in untreated hypertensive and normotensive subjects. J. jap. soc. clin. nutr. 23: 26–31.

[92] Kajimoto O, Nakamura Y, Yada H, Moriguchi S, Hirata H, Takahashi T (2001) Hypotensive effects of sour milk in subjects with mild or moderate hypertension. J. jpn. soc. nutr. food sci. 54: 347–354.

[93] Yasuda K, Aihara K, Komazaki K, Mochii M, Nakamura Y (2001) Effect of large high intake of tablets containing 'lactotripeptides (VPP, IPP)' on blood pressure, pulse rate and clinical parameters in healthy volunteers. J. nutr. food 4: 63–72.

[94] Law MR, Wald NJ, Morris JK, Jordan JE (2003) Value of low dose combination treatment with blood pressure lowering drugs: analysis of 354 randomised trials. Br. med. j. 326: 1427–1431.

[95] Pripp AH (2008) Effect of peptides derived from food proteins on blood pressure: a meta-analysis of randomized controlled trials. Food nutr. res. 5: 1–9.

[96] Xu JY, Qin LQ, Wang PY, Li W, Chang C (2008) Effect of milk tripeptides on blood pressure: a meta-analysis of randomized controlled trials. Nutrition 24: 933–940.

[97] Engberink MF, Schouten EG, Kok FJ, van Mierlo LA, Brouwer IA, Geleijnse JM (2009) Lactotripeptides show no effect on human blood pressure: results from a double-blind randomized controlled trial. Hypertension 51:399-405.

[98] Usinger L, Jensen LT, Flambard B, Linneberg A, Ibsen H (2010) The antihypertensive effect of fermented milk in individuals with prehypertension or borderline hypertension. J.hum. hypertens. 24: 678–683

[99] Cicero AFG, Gerocarni B, Laghi L, Borghi C (2011) Blood pressure lowering effect of lactotripeptides assumed as functional foods: a meta-analysis of current available clinical trials. J. hum. hypertens. 25: 425–436.

[100] Vaidyanathan S, Jermany J, Yeh C, Bizot MN, Camisasca R (2006) Aliskiren, a novel orally effective renin inhibitor, exhibits similarpharmacokinetics and pharmacodynamics in Japanese and Caucasian subjects. Br. j. clin. pharmacol. 62: 690–698.

[101] Arsenault J, Lehoux J, , Lanthier L Cabana J, Guillemette G, Lavigne P, Leduc R, Escher E (2010) A single-nucleotide polymorphism of alanine to threonine at position 163 of the human angiotensin II type 1 receptor impairs losartan affinity. Pharmacogenet Genomics 20: 377–388.

[102] Boelsma E, Kloek J (2010) IPP -rich milk protein hydrolysate lowers blood pressure in subjects with stage 1 hypertension, a randomized controlled trial. J. nutr. 9:52 doi:10.1186/1475-2891-9-52.

[103] Hernández-Ledesma B, Amigo L, Ramos M, Recio I (2004) Application of high-performance liquid chromatography–tandem mass spectrometry to the identification of biologically active peptides produced by milk fermentation and simulated gastrointestinal digestion. J. chromatogr. A 1049: 107–114.

[104] Mäkinen S, Johansson T, Vegarud G, Pihlava JM, Pihlanto A (2012) Angiotensin I-converting enzyme inhibitory and antioxidant properties of rapeseed hydrolysates. J. funct. foods (in press)

[105] Matsui T, Li CH, Osajima Y (1999) Preparation and characterization of novel bioactive peptides responsible for angiotensin I-converting enzyme inhibition from wheat germ. J. pept. sci. 5: 289–297.

[106] Vermeirssen V, Van Camp J, Verstraete W (2004) Bioavailability of angiotensin I converting enzyme inhibitory peptides. Br. j. nutr. 92: 357–366.

[107] Maeno M, Yamamoto N, Takano T (1996) Identification of an antihypertensive peptide from casein hydrolysate produced by a proteinase from Lactobacillus helveticus CP790. J. dairy sci. 79: 1316–1321.

[108] Eriksen EK, Holm H, Jensen E, Aaboe R, Devold TG, Jacobsen M, Vegarud GE (2010) Different digestion of caprine whey proteins by human and porcine gastrointestinal enzymes. Br. j. nutr. 104:374-381.

[109] Pauletti GM, Gangwar S, Knipp GT, Nerurkar MM, Okumu FW, Tamura T, Siahaan TJ (1996) Structural requirements for intestinal absorption of peptide drugs. J. control release. 41:3–17.

[110] Daniel H (2004) Molecular and integrative physiology of intestinal peptide transport. Annu. rev. physiol. 66:361–384.

[111] Foltz M, Cerstiaens A, van Meensel A, Mols R, van der Pijl PC, Duchateau GSMJE, Augustijns P (2008) The angiotensin converting enzyme inhibitory tripeptides Ile-Pro-Pro and Val-Pro-Pro show increasing permeabilities with increasing physiological relevance of absorption models. Peptides 29: 1312–1320.

[112] Satake M, Enjoh M, Nakamura Y, Takano T, Kawamura Y, Arai S, Shimizu M (2002) Transepithelial transport of the bioactive tripeptide Val-Pro-Pro, in human intestinal Caco-2 cell monolayers. Biosci. biotech. bioch. 66: 378–384.

[113] Quiros A, Davalos A, Lasuncion MA, Ramos M, Recio I (2008) Bioavailability of the antihypertensive peptide LHLPLP: Transepithelial flux of HLPLP Int. dairy j. 18: 279-286.

[114] Deacon CF, Nauck MA, Toft-Nielsen M, Pridal L, Willms B, Holst JJ (1995) Both subcutaneously and intravenously administered glucagon-like peptide I are rapidly degraded from the NH2-terminus in type II diabetic patients and in healthy subjects. Diabetes. 44:1126–1131.

[115] Jauhiainen T, Wuolle K, Vapaatalo H, Kerojoki O, Nurmela K, Lowrie C, Korpela R (2007) Oral absorption, tissue distribution and excretion of a radiolabelled analog of a milk-derived antihypertensive peptide, Ile-Pro-Pro, in rats. Int. dairy j. 17:1216–1223.

[116] van der Pijl PC, Kies AK, Ten Have GA, Duchateau GS, Deutz NE (2008) Pharmacokinetics of proline-rich tripeptides in the pig. Peptides. 29:2196–2202.

[117] Foltz M, Meynen EE, Bianco V, van Platerink C, Koning TMMG, Kloek J (2007) Angiotensin converting enzyme inhibitory peptides from a lactotripeptide-enriched milk beverage are absorbed intact into the circulation. J. nutr. 137:953–958.

[118] Shaji J, Patole V (2008) Protein and peptide drug delivery: Oral approaches. J. pharm. sci. 70:269-277.

[119] Kies AK, Van Der Pijl P (2012) Peptide availability USA Patent Application 20120040895.

[120] Ko WC, Cheng ML; Hsu KC, Hwang YS (2006) Absorption-enhancing treatments for antihypertensive activity of oligopeptides from tuna cooking juice: In vivo evaluation in spontaneously hypertensive rats. J. food sci. 71:13-17.

[121] Fujita H, Sasaki R, Kurahashi K, Yoshikawa M (1995) Potentiation of the antihypertensive activity of orally administered ovokinin, a vasorelaxing peptide derived from ovalbumin, by emulsification in egg phosphatidyl-choline. Biosci. biotech. bioch.59:2344–2345.

Chapter 11

Bifidobacterium in Human GI Tract: Screening, Isolation, Survival and Growth Kinetics in Simulated Gastrointestinal Conditions

Nditange Shigwedha and Li Jia

Additional information is available at the end of the chapter

1. Introduction

Many species of lactic acid bacteria (LAB), *Bacillus,* and fungi such as *Saccharomyces* and *Aspergillus* have been used over the years in the food industry. A few have gained the probiotic status – defined as live microorganisms, which when administered in adequate amounts confer a health benefit on the host (Joint FAO/WHO, 2002) – and most of this belong to *Lactobacillus* (e.g., *L. bulgaricus, L. acidophilus, L. rhamnosus, L. casei, L. johnsonii, L. reuteri,* etc.), *Streptococcus* (e.g., *S. thermophilus,* etc.), and *Bifidobacterium* (e.g., *B. bifidum, B. longum, B. breve, B. infantis*) genera. Bifidobacteria is the predominant species of bacteria in the normal intestinal flora of healthy breast-fed newborns where they constitute more than 95% of the total population (Yildirim & Johnson, 1998). Numerous *Bifidobacterium* strains have gained recognition as probiotics because of their various therapeutic health benefits, including resistance to enteric pathogens (*Clostridium spp., Salmonella spp., Candida spp., Escherichia coli spp. and Listeria monocytogenes*), aid in lactose digestion and/or help to regulate digestion, anti–colon cancer effect, the immune system modulation, anti-allergy, and hepatic encephalopathy (Jia *et al.,* 2010), and also for having a protective effect against acute diarrhoea (Liepke *et al.,* 2002). The food industry recognized the market potential of the numerous strain-specific positive health benefits of the bifidobacteria cultures, namely in beverages. Bifidobacteria can also be administered as capsules or tablets or incorporated into food as dietary adjuncts and into baby foods (Lourens-Hattingh & Viljoen, 2001; Patrignani *et al.,* 2006). In addition, bifidobacteria lower inositol phosphate content during bread making (Palacios *et al.,* 2008).

Several investigators have speculated that the survival of most bifidobacteria is not exceptionally high in most dairy products due to low pH and/or exposure to oxygen.

Nevertheless, problems may arise as a consequence of the difficulties of isolation and cultivation of bifidobacteria. Only a few studies have been published concerning the isolation and characterization of plasmids from bifidobacteria. The human gastrointestinal (GI) tract is the largest tube, running through the body and which include mouth and/or oral cavity, oesophagus, stomach, small intestine and large intestine. (Figure 1).

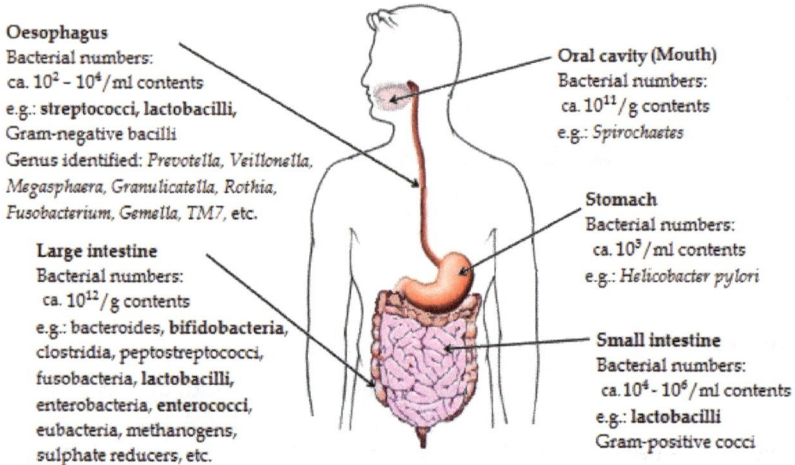

Oesophagus
Bacterial numbers:
ca. 10^2 - 10^4/ml contents
e.g.: streptococci, lactobacilli,
Gram-negative bacilli
Genus identified: *Prevotella, Veillonella, Megasphaera, Granulicatella, Rothia, Fusobacterium, Gemella, TM7*, etc.

Large intestine
Bacterial numbers:
ca. 10^{11}/g contents
e.g.: bacteroides, bifidobacteria,
clostridia, peptostreptococci,
fusobacteria, lactobacilli,
enterobacteria, enterococci,
eubacteria, methanogens,
sulphate reducers, etc.

Oral cavity (Mouth)
Bacterial numbers:
ca. 10^{11}/g contents
e.g.: *Spirochaetes*

Stomach
Bacterial numbers:
ca. 10^3/ml contents
e.g.: *Helicobacter pylori*

Small intestine
Bacterial numbers:
ca. 10^4 - 10^6/ml contents
e.g.: lactobacilli
Gram-positive cocci

Figure 1. The human gastrointestinal tract and its microbiota.

1.1. The oral cavity

Ingested foodstuff first comes into contact with the oral cavity, which is composed of different niches of microbial population. In the oral cavity, bacteria are the main group of microorganisms, although viruses and yeasts can also be found. The main ecological habitants of the mouth are the mucosa of lips, cheeks and palate, the tongue, the tooth surface, the saliva, and the tonsillar area. The population of microorganisms in each section is mainly dependent on the presence of oxygen and nutrients as well as the flow rate of the saliva (see Figure 2). The major species in the oral cavity are lactic acid bacteria of the genera *Streptococcus, Lactobacillus* and *Bifidobacterium*. In dental plaque and oral infections, many anaerobic species have been isolated, mainly *Prevotella* and *Porphyromonas* species, as well as *Eubacterium, Actinomyces* and *Veillonella* (Hartemink, 1999).

The main source of nutrients and energy for oral bacteria is the ingested food, especially carbohydrates, which are rapidly metabolized to lactic and acetic acids by the predominant LAB, leading to a rapid drop in the pH of the saliva after ingestion of carbohydrates. The surplus carbohydrates can be incorporated into exopolysaccharides by a large number of bacteria and be used as energy storage compounds, or as attachment factors (Hartemink, 1999).

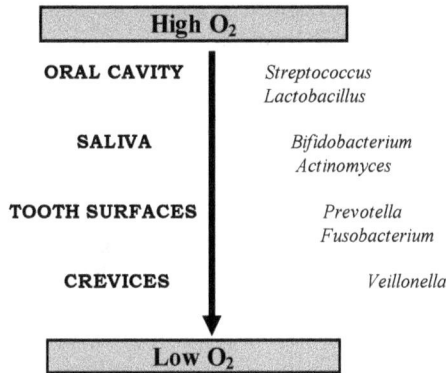

Figure 2. Relationship between bacterial species, oxygen tension and habitat in the oral cavity.

1.2. The oesophagus

In quantitative terms, the oesophagus and stomach carry the lightest microbial loads in the human GI tract. The predominant culturable bacteria are facultative anaerobes, originating in the oral cavity, such as streptococci and lactobacilli, which occur in relatively small numbers (*ca.* $10^2 - 10^3$ cm^{-2} or ml^{-1} of the mucosal surface or lumenal aspirate, respectively) (Macfarlane & Dillon, 2007). The majority of oesophageal bacteria (including the largely α-haemolytic *Streptococcus* species) are cultivable and are almost 10^4 bacteria per mm^2 mucosal surface of the distal oesophagus (Pei *et al.*, 2004). While the bacterial biota in the distal oesophagus is likely to be similar to that of the oropharynx (Kazor *et al.*, 2003), many other species of *Pseudomonas tolaasii, Pseudomonas influorensces, Pseudomonas syringae, Pseudomonas putida,* uncultured *Duganella, Stenotrophomonas maltophilia, Janthinobacterium lividum, Lactobacillus paracasei, Propionibacterium acnes, Pseudomonas Antarctica / meridiana,* and *Brevundimonas bulata* exist in the oesophagus (Pei *et al.*, 2004). Other selected members of the bacterial genera found in human distal oesophagus are given in Figure 1.

1.3. The stomach

In general, the human stomach has a remarkably low pH. The normal resting gastric juice's pH is below 3.0, which prevents virtually all bacterial growth, and which is bactericidal for most transient species, especially the LABs. During and shortly after a meal, the pH may increase to values around 6.0. This will allow passing bifidobacteria to survive the gastric juice prior to proceeding onto the small intestine (to battle the bile salts). The resident flora of the gut lumen is highly acidic tolerant and consists mainly of lactobacilli and streptococci.

In the stomach mucosa, the pH is much higher, and bacterial populations may be higher, as well. In addition to lactobacilli and streptococci, some other bacterial species and yeasts may be present (Hartemink, 1999). The gastric juice plays a significant role in digestion of proteins, by activating digestive enzymes, making ingested proteins unravel so that digestive enzymes can alter protein down to individual amino acids. Fermentation of ingested carbohydrates in the stomach hardly occurs.

1.4. The small intestine

When the partially digested food enters the small intestine, it is mixed with intestinal secretions, such as bile, pancreatic enzymes and bicarbonates. The bile in particular has a strong bactericidal effect. Together with a strong-fluid secretion by the intestinal mucosa, this also prevents extensive colonization of the small intestine. Colonization usually takes place in crypts and blind loops. In this lower part of the small intestine, the movement is slightly reduced, the bile is diluted, the pH becomes more neutral, and the oxygen tension drops rapidly. This favours the growth and/or transit of different bacteria, initially mainly aerotolerant species, and in the ileum also strict anaerobes as revealed in Figure 3 (Hartemink, 1999). There is not much carbohydrate fermentation in the small intestine in healthy humans, due to the flow rate and the little bacterial mass.

In studies undertaken in pigs, it has been reported that the conditions in the small intestine differed widely. The pH is much higher, and the bile secretion is less abundant, which results in an extensive bacterial growth in the small intestine. This also results in substantial fermentation of ingested carbohydrates. The human body is projected to produce between 20 to 30 g of bile salts per day to replace the loss occurring in the excreta (250 to 500 mg), and these are typically stored in the gall bladder (Glickman, 1980).

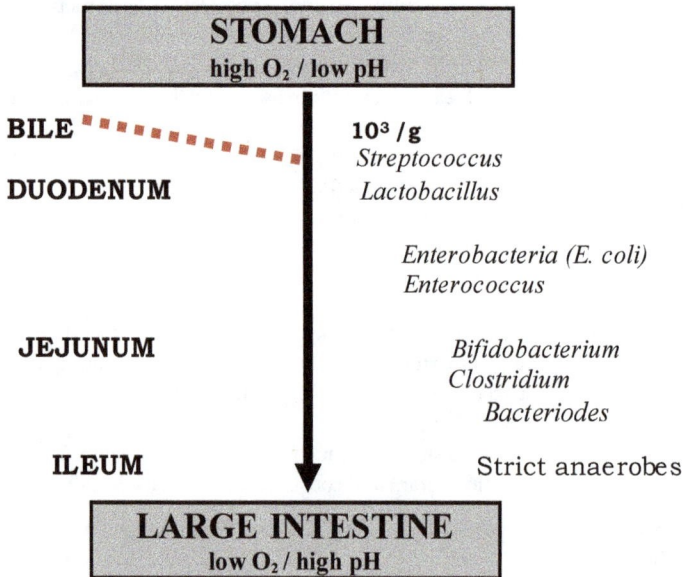

Figure 3. Appearance of bacterial species, oxygen tension and habitat in the small intestine.

1.5. The large intestine

In the large intestine, the flow rate of the digesta decreases considerably. In addition, the bile is even more diluted, and the pH is close to neutral. Total logarithmic counts may reach up to 10^{11} bacteria/gram contents. Higher numbers have been reported, but it is physically impossible to achieve a number over 10^{12} bacteria/gram faecal dry weight, taking into account the average balance of faeces and the dimensions of an average bacterium. It is estimated that over 400 different bacterial species reside in the human large intestine. Of these, about 200 have been validly described, but often non-identifiable strains are reported.

In addition to the resident bacteria, transient bacteria are often isolated. The dominant floras in the large intestine are relatively stable, and they include *Bifidobacterium*, *Bacteriodes* and anaerobic cocci. Large variations also exist in the less dominant species, especially among the facultative or aerotolerant species like *E. coli* and lactobacilli. The numbers of the dominant species are also comparable in different population. Differences in counts are more often due to the methodology used, rather than actual differences. As in individuals, the counts of less dominant species differ widely between different populations. Among the dominant bacterial groups are members of the genera *Bacteriodes*, *Bifidobacterium*, *Coprococcus*, *Peptostreptococcus*, *Eubacterium* and *Ruminococcus*. Members of the following genera are often isolated and are available in lower numbers: *Fusobacterium*, *Streptococcus*, *Lactobacillus*, *Enterococcus*, *Veillonella*, *Megasphaera*, *Propionibacterium* and *Enterobacteriaceae*.

It is indispensable to emphasize here that the principal function of the GI tract includes breakdown and absorption of food components and water. In general, degradation takes place in the upper part of the GI tract, whereas the major sites of absorption are in the lower part of small intestines and the large intestines. Degradation and absorption are enhanced by the excretion of the large number of digestive enzymes, such as glycosidases, lipases, peptidases and proteinases. The colon receives digesta from the intestinal ileum approximately 5 h after food ingestion. Thereafter, rate of motion slows progressively from the caecum towards the distal colon. Concurrent with this is an increase in water absorption; thus gut contents in the proximal colon are more or less liquid in nature but have a faecal like appearance distally (Hartemink, 1999).

For most of the world's population, the standard gut transit time is 60 h, with a variety of 23 – 168 h. The colon itself has a capacity of approximately 500 ml with about 220 g contents. In general, stools weight correlates inversely with transit time. Studies with healthy volunteers have indicated that speeding up colonic transit times from 67 to 25 h resulted in an increase in stool weight from 148 to 285 g/day. Conversely, when transit time increased, stool weight decreases from 182 to 119 g/day (Hartemink, 1999). The differences are mainly due to changes in the water content of the faecal mass.

The structure of faeces is highly variable. Bacteria may constitute up to 55% of the total solids, whilst fibre and other non-digested, non-fermented compounds represent less than 17% of the weight of which about 24% is soluble material. Faecal water content may be as high as 70% of the total weight. Stool size is influenced by both dietary and endogenous factors. Endogenous factors mainly operate through hormones on the intestinal motility. The well-known endogenous factors include: decreased peristaltic movements during exercises and menstrual cycle. Dietary factors, like non-digestible fibres and polyalcohols (sorbitol), may retain water and thus increases stool bulk. High amounts of these factors may cause diarrhoea, due to the increased osmotic pressure.

2. Screening and isolation procedures of bifidobacteria strains

Molecular methods have shown that the average percentage of bifidobacteria in the GI tract of humans is approximately 3% of total microbiota, or they occur at a concentration of $10^9 - 10^{10}$ CFU/g of faeces (Jia *et al.*, 2010). As to achieve intestinal colonization in humans or animals, bifidobacteria have to endure inhibitory substances secreted by the host, such as gastric acid in the stomach and bile salts (in the small intestine). Although, both the gastric pH (pH < 3) and bile salts are strongly bacteriocidal, some resistant bifidobacteria can handle the low pH's ranges of the stomach and also survive the effects of bile salts in the small intestine of humans. These can be isolated and screened for their leading roles as probiotics.

2.1. Isolation and cultivation of bifidobacteria resistant to acidic pH and bile salts

2.1.1. Isolation via stress-shock procedure

Selection of acid and bile resistant bifidobacteria has been based on the stressing isolation method developed by Chung *et al.*, (1999). Faecal samples are collected from infants and/or adults. The tube containing the faecal sample is promptly screened for the isolation of resistant strains, as follows: Faecal samples (0.8 g each) are inoculated into 8 ml of Transga-lactooligosaccharide-propionate (TP) medium as an enrichment medium for the bifidobacteria. After an anaerobic incubation for 12 h at 37 °C, 0.8 ml of the incubated cultures is transferred into fresh TP medium with pH adjusted to 2.0 and incubated anaerobically for another 12 h at 37 °C. After the acid exposure, an aliquot (0.8 ml) of the incubation medium is transferred into fresh TP medium supplemented with 1.5% ox-gall, and the incubation continued for another 2 h at 37 °C. The resulting incubation medium is serially diluted and plated on TP agar, to select colonies of the resistant bifidobacteria strains. To isolate reference strains, serially diluted *Bifidobacterium* cells grown in the regular TP medium are plated on TP agar medium. In most of the isolation studies, *B. adolescentis, B. longum, B. infantis, B. bifidum* and owner identified *Bifidobacterium* strains (commonly called "own isolates" in microbiology) are used as reference strains. The reference strains are utilized for the convenience of comparison to the resistant strains. Microscopic analysis (1000 × with immersion oil) is routinely performed to confirm *Bifidobacterium* morphology.

In addition, *Bifidobacterium* cells are examined for their biochemical and morphological characteristics according to the Bergey's Manual of Determinative Bacteriology. The cultures are grown in Man, Rogosa, and Sharpe (MRS) medium under anaerobic conditions, in a microprocessor-controlled anaerobic chamber. Cultures are incubated for 18 h at 37 °C and stored at 3 – 5 °C between transfers. For the fermentation test, 0.5 ml of 10% substrate solutions (which were membrane filtered through 0.45 µM filter), are added to 9.5 ml of Peptone Yeast-extract Fildes (PYF) basal medium (Mitsuoka, 1990). After 2.5 days of strictly anaerobic incubation, the pH of the growth medium is measured. Tubes showing pH values below 5.5 are considered to be positive for fermentation. The presence of acetate and lactate in the fermented PYF containing glucose medium is assayed by using gas chromatography (GC) or high performance liquid chromatography (HPLC).

2.1.2. Isolation and screening via stress-shock

Briefly, faecal samples of 3 to 5 days old new-born babies are collected and taken to the laboratory for immediate analysis and isolation of bifidobacteria. About 2 g of each faeces sample is placed in a sterile test-tube (30 ml) and closed tightly with a rubber-stopper. For optimal survival of these highly sensitive anaerobic bacteria, the samples are treated within 15 min after faeces emission, or else the samples are kept in an anaerobic environment until analysis (maximum of 10 h). Screening for the isolation of resistant strains is as follows: faecal samples (2 g each) are inoculated into 10 ml test-tubes of Raffinose-Bifidobacterium (RB) broth (pH 6.8). After an anaerobic incubation for 12 h at 38.5 ºC, 1 ml of the incubated culture is transferred into 10 ml of fresh RB medium with pH adjusted to 3.0 and incubated anaerobically for 2 h at 38.5 ºC. After the acid exposure, an aliquot (1 ml) of the incubation medium is transferred into 10 ml of fresh RB medium supplemented with 1% ox-gall, and the incubation continues for another 2 h at 38.5 ºC. The resulting incubation medium is serially diluted (10-folds) in a pre-reduced Ringer solution with 5 – 10% glycerol for the inhibition of the cellulolytic activity of the fungus. An aliquot of 100 µl from each dilution is plated directly on RB and MRS agars using the surface streak method and incubated anaerobically at 38.5 °C for 3 – 4 days to determine colonies of the resistant *Bifidobacterium* strains.

Likewise, the isolate designated *B. longum* GB-03 was isolated from a pharmaceutical product called Golden Bifid (containing a combination of unspecified *Bifidobacterium* spp., *Streptococcus thermophilus* and *Lactobacillus bulgaricus*) using a similar approach. The first step is crucial to reveal that a single piece (0.5 g) has to be dissolved in 0.2 ml test-tube of sterilized distilled water before being inoculated into 10 ml test tube of fresh RB-medium.

3. Morphological identification of bifidobacteria by phase contrast microscopy (PCM)

In the morphological analysis of bifidobacteria population, *in situ*, in human faeces and/or other foods products, microscopes have been used to determine the degree of heterogeneity of these probiotic's populations. The morphology of bifidobacteria determined microscopically has been used as an aid to phenotypic differentiation within the group, while the effect of medium type, low pH and high bile salt concentrations on the bifidobacterial cell morphology has also been studied by this method. Individual

Bifidobacterium strains are characterized phenotypically, including morphology identification by phase contrast microscopy (PCM).

Bifidobacteria are gram-positive, anaerobic, rods of various shapes (short, regular, thin cells with pointed ends, coccoidal regular cells, long cells with slight bends or protuberances) or a variety of branching (pointed, slightly bifurcated, club-shaped or spatulated extremities), single or chains of various arrangements (in star-like aggregates or disposed in "V" or "Y" or else "palisade" arrangements) (Scardovi, 1986).

As a pattern to characterize the heterogeneous population of bifidobacteria associated with human origin and other sources, the PCM examinations and two different media (RB & modified MRS) were used to demonstrate a better phenotypical correlation of the natural isolates to the reference strains on RB, MRS and modified MRS media as shown in Figures 4.1 – 4.12). These media are unique and appear to be still the most predominant in culturing the bifidobacteria strains.

Isolates of bifidobacteria are normally cultured anaerobically on appropriate agars at 38 ºC for 3 – 4 days. For gram-staining, a loopful of the culture is streaked on microscope slides (46 × 25 mm) and the staining technique followed thoroughly. Subsequently, the slide is observed under phase contrast microscopy, preferably at 1000 × magnification by oil immersion and can be photographed as well, using the images advanced software package if available.

3.1. Morphological characterization of *Bifidobacterium* reference strains

The basic morphologies, namely short, regular, thin cells with pointed ends, coccoidal regular cells, and long cells with slight bends or protuberances are discernible among the 2 *Bifidobacterium* reference strains (*B. adolescentis* and *B. infantis*) shown in Figures 4.1, 4.2, 4.3, 4.4 and 4.11) on modified MRS and RB media. From these micrographs alone, it is obvious to validate that individual variations of the average phenotypic morphologies of bifidobacteria are present as described earlier. The PCM also provided a rapid and clear visualization of the basic bifidobacteria cell morphology, while at the same time, allowed only broad comparisons amongst the bifid structures within a mixture of 2 other LABs (*Streptococcus thermophilus* & *Lactobacillus bulgaricus*) (see Figure 4.12).

The typical colonies of bifidobacteria are altogether round and white on RB and modified MRS media. Colonies are usually picked off of a suitable plate and may be kept sub-cultured 2 – 3 times on a freshly prepared agar as to obtain pure culture without contamination. The morphologies of the 2 reference strains and their relationship to each will now be discussed separately. When the strain of *B. adolescentis* is resuscitated and cultured on modified MRS medium (Figure 4.1) or RB medium (Figure 4.2), it may be differentiated clearly from the *B. infantis* (Figures 4.3 & 4.4) on the basis of morphology. The *B. infantis* was also resuscitated and cultured under the similar conditions. As it can be observed from Figure 4.2, *B. adolescentis* on RB displayed long and thick rod-shaped and regular coccoidal cells. The cells of *B. adolescentis* strain on RB were almost paired and assembled, a feature which was highlighted by PCM. The existence of distinct "V"- and/or "Y"-shapes and some long cells with protuberances or slight curvature of this isolates when grown on RB agar, is a powerful diagnostic feature, particularly when distinguishing this specie from closely related *B. minimum* when grown on Trypticase-Phytone-Yeast extract (TPY) agar stabs (Biavati *et al.*, 1982). In addition, curved cells with smooth and rounded ends are the most one dominating in the micrograph. These features were not compatible with descriptions of this particular species' morphology as described by Reuter (1963), but were common to other species of the genus.

The *B. infantis* strain displayed slender, often short rod-shaped and of the typical club-shaped extremities, which cells of these species are reported to exhibit (see Figures 4.3 & 4.4). The morphology of this strain is almost the same when grown on both the MRS and RB solid growth media. Furthermore, *B. infantis* showed a distinct tendency for chain formation on RB medium. These cells often occurred in "V" and "Y"-shapes and were similar to that of many other species of the genus. Nevertheless, it was also possible to differentiate between this strain and the closely related *B. longum* GB-03 (own isolate, Fig. 4.6) on the basis of small variety of club-shaped extreme morphology.

3.2. Morphological differentiation of isolates of bifidobacteria

Morphological consistency is greater among the *Bifidobacterium* isolate (*B. longum* GB-03 and *B. bifidum* WN-04) as shown in Figures 4.5 to 4.11) than the *Bifidobacterium* reference strains. Cell shapes ranged from long and thick–rods with protuberances to long and thin–rods with blunted ends and slightly bifurcated club-shaped extremities, with a number of variations on these basic shapes. Two morphological groups and their potential significance are discussed separately below.

3.2.1. Long and thick–rods with protuberances cell morphology

Figures 4.6 and 4.8 display both isolates of *B. longum* GB-03 and *B. bifidum* WN-04 on RB medium, which consisted of long and thick cells with slight bends. The regular morphology of these cells and the star-like aggregates arrangement (Figure 4.6) was evident under the PCM when grown on RB agar. Also, the presences of sparsely distributed single cells were also evident under the PCM (Figure 4.8). The morphology of these cells was consistent with any of the *Bifidobacterium* reference strains discussed previously. The isolates' morphologies resembled the reference strain of *B. infantis* which are never elongated but have a penchant for group formation (Figure 4.4).

Although no conclusions could be drawn on the basis of morphology alone, the presence of "V"-shaped rods, protuberances with a large variety of bending in *B. bifidum* WN-04 isolate appeared to resemble the reference strains of *B. bifidum*, especially the "amphora-like" cells that are characteristic (Sundman & Bjorksten, 1959). On the RB media, PCM analysis allowed a better correlation of the natural isolates to the reference strains. Speciation of *B. longum* GB-03 (in Figure 4.6) conversely appeared to favour the reference strain of *B. longum*, especially the ultra-elongated and relatively thin cellular elements with slightly irregular contours (Reuter, 1963).

B. *adolescentis* on modified MRS medium   B. *adolescentis* on RB medium

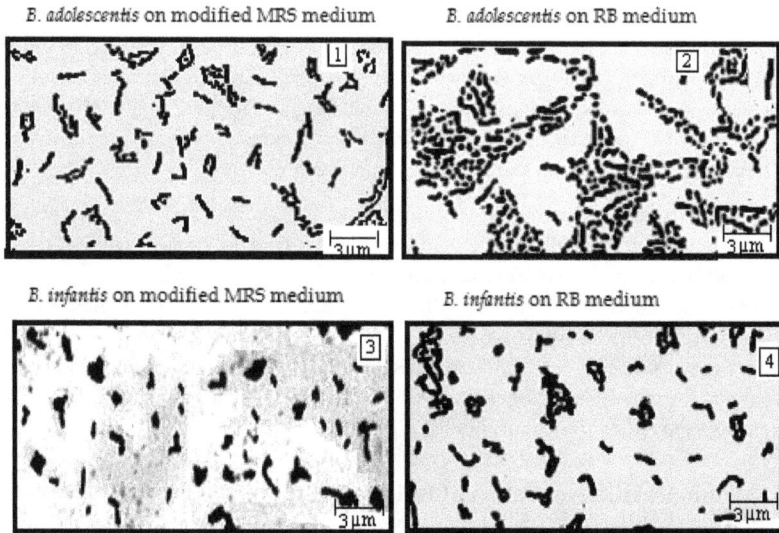

Phase Contrast Micrographs of *Bifidobacterium* reference strains: **Fig. 4.1**, *B. adolescentis* on modified MRS; **Fig. 4.2**, *B. adolescentis* on RB; **Fig. 4.3**, *B. infantis* on modified MRS and **Fig. 4.4**, *B. infantis* on RB, taken at 1000 × magnifications.

B. *longum* GB-03 on modified MRS medium   B. *longum* GB-03 on RB medium

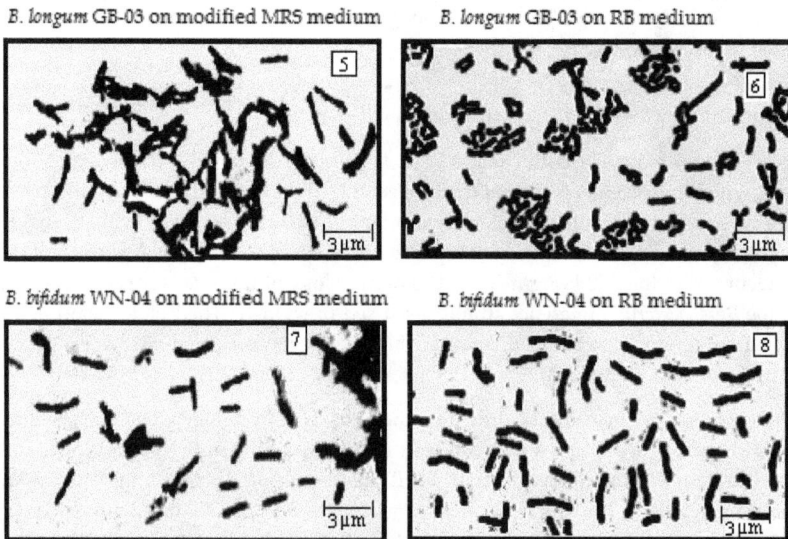

Phase Contrast Micrographs of the isolate strains: **Fig. 4.5**, *B. longum* GB-03 on modified MRS; **Fig. 4.6**, *B. longum* GB-03 on RB; **Fig. 4.7**, *B. bifidum* WN-04 on modified MRS and **Fig. 4.8**, *B. bifidum* WN-04 on RB, taken at 1000 × magnifications.

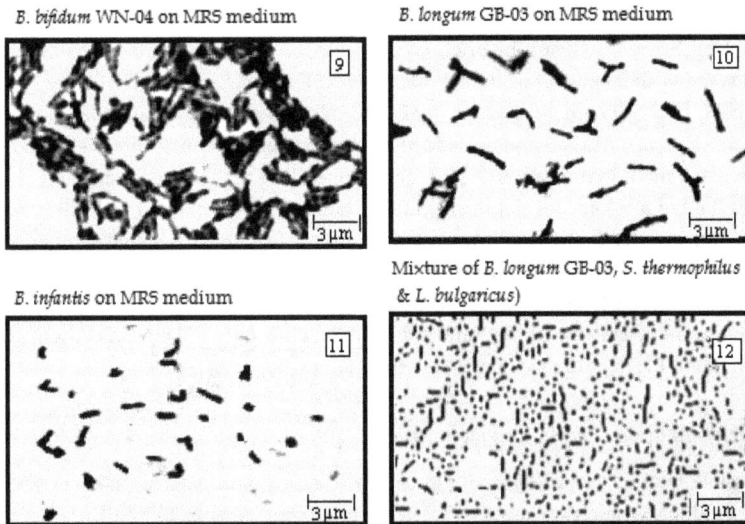

Figure 4. Phase Contrast Micrographs of *Bifidobacterium* strains: **Fig. 4.9**, *B. bifidum* WN-04 on unmodified MRS; **Fig. 4.10**, *B. longum* GB-03 on unmodified MRS; **Fig. 4.11**, *B. infantis* on unmodified MRS and **Fig. 4.12**, *B. longum* GB-03 and an assortment of other 2 Lactic Acid Bacteria (*Streptococcus thermophilus* & *Lactobacillus bulgaricus*), taken at 1000 × magnification.

3.2.2. Long and thin–rods with blunted ends cell morphology

This was the most common type of morphology encountered among the *Bifidobacterium* isolates of *B. longum* GB-03 and *B. bifidum* WN-04 on the unmodified MRS agar (Figures 4.5, 4.7, 4.9 & 4.10). Since only the general cell structure was used to differentiate this species from the other bacteria, PCM proved sufficient for this purpose. Variations of morphology within these small groups were visible under PCM as indicated by the following examples. The isolate of *B. longum* GB-03 in Figure 4.12 exemplified the diversity of rods and coccus cells morphology including bifid structures also; with the absence of any coccus build cells when grown on RB agar in Figure 4.6. By comparison with the reference strains, the cells morphology of *B. longum* GB-03 isolate is more peculiar to that displayed by *B. infantis* (Figure 4.4) and the isolate of *B. bifidum* WN-04 (Figures 4.7 & 4.8). All the *Bifidobacterium* isolates displayed long and short club-shaped rods, most of which were long and thin with blunted ends and of conventional "V" and/or "Y"-shaped cells.

3.3. Confirmation of identity of *Bifidobacterium* strains

3.3.1. Fructose-6-Phosphate Phosphoketolase (F6PPK) verification test

F6PPK is certainly a key enzyme in the "bifidus pathway" and it allows the discrimination of the specific feature on expression of fructose-6 phosphate in cellular extracts that assigned the bifidobacteria to the genus level (Sgorbati, 1979).

The procedure to test for the F6PPK activity in the *Bifidobacterium* strains is still practised as described by Scardovi (1986). In brief, cells harvested from 10 ml RB or MRS broth are washed twice with 50 mM phosphate buffer (pH 6.5). The cells are disrupted by sonication in the cold, and 0.25 ml of each of NaF and Na iodoacetate solution and fructose-6-phosphate (Na Salt: 70% purity) are added to the sonicate. The reaction is stopped by the addition of 1.5 ml of hydroxylamine HCl, and 1 ml each of trichloroacetic acid and 4 M HCl. Finally, 1.0 ml of a colour-developing agent (FeCl₃.6H₂O 5% (w/v) in 0.1 M HCl) is added. A tube without fructose-6-phosphate serves as a blank, to facilitate the visual comparison. The formation of acetyl phosphate from fructose-6-phosphate, shown by the reddish violet colour formed by the ferric chelate of its hydroxamate is an indicator for F6PPK. This is the distinctive and key enzyme of the "bifid shunt" that characterizes the genus. There are three subtypes of F6PPK in bifidobacteria as shown in Figure 5.

3.3.2. Determination of acetic and lactic acids

One possible method of validating the presence of acetic and lactic acids in the fermented milk by bifidobacteria can be assayed by using High Performance Liquid Chromatography (HPLC). Samples for this analysis are prepared by using a modified method described by Dubey & Mistry, (1996).

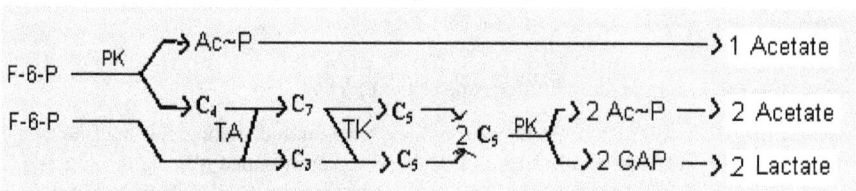

Figure 5. Fermentation of hexose for carbohydrate metabolism (the "bifid shunt"), based on Schlegel (1993), where PK, phosphoketolase; TA, transaldolase; TK, transketolase; Ac~P, acetyl phosphate; GAP, glyceraldehydes-3-phosphate.

The strains were maintained anaerobically by propagation in MRS broth (peptone: 10 g/l; meat extract: 8 g/l; yeast extract 5 g/l; D(+)glucose: 20 g/l; di-potassium hydrogen phosphate: 2 g/l; di-ammonium hydrogen citrate: 2 g/l; Tween-80: 1 ml/l; sodium acetate: 5 g/l; magnesium sulfate: 0.2 g/l; manganese sulfate: 0.04 g/l, supplemented with 0.05% (w/v) cysteine-hydrochloride).

The production of acetic and lactic acids, spore formation, aerobic and anaerobic growth, gram reactions, motility, gas production from lactose and carbohydrates fermentation tests are some of the confirmation tests that proves highly diagnostic personality characteristics of different *Bifidobacterium* spp as summarized in Table 1. Furthermore, the taxonomy of bifidobacteria has changed ever since they were first isolated. They had been assigned to the genera *Bacillus*, *Bacteroides*, *Nocardia*, *Lactobacillus* and *Corynebacterium* among others, before being recognized as a separate genus in 1974.

Many of the *Bifidobacterium* species groupings are heterogeneous and the entire genera have been re-examined using DNA-DNA hybridization. A point is made here that, instant phenotypic characterization of most bacteria within their respective genera relies on biochemical tests such as the proportion of acetic and lactic acid relative to the end product of metabolism; the ratio of acetic and lactic acid produced; some key carbohydrate fermentations; colonies and phenotypic morphologies; and the presence of fructose-6-phosphate phosphoketolase (F6PPK), a key enzyme in the bifidus pathway.

Characteristics	*Bifidobacterium* Strains			
	B. bifidum[1]	*B. longum*[1]	*B. infantis*[1]	*B. adolescentis*[2]
Spore forming	−	−	−	−
Motility	−	−	−	−
Gram reaction	+	+	+	+
Morphology: rods, pleiomorphic	+	+	+	+
Anaerobic growth	+	+	+	+
Aerobic growth	−	−	−	−
Gas from lactose	−	−	−	−
Catalase	−	−	−	−
F6PPK	+	+	+	+
Acetic and lactic production (ratio 3:2)	+	+	+	+
Carbohydrates Fermentation Test				
Cellobiose	−	+	−	+
Fructose	+	+	+	+
Fructooligosacharides	−	+	+	+
Galactose	+	+	+	+
Glucose	+	+	+	+
Isomaltose	−	+	+	+
Lactose	+	+	+	+
Maltitol[3]	−	−	−	−
Mannose	−	−	−	−
Melezitose	−	−	−	−
Raffinose	+	+	+	+
Stachyose	+	+	+	+
Trehalose	−	−	−	−
Xylose	−	+	−	+

Legends on Table 1: [1] Obtained from American Type Culture Collection, Rockville, USA. [2] Obtained from China General Microorganisms Culture Collection Center, Beijing, China. [3] Maltitol is still widely used as a non-cariogenic sweetener and sugar substitute but is as yet not used as a possible prebiotic. + positive results or fermentation; − negative results or no fermentation observed. F6PPK (fructose-6-phosphate phosphoketolase).

Table 1. Phenotypic characteristics of some of the pH- and bile salts-resistant bifidobacteria tested.

4. Common media used in isolation and detection of bifidobacteria

Many different media for bifidobacteria are outlined in Table 2.

Medium	Selectivity based on*	Used for
Acetylglucosamine-Lactose (AL) agar	lactose, acetylglucosamine	faeces
AMC-agar	nal, polymyxin B, kan, iac, TTC, LiCl, prop	B. longum
Bifidobacterium selective (BS) agar	LiCl, neo, paro, prop	faeces
Bifidobacterium selective medium (BBM-agar)	nal, rifampicine, raffinose	faeces
Bifidus Blood Agar	aniline blue, blood	faeces
Bif-medium	human whey, nal, paro, aztreonam, netilmycin	dairy products
Bifidobacterium Iodoacetate Medium (BIM-25 agar)	kan, nal, iac, neo, polymyxin B	sewage
BS-agar	LiCl, neo, paro, prop	faeces
China Blue (CB) agar	specific impact of china blue	faeces
GL-agar	galactose, LiCl	dairy products
Liver Cysteine Lactose (LCL) agar	lactose, liver infusion	faeces
LP agar	lactose, LiCl, prop	dairy products
Modified Rogosa agar	neo, paro, prop, LiCl	dairy products
MPN-agar	lactose, nal	faeces
MRS-LP-agar	prop, LiCl	dairy products
Neomycin Paromomycin Lithium Nalidixic acid (NPLN) agar	LiCl, nal, neo, paro, prop	faeces, dairy products
Propionate or Beerens agar	propionic acid, pH 5.0	faeces
Raffinose-Bifidobacterium (RB) Agar	raffinose, LiCl, propionate	faeces, dairy products
RCM (modified)	low pH	dairy products
RCM + stain	Loeffler's methylene blue stain	dairy products
Rogosa agar	low pH	faeces, dairy products
Rogosa (modified)	neo, paro, prop, LiCl	dairy products
Rogosa-N	low pH, nal	faeces
Tomato Casein Peptone Yeast Agar (TCPY)	tomato juice	faeces
Transgalactosyloligosaccharide (TOS-Agar)	TOS	faeces, dairy products
TOS-Agar (modified)	TOS, nal, neo, paro	dairy products
TPYd-agar	dicloxacillin	dairy products
TTC-agar	TTC	faecal contamination
VF-agar (modified)	LiCl, prop, neo, sodium lauryl sulfate	dairy products
YN-6 agar	lactose, nal, neo, bromocresol green	faeces, sewage

Legends on Table 2: *iac = iodoacetic acid, kan = kanamycin, LiCl = lithiumchloride, nal = nalidixic acid, neo = neomycin, paro = paromomycin, prop = propionate, TOS = transgalactosyl oligosaccharides, TTC = 2,3,5-triphenyl-tetrazoliumchloride

Table 2. Popular media used for the enumeration of bifidobacteria from faeces, dairy- and pharmaceutical products, (Adapted from prebiotic effect on non-digestible oligo- and polysaccharides by Hartemink, 1999).

Media used for the detection of bifidobacteria can be classified in 5 different groups. These are non-selective medium (such as MRS and Rogosa), medium without antibiotics but with elective carbohydrate, medium with antibiotics, medium with propionate, and medium with elective substance and/or low pH (Table 3).

Medium	Group
Acetylglucosamine - Lactose (AL) agar	2
Bifidobacterium selective (BS) agar	3, 5
Bifidobacterium selective medium (BBM) agar	2, 3
Bifidus Blood agar	5
Bifidobacterium Iodoacetate Medium (BIM-25) agar	3
China Blue agar	5
Liver Cysteine Lactose (LCL) agar	2
Rogosa agar	1
Modified Rogosa agar	3, 5
MPN-agar	2, 3
MRS	1
MRS agar with LiCl and antibiotics (MRS-NN)	3, 5
Neomycin Paromomycin Lithium Nalidixic acid (NPLN) agar	3, 4
Propionate agar or Beerens agar	4
Raffinose-Bifidobacterium (RB) agar	2, 4
Reinforced Clostridial agar with Cephalothin and blood (RCB)	3
Tomato Casein Peptone Yeast agar (TCPY)	5
Tomato Casein Peptone Yeast agar (TPCY) with azide	5
Tomato Casein Peptone Yeast agar (TPCY with sorbic acid	5
Tomato Casein Peptone Yeast agar (TPCY with antibiotics	3
Transgalactosyloligosaccharide (TOS - agar)	2
TTC-agar	5
x-Gal medium	5
YN-6 agar	2, 3, 5

Legends on Table 3: group: 1 = non selective medium, 2 = medium without antibiotics but with elective carbohydrate, 3 = medium with antibiotics, 4 = medium with propionate, 5 = medium with elective substance and/or low pH

Table 3. Media used for the detection of bifidobacteria from faeces (Source: Hartemink, 1999).

Combinations and media belonging to more than one group are also used. From the large number of media used, it can be concluded that there is no standard medium for the detection of bifidobacteria. *Bifidobacterium* spp. in the GI tract of humans are normally present in an adequate amounts and estimated to be between 10^9 and 10^{10} colony forming units (CFU) per gram wet weight or around 3% of total microbiota (Jia *et al.*, 2010). However, the selectivity of independent media for the quantification of bifidobacteria is thoroughly examined and tested with different baby faeces.

The experimental results of 3 media (PROP, RB and NPLN) tested on bifidobacteria show a wide variation in counts for the different samples (see Figure 6). Absolute counts are highest for the faecal samples on NPLN, followed by RB in 8 of 9 samples. PROP showed the lowest counts. However, as it can be observed from the same Figure 6, the principal difference between these 3 media is exceedingly little, actually less than one log unit.

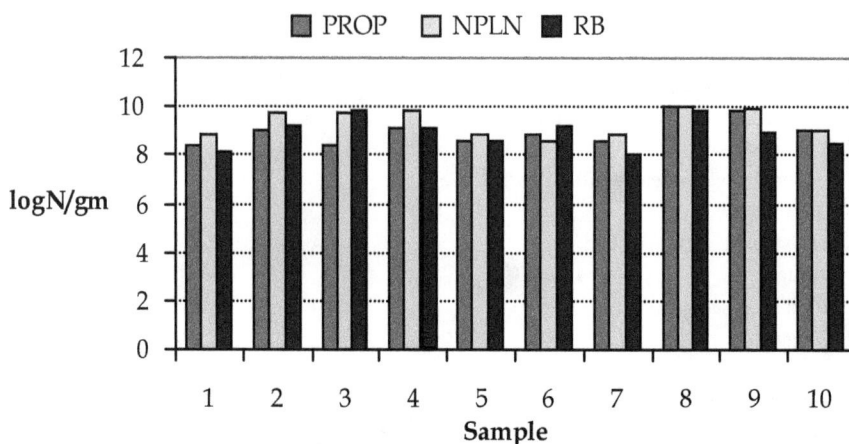

Figure 6. Counts (log CFU/gm wet weight) on PROP, NPLN and RB media in babies' faeces.

Selectivity is also determined by microscopic observations of all different colony morphologies on all countable (between 10 and 150 colonies/plate) plates (see Table 4). Based on morphologies, selectivity is highest for babies' faeces with NPLN with 29% false positive colonies (growth, but no bifidobacterial morphology). PROP showed 39% false positive and RB with 50% false positives. False negatives (non-typical colonies, but bifid morphology) can be determined on RB, as this is the only medium for which typical colonies are described. However, no false negatives were observed in this work.

Medium	Babies' faeces		
	morphology		
	n	typical	non-typ. [b]
RB pos [c]	24	12	12 (50)
RB neg	4	0	4 (0)
PROP	18	11	7 (39)
NPLN	28	20	8 (29)

Legends on Table 4: [b] number in brackets is the percentage of false positive (typical colony, non-typical morphology) or false negatives (non-typical colony, typical morphology) of the colonies tested. [c] pos = colonies showing characteristics for bifidobacteria, neg = colonies not showing characteristics for bifidobacteria. Bifidobacteria characteristics were defined as yellow-green colonies with a yellow halo. This attribute could only be determined on RB, as no characteristics were defined for other media.

Table 4. Selectivity of media for bifidobacteria.

Most false positive colonies are reported to be different cocci (mono-, diplo- or streptococci), spore-forming rods and short rods. No yeast is observed on any of the media tested. Based on the actual counts, selectivity can only be determined for RB, as the colonies of bifidobacteria and non-bifidobacteria cannot be determined for the other media and not all colonies are tested for their morphology. Selectivity as percentages of non-typical colonies ranges from around 5 – 7%.

Colonies of different shapes can be tested microscopically. Bacterial morphology is determined, and typical and non-typical morphology is also determined. Typical morphology of bifidobacteria is branched or bifid-shaped rods. For the determination of bifidobacteria, none of the 3 media tested was decidedly selective. In this study, the occurrence of false positive or false negative colonies was determined. The lowest incidence of potential false positive colonies was observed on NPLN, but in all 3 media, the number of non-bifidobacteria capable of growing on the selective media was remarkably high. When many different species are capable of growing on the medium, an increase of one of these species may result in serious mistakes in calculating bifidobacteria. NPLN and RB gave slight higher counts than PROP. The incidence of false positive, based on morphologies on RB was comparable with that on the PROP and slightly higher than that on NPLN. The incidence of competitive flora was relatively low (less than 10% of the total colonies on the plates), as bifidobacteria are one of the main groups of intestinal bacteria in humans.

PROP medium has been described as the best medium for the determination of bifidobacteria by Silvi et al., (1996), but they also concluded that the total bifidobacterial counts were significantly lower on PROP than on the other media tested. Similarly, Favier et al., (1997) concluded that PROP underestimated bifidobacteria in some of their samples. Both studies used human faeces as the test substrate. Several other studies, in which PROP agar is used, also show significantly lower bifidobacterial counts than most other studies (Favier et al., 1997).

NPLN, which has been described as the medium of choice to choose bifidobacteria in dairy products, showed many cocci. This was in accordance with results observed by Silvi et al., (1996). In the same study, BIM-25 was tested, and this medium was found to be non-specific. All these 3 media performed reasonably well for human faeces and bifidobacteria can reliably be counted. The typical colonies morphological trait and the basic cellular-morphology of bifidobacteria were demonstrated well by RB media, with reference to NPNL and PROP medium. The RB medium presented strains with double thickness diameter and more bifurcated cellular morphology under phase contrast microscopy.

5. Experimental procedures for the enumeration of bifidobacteria and determining microbial inactivation by low acidic pH or bile salts

LAB or bifidobacteria strains can be selected or isolated from commercial or alleged "own isolates" strains, from freeze-dried cultures which are resuscitated to stationary phase in MRS broth at a ratio of 2% of the volume of the fresh broth. Decimal dilutions are put onto Raffinose–Bifidobacterium (RB) agar plates whose pH had to be adjusted to 6.8 – 7.0 with 2

N NaOH. The agar plates are then incubated anaerobically at 38.5 ºC for 3 – 4 d and number of colony forming units (CFU)/mL are determined. Two hundred microliters of each strain containing about 10^8 CFU/mL is aseptically transferred into test tubes containing 9 mL of diluted MRS medium with pH adjustments of 3.0, 3.5, 4.0, or 4.5, using 2 N HCl. These suspensions are incubated anaerobically at 38.5 ºC and numbers of survivors are determined after various times as shown in Figure 7 (A). Cells were harvested by centrifugation at 5 ºC, were washed with phosphate-buffered saline (PBS) and were re-suspended in diluted MRS medium without pH adjustment. After thorough mixing on a vortex mixer, the concentration of surviving cells is determined by anaerobic pour plate counts, using 2 plates of RB agar per dilution, and incubated at 38.5 ºC for 3 – 4 days.

Similarly, treatments for the bile salts are carried out at the final concentrations of 0.15%, 0.30%, 0.45%, and 0.60% ox-gall in diluted MRS medium (pH 6.8), exposed to appropriate times as to low pH and incubated anaerobically at 38.5 ºC (see Figure 7 (B). The cells are harvested by centrifugation, washed with PBS, re-suspended in diluted MRS medium without pH adjustment, and mixed using a vortex mixer as described for acidic pH conditions before. Numbers of CFU of bifidobacteria surviving the lytic effect of bile salts are also determined by anaerobic pour plate counts on RB agar after anaerobic incubation for 3 – 4 d at 38.5 ºC.

5.1. Characterizations for $D_{(acid)}$-, $D_{(bile)}$-, $z_{(acid)}$, and $z_{(bile)}$-values

$D_{(acid)}$-value is defined as the time (in min) required at a specified acidic pH to reduce the number of cells by 90%, while $D_{(bile)}$-value is defined as the time (in min) required at a specified concentration of bile salts to reduce the number of cells by 90%. In fact, the $D_{(bile)}$-value of any LAB or bifidobacterial strain is directly proportional to the bile salt concentrations, while the $D_{(acid)}$-value is inversely proportional to the acidic pH.

The $z_{(acid)}$ and/or $z_{(bile)}$-values, on the other hand, is defined as a decrease in pH value (pH < 4.5) or an increase in bile salt concentration (% ox-bile) required to reduce the D-values by 1 log cycle, however, respectively. The $D_{(acid)}$- and $D_{(bile)}$-values can be directly calculated from the absolute values of the reciprocal of the slopes of the linear-regression equations, using a Microsoft Office–Excel software. It is essential to emphasize that, the regression lines must be applied to all the treatments, for which restriction of the R-squared (R^2) value is pragmatic above 0.8920. Moreover, the $D_{(acid)}$- and/or $D_{(bile)}$-values can also be calculated algebraically from the regression equation derived using the method of least-squares to be able to produce the $z_{(acid)}$- and/or $z_{(bile)}$-values for the probiotic strains.

In order to determine the $z_{(acid)}$- and/or $z_{(bile)}$-values, the formula is exactly the same as that for heat resistance, replacing T (temperature) with pH values or bile salts (BS) concentrations as described by Equations (1) and (2), respectively. In both of these cases, the effect of acidic conditions and bile salts is determined from the reduction in concentration of colony-forming units. One has to pay attention that the dynamic $z_{(acid)}$- and/or $z_{(bile)}$-values are calculated for a period of exponential destruction of microbial cells (following the logarithmic order of death), using both Equations (1) and (2).

A)   **Stomach pH Treatments: 1st Simulated Stressing Method**

Bifidobacterium spp.
Suspended in Simulated Gastric Juice (SGJ) @ ambient temp. (38.5°C).

Treatment - pH 3.0	**Treatment - pH 3.5**	**Treatment - pH 4.0**	**Treatment - pH 4.5**
Incubate anaerobically. Hold for 50, 100, 150, 200, 250, 300, 350 min, respectively.	Incubate anaerobically. Hold for 150, 300, 450, 600, 750, 900, 1050 min, respectively.	Incubate anaerobically. Hold for 250, 500, 750, 1000, 1250, 1500, 1750 min, respectively.	Incubate anaerobically. Hold for 350, 700, 1050, 1400, 1750, 2100, 2450 min, respectively.

Centrifuge @ 3000 rmp for 15 min at 5°C.
Wash with Phosphate-Buffered Saline.
Re-suspend in SGJ @ pH 6.8.
Vortex well.
Pour and/or spread plate count on RB agar @ 38.5°C / 72 hours.

B)   **Bile Salt Treatments: 2nd Simulated Stressing Method**

Bifidobacterium spp.
Suspended in SGJ @ ambient (38.5°C).

Treatment - 0.15% oxgall	**Treatment - 0.30% oxgall**	**Treatment - 0.45% oxgall**	**Treatment - 0.60% oxgall**
Incubate anaerobically. Hold for 350, 700, 1050, 1400, 1750, 2100, 2450 min, respectively.	Incubate anaerobically. Hold for 250, 500, 750, 1000, 1250, 1500, 1750 min, respectively.	Incubate anaerobically. Hold for 150, 300, 450, 600, 750, 900, 1050 min, respectively.	Incubate anaerobically. Hold for 50, 100, 150, 200, 250, 300, 350 min, respectively.

Centrifuge @ 3000 rmp for 15 min at 5°C.
Wash with Phosphate-Buffered Saline.
Re-suspend in SGJ @ pH 6.8.
Vortex well.
Pour and/or spread plate count on RB agar @ 38.5°C / 72 hours.

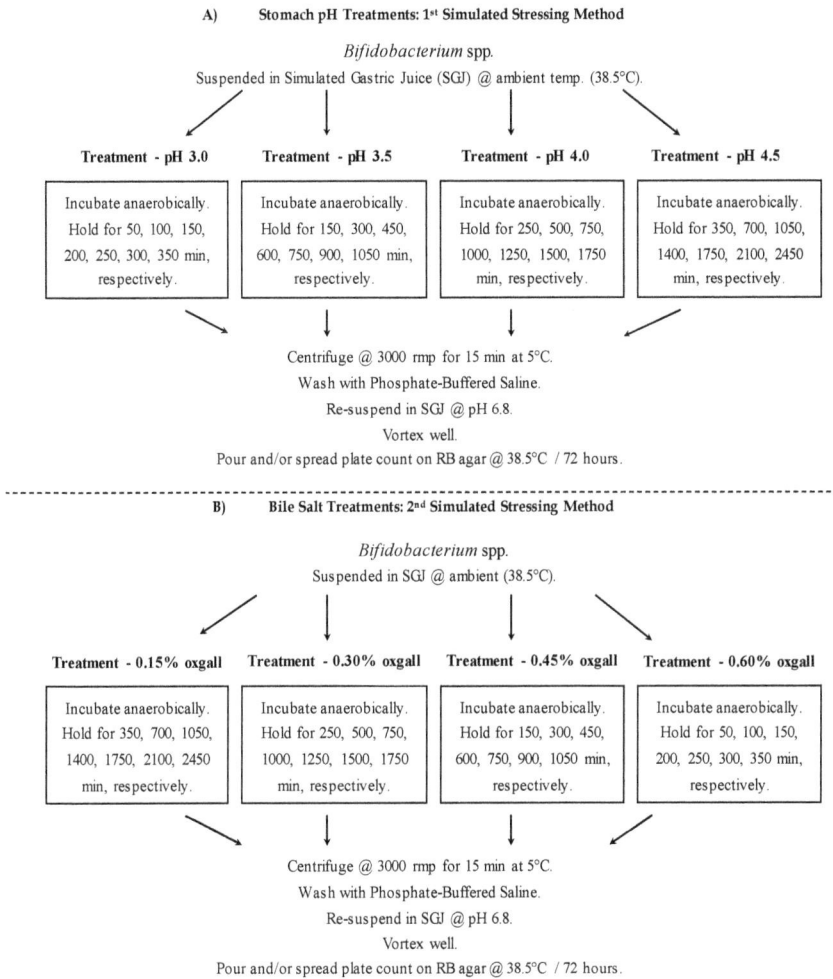

Figure 7. Schematic diagram of treatment groups for the selected bifidobacteria. A): Influence of gastric acidity and its residence time. B): Influence of bile salt(s) and its residence time.

$$z_{(acid)} = \frac{-(acid\ pH_2 - acid\ pH_1)}{Log_{10}\left(\dfrac{D_{(acid\ 1)}}{D_{(acid\ 2)}}\right)} \tag{1}$$

$$z_{(bile)} = \frac{BS_2 - BS_1}{Log_{10}\left(\dfrac{D_{(bile\ 1)}}{D_{(bile\ 2)}}\right)} \tag{2}$$

where z is the acid pH value (in Equation 1) or bile salts value (in Equation 2) required for a ten-fold reduction in D-values; pH_1 is the acidic value of pH 1; pH_2 is the acidic value of pH 2; BS_1 is the concentration of bile salts 1 (%); BS_2 is the concentration of bile salts 2 (%); $D_{(acid1)}$ or $D_{(bile1)}$ is a D-value obtained at either pH_1 or BS_1; and $D_{(acid2)}$ or $D_{(bile2)}$ is a D-value obtained at either pH_2 or BS_2.

5.2. Survival of bifidobacteria in simulated acidic pH of human stomach

Figure 8 shows the survival of selected *Bifidobacterium* strains exposed to various acidic pH levels. The bifidobacterial counts (range: 5.5 – 6.7 log CFU/ml) of all four strains at pH 3.5 after 5 h are an indication of resistance that may perhaps simulate gastric conditions. In fact, with *B. bifidum*, *B. infantis*, and *B. longum* the counts were > 2.0 log CFU/ml after exposure for 5 h, which indicates that these strains are relatively resistant at pH 3.0. However, as observed from all the experimental results in Figure 8, the *B. adolescentis* strain is more sensitive than the other three strains to all the acid treatments. For instance, numbers of *B. adolescentis* are 3.4 log CFU/ml after 10 h at pH 3.5, but below the level of exposure thereafter, while a similar count (~3.4 log CFU/ml) is observed with the other three strains after 12.5 h. This 2.5 h difference in survival at pH 3.5 is approximately the time chosen by Olejnik *et al.*, (2005) to control acid resistance, as these times simulate residence time in the stomach. On this basis, this specific strain of *B. adolescentis* is considered a less-resistant strain with respect to gastric acidity.

Many other researchers have found pH 2.0 and pH 3.0 to be lethal and sublethal pH values respectively for lactic acid bacteria (LAB), including bifidobacteria (Khalil *et al.*, 2007). It is vital to stress that probiotics are able to confer health benefits despite the brief exposure to exact acidic conditions following ingestion. Although a log-scale reduction of viability may occur, it may still mean that a sufficient number of bifidobacteria survive the gut, depending on the dose. Moreover, the exposure to acid does not mean that the potential health benefits are lost. Some cells may die, and some may be injured. However, these cells may recover later, and they may also have beneficial effects on health. The mechanism may be mediated, for example, through the components of the cell wall of the probiotics which will then be available in both dead and living cells. It should be also borne in mind that, the food matrix in which the probiotics are consumed is likely to have a strong effect on the survival of the bacteria in the gut.

In Figure 8, for example, it is possible to say that the high survival counts of *B. bifidum*, *B. infantis*, and *B. longum* exposed to pH 4.5 after 41 h is a representation of culture stability curves of the acid resistant strains (log CFU/ml) in a food matrix. While the pH of 4.5 does not represent gastric acid conditions, it is a typical representation of fermented products, and in such products, the survival counts are expected to last for much longer periods during shelf life, especially at the refrigeration temperature of 4 °C.

5.3. Survival of bifidobacteria in simulated bile salts nature

Resistance to bile salts is considered an intrinsic property for probiotic strains to survive the conditions in the small intestine. The physiological bile salt concentration in the GI tract of humans is estimated to be 0.3 – 0.4% w/v (Jia *et al.*, 2010). As shown in Figure 9, the linear regressions of the loss of CFU did express satisfactorily that *B. bifidum*, *B. infantis*, and *B. longum* are more-resistant strains to the bile salts. These 3 strains survived well in 0.45% (w/v) bile salts, with more than 4.5 log CFU/ml present after 10 h. Their capacity to survive high bile salt concentrations suggests the existence of defence mechanisms and confirms that strains showing antagonistic effect against enteric pathogens should be able to compete successfully with the pathogens in the GI tract. It is clear that *B. adolescentis* is again the less-resistant strain encountered with only 2.8 log CFU/ml surviving after 10 h in 0.45% bile salts (see Figure 9). Therefore, *B. adolescentis* is considered the less-resistant strain, while *B. bifidum*, *B. infantis*, and *B. longum*, in that order, are considered the more-resistant. It is well known that, the bile salt hydrolytic (BSH) activity may be the contributing factor towards the resistance of the LABs and to the toxicity of conjugated bile salts in the duodenum, and therefore, is an essential colonization factor.

5.4. The feasibility of $D_{(acid)}$-, $D_{(bile)}$-, $z_{(acid)}$- and $z_{(bile)}$-values for selection of probiotic strains and for determining the mechanisms of resistance to acid and bile salts stress

Table 5, shows that accurate tabulation of the $D_{(acid)}$- or $D_{(bile)}$-values and their respective $z_{(acid)}$- or $z_{(bile)}$-values is tremendously helpful in evaluating the resistance and susceptibility of probiotics to acidic pH and high bile salt concentrations, respectively. Both the estimated $D_{(acid)}$- and $D_{(bile)}$-values validated that the most acid- and bile-resistant strain is *B. bifidum* followed by *B. infantis*, *B. longum*, and final *B. adolescentis*. It is also possible to observe in Table 5, that, increasing the bile salt concentration from 0.15 to 0.60% had a greater impact on survival than decreasing the pH values from 4.5 to 3.0, with the $D_{(bile)}$-values of *B. bifidum* decreasing from 17.40 to 1.40 min and the $D_{(acid)}$-values decreasing from 23.80 to 1.10 min. Similar trends are observed with all other *Bifidobacterium* strains. However, decreases of depicted $z_{(acid)}$-value in the pH value (pH<4.5) or increases of depicted $z_{(bile)}$-value in the bile salt concentrations (% ox-bile) are expected to cause a 1-log reduction in their respective D-values. In practice, $z_{(acid)}$- or $z_{(bile)}$-value measures how the sensitivity of probiotic strains is to small changes in [H⁺] and/or [OH⁻] or bile salts. As for probiotics to gain intestinal colonization in humans or animals for their proclaimed therapeutic health benefits, obviously, they have to tolerate inhibitory substances secreted by the host, such as gastric acids (in the stomach) and bile salts (in the small intestine).

Of all ions, H⁺ and OH⁻ are the most mobile, and minor changes in their concentrations show significant effects on microorganisms. Most organisms survive better when these ions are present in approximately equal concentrations, that is, pH 7.0. Although many bacteria tolerate higher pH values, only a few are acid tolerant or acidophilic. In addition, many other bacteria are tolerant of small pH variations, especially in the pH range of 6.0 to 9.0. For instance, if the pH of the medium changes rapidly, there may be a transient change in the intracellular pH, and this is usually readjusted to the original pH within 30 min. Consequently, any damage produced by adverse pH is not actually due to the H⁺ and/or

OH⁻, but to the effect of these ions on the proportion of undissociated weak acids or bases, which penetrate more readily into the bacterial cell than the ionized forms. In contrast, bile salts are biological detergents synthesized in the liver from cholesterol, conjugated to either glycine or taurine, and are then secreted into the intestine where they facilitate fat absorption. Bile salts are well known to be toxic for many cells as they disrupt the lipid bilayer structure of the cellular membranes. Many earlier studies revealed that the autochthonous gastrointestinal microbiota must develop strategies to protect themselves against bile salts.

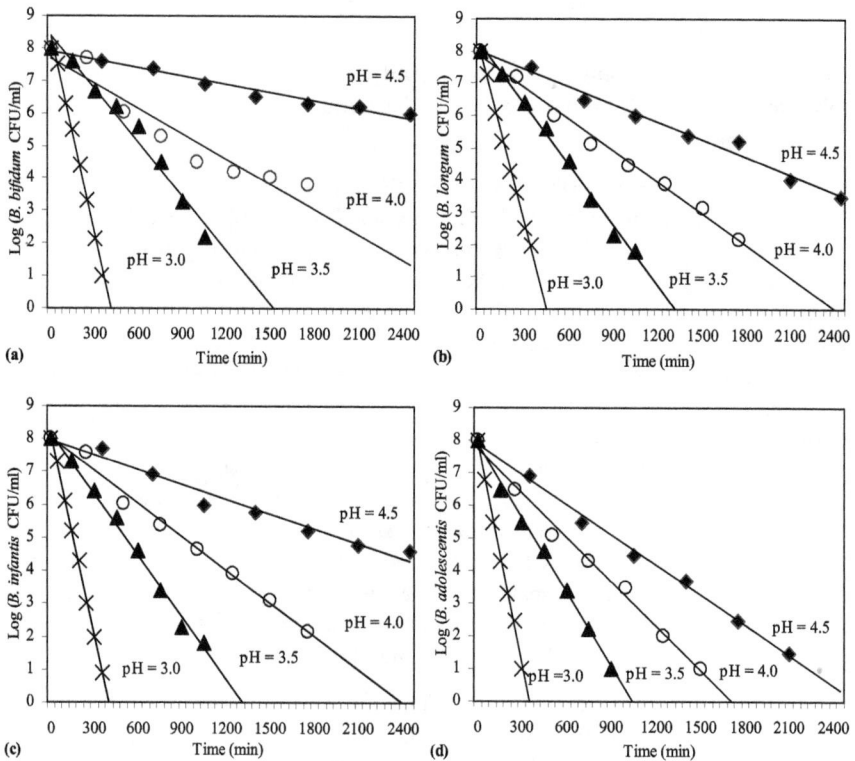

Figure 8. Linear regressions of the loss of CFU for the selected bifidobacteria strains when exposed to simulated gastric acidity of pH 3.0, pH 3.5, pH 4.0 and pH 4.5, respectively: (a) *B. bifidum*, (b) *B. longum*, (c) *B. infantis*, (d) *B. adolescentis*.

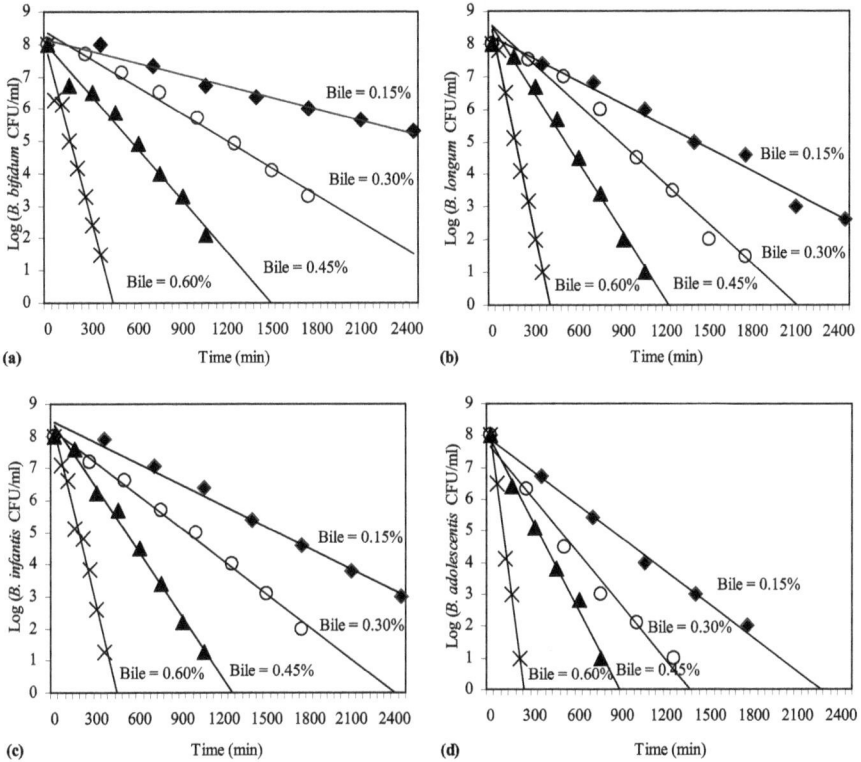

Figure 9. Linear regressions of the loss of CFU for the selected bifidobacteria strains when exposed to high bile salt (oxgall) concentrations of 0.60%, 0.45%, 0.30% and 0.15%, respectively: (a) *B. bifidum*, (b) *B. longum*, (c) *B. infantis*, (d) *B. adolescentis*.

The individual $z_{(acid)}$- or $z_{(bile)}$-values calculated from their $D_{(acid)}$- and $D_{(bile)}$-values ranged from 1.11 – 1.55 pH units and 0.40 – 0.49%, respectively (Table 5). Although the combination of both the low acidic pH and bile salts is not assessed, it is assumed that at pH < 3.0, and 0.60% of ox-bile, the combined effects could be more synergistic and even greater in magnitude for probiotic bacteria to survive. Additionally, the $D_{(acid)}$- and $D_{(bile)}$-values reveal a modern and efficient sorting order of the more-resistant probiotic strains to these two distinct hostile GI tract conditions in humans. Many authors have investigated the effect of bile on survival of LAB. For example, Kim *et al.*, (1999) examined the effect of bile concentration in the range of 0 – 0.4% on survival of *Lb. lactis* and found bile to be toxic at concentrations over 0.04%. Shimakawa *et al.*, (2003) reported that 0.2% oxgall in the growth medium inhibited growth of *B. breve* strain Yakult. Others detected that all bacterial cells were killed by 0.2% bile and higher (Olejnik *et al.*, 2005). However, Khalil *et al.*, (2007) reported higher resistance to bile salts, with viability of strains apparently increasing when exposed to high levels of oxgall (0.4%).

pH	$D_{(acid)}$-value (min)			
	B. bifidum	*B. infantis*	*B. longum*	*B. adolescentis*
4.5	23.80	14.10	12.00	7.60
4.0	8.40	6.00	5.70	3.98
3.5	3.00	2.70	2.60	2.05
3.0	1.10	1.10	1.20	1.10
	$z_{(acid)}$-value (in pH units)			
	1.11	1.55	1.35	1.55
Bile Salts	$D_{(bile)}$-value (min)			
0.15%	17.40	10.50	9.60	6.80
0.30%	7.40	5.20	4.70	3.20
0.45%	3.20	2.55	2.30	1.58
0.60%	1.40	1.30	1.10	0.75
	$z_{(bile)}$-value (% ox-bile concentration)			
	0.40	0.48	0.49	0.46

Table 5. Selected *Bifidobacterium* strains and their calculated $D_{(acid)}$-, $D_{(bile)}$-, $z_{(acid)}$- and $z_{(bile)}$-values.

As compared to previous studies, the practicality of $D_{(acid)}$-, $D_{(bile)}$-, $z_{(acid)}$- and $z_{(bile)}$-values as new kinetic-measurements applied in this study, are indeed, quick to identify comparably higher survival of bifidobacteria cells (> 4.1 log CFU/ml after 2.5 h) at elevated bile salt concentrations of 0.6% (w/v), thereby confirm also that the individual *Bifidobacterium* strains are resistant to harsh intestinal conditions in the following order: *B. bifidum* > *B. infantis* > *B. longum* > *B. adolescentis*. A number of researchers reported that *B. infantis* had the highest survival rates followed by *B. bifidum, B. breve and B. longum,* when exposed to bile salt at concentrations ranging from 0 to 3 g/l. In contrast, the literature contains also one preliminary report that *B. longum* exhibited the highest tolerance to bile salts followed by *B. bifidum* and *B. infantis,* which was almost the exact opposite in order of their tolerance to acidic pH. These contrasting observations may reflect the strain-specific resistance to acid or bile salts stress. It also indicates that tolerance is strain- rather than species-specific. Likewise, the source of isolation of the probiotic strains is particularly influential too.

6. Conclusion

Apart from the isolation, enumeration, unequivocal taxonomical characterization, screening and selection of tolerant strains of bifidobacteria to gastric acid and bile salts studies, the assessment of the tolerant bifidobacteria to bile salts and low pH has been made possible by use of D- and z-value concept. After log-conversion, inactivation followed first-order kinetic law whereby validating the kinetic assumptions of the latter concept. The projected $z_{(acid)}$- and $z_{(bile)}$-values were all fairly similar for the bifidobacteria strains and suggested the effect of increasing the bile salt concentration or decreasing the pH on the $D_{(acid)}$- and $D_{(bile)}$-values. This approach is useful for measuring the resistance and sensitivity of lactic acid bacteria or bifidobacteria to these two hostile gastrointestinal conditions. The approach pursued in this chapter would be extremely useful for predicting the suitability of bifidobacteria and/or other LAB as probiotics for use in real life situations. While the mechanisms of probiotic survival in the GI tract could be more complex, the practical utility of the $D_{(acid)}$- and/or $D_{(bile)}$- and their $z_{(acid)}$- and $z_{(bile)}$-values is significant.

Author details

Nditange Shigwedha* and Li Jia
University of Namibia & Meat Corporation of Namibia, Namibia

Acknowledgement

We are grateful to Ms. Masa Vidovic and the entire staff of the publisher for helpful advices and the opportunity given to us. We would also like to thank Mr. Richard Shigwedha for the good cooperation.

* Corresponding Author

7. References

Biavati, B.; Scardovi, V. & Moore, W.E.C. (1982). Electrophoresis Patterns of Protein in the Genus *Bifidobacterium* and Proposal of Four New Species. *International Journal of Systematic Bacteriology*, Vol.32, No.3, (July 1982), pp. 358-373

Chung, H.S.; Kim, Y.B.; Chun, S.L. & Ji, G.E. (1999). Screening and Selection of Acid and Bile Resistant Bifidobacteria. *International Journal of Food Microbiology*, Vol.49, No.1-2, (March 1999), pp. 25-32

Dubey, U.K. & Mistry, V.V. (1996). Effect of Bifidogenic Factors on Growth Characteristics of Bifidobacteria in Infant Formulas. *Journal of Dairy Science.* Vol.79, No.7, (July 1996), pp. 1156-1163

Favier, C.; Neut, C.; Mizon, C.; Cortot, A.; Colombel, J.F. & Mizon, J. (1997). Fecal β-D-Galactosidase Production and Bifidobacteria are Decreased in Crohn's Disease. *Digestive Diseases and Sciences*, Vol.42, No.4, (April 1997), pp. 817-822

Glickman, R.M. (1980). Intestinal Fat Absorption, In: *Nutrition and Gastroenterology*, M. Winick, (Ed.), 29-41, Wiley, New York, USA

Hartemink, R. (1999). *Prebiotic Effects of Non-Digestible Oligo- and Polysaccharides*, Ponsen & Looijen, ISBN 90-5808-051-X, Wageningen, The Netherlands

Jia, L.; Shigwedha, N. & Mwandemele, O.D. (2010). Use of D_{acid}-, D_{bile}-, z_{acid}- and z_{bile}-Values in Evaluating Bifidobacteria with Regard to Stomach pH and Bile Salt Sensitivity. *Journal of Food Science*, Vol.75, No.1, (January/February 2010), pp. M14-M18

Joint FAO/WHO. (2002). Guidelines for the Evaluation of Probiotics in Food. Ontario, Canada: 30 April - 1 May 2002, Available from
http://www.who.int/foodsafety/publications

Kazor, C.E.; Mitchell, P.M.; Lee, A.M.; Stokes, L.N.; Loesche, W.J.; Dewhirst, F.E. & Paster, B.J. (2003). Diversity of Bacterial Populations on the Tongue Dorsa of Patients with Halitosis and Healthy Patients. *Journal of Clinical Microbiology*, Vol.41, No.2, (February 2003), pp. 558-563

Khalil, R.; Mahrous, H.; El-Halafawy, K.; Kamaly, K.; Frank, J. & El Soda, M. (2007). Evaluation of the Probiotic Potential of Lactic Acid Bacteria Isolated from Faeces of Breast-Fed Infants in Egypt. *African Journal of Biotechnology*, Vol.6, No.7, (April 2007), pp. 939-949

Kim, W.S.; Ren, J. & Dunn, N.W. (1999). Differentiation of *Lactococcus lactis* Subspecies *lactis* and Subspecies *cremoris* Strains by their Adaptive Response to Stresses. *FEMS Microbiology Letters*, Vol.171, No.1, (February 1999), pp. 57-65

Liepke, C.; Adermann, K.; Raida, M.; Magert, H-J.; Forssmann, W-G. & Zucht, H-D. (2002). Human Milk Provides Peptides Highly Stimulating the Growth of Bifidobacteria. *European Journal of Biochemistry*, Vol.269, No.2, (January 2002), pp. 712-718

Lourens-Hattingh, A. & Viljoen, B.C. (2001). Growth and Survival of a Probiotic Yeast in Dairy Products. *Food Research International*, Vol.34, No.9, (November 2001), pp. 791-796

Macfarlane, S. & Dillon, J.F. (2007). Microbial Biofilms in the Human Gastrointestinal Tract. *Journal of Applied Microbiology*, Vol.102, No.5, (May 2007), pp. 1187-1196

Mitsuoka, T. (1990). Bifidobacteria and their Role in Human Health. *Journal of Industrial Microbiology & Biotechnology*, Vol.6, No.4, (April 1990), pp. 263-267

Olejnik, A.; Lewandowska, M.; Obarska, M. & Grajek, W. (2005). Tolerance of *Lactobacillus* and *Bifidobacterium* Strains to Low pH, Bile Salts and Digestive Enzymes. *Electronic Journal of Polish Agricultural Universities*, Vol.8, No.1, Available from http://www.ejpau.media.pl/volume8/issue1/art-05.html

Palacios, M.C.; Haros, M.; Rosell, C.M. & Sanz, Y. (2008). Selection of Phytate-Degrading Human Bifidobacteria and Application in Whole Wheat Dough Fermentation. *Food Microbiology*, Vol.25, No.1, (February 2008), pp. 169-176

Patrignani, F.; Lanciotti, R.; Mathara, J.M.; Guerzoni, M.E. & Holzapfel, W.H. (2006). Potential of Functional Strains, Isolated from Traditional Maasai Milk, as Starters of the Production of Fermented Milks. *International Journal of Food Microbiology*, Vol.107, No.1, (March 2006), pp. 1-11

Pei, Z.; Bini, E.J.; Yang, L.; Zhou, M.; Francois, F. & Blaser, M.J. (2004). Bacterial Biota in the Human Distal Esophagus. *Proceedings of the National Academy of Sciences (PNAS)*, Vol.101, No.12, (March 2004), pp. 4250-4255.

Reuter G. 1963. Vergleichede Untersuchungen über die Bifidus-Flora im Säuglings- und Erwachsenenstuhl. *Zentralbl Bakteriol Parasitenkd Infektionskr Hyg Abt I Orig*, Vol.191, (1963), pp. 486-507

Scardovi, V. (1986). Genus *Bifidobacterium*, In: *Bergey's Manual of Systematic Bacteriology*, P.H.A. Sneath, N.S. Mair, M.E. Sharpe, & J.G. Holt (Eds.), 1418-1434, Williams & Wilkins, Baltimore, USA

Schlegel, H.G. (1993). *General Microbiology*, Cambridge University Press, ISBN 0-521-43372-X, Melbourne, Australia

Sgorbati, B. (1979). Preliminary Quantification of Immunological Relationships among the Transaldolases of the Genus *Bifidobacterium*. *Antonie van Leeuwenhoek*, Vol.45, No.4, (April 1979), pp. 557-564

Shimakawa, Y.; Matsubara, S.; Yuki, N.; Ikeda, M. & Ishikawa, F. (2003). Evaluation of *Bifidobacterium breve* Strain Yakult-Fermented Soymilk as a Probiotic Food. *International Journal of Food Microbiology*, Vol.81, No.2, (March 2003), pp. 131-136

Silvi, S.; Rumney, C.J. & Rowland, I.R. (1996). An Assessment of the Three Selective Media for Bifidobacteria in Faeces. *Journal of Applied Microbiology*, Vol.81, No.5, (November 1996), pp. 561-564

Sundman, V.; Björksten, K.A.F. & Gyllenberg, H.G. (1959). Morphology of the Bifid Bacteria (Organisms Previously Incorrectly Designated *Lactobacillus bifidus*) and Some Related Genera. *Journal of General Microbiology*, Vol.21, No.2, (October 1959), pp. 371-384

Yildirim, Z. & Johnson, M.G. (1998). Characterization and Antimicrobial Spectrum of Bifidocin B, a Bacteriocin Produced by *Bifidobacterium bifidum* NCFB 1454. *Journal of Food Protection*, Vol.61, No.1, (January 1998), pp. 47-51

Lactic Acid Bacteria as Probiotics: Characteristics, Selection Criteria and Role in Immunomodulation of Human GI Muccosal Barrier

Daoud Harzallah and Hani Belhadj

Additional information is available at the end of the chapter

1. Introduction

As it was reported by Chow (2002), the notion that food could serve as medicine was first conceived thousands of years ago by the Greek philosopher and father of medicine, Hippocrates, who once wrote: 'Let food be thy medicine, and let medicine be thy food'. However, during recent times, the concept of food having medicinal value has been reborn as 'functional foods'. The list of health benefits accredited to functional food continues to increase, and the gut is an obvious target for the development of functional foods, because it acts as an interface between the diet and all other body functions. One of the most promising areas for the development of functional food components lies in the use of probiotics and prebiotics which scientific researches have demonstrated therapeutic evidence. Nowadays, consumers are aware of the link among lifestyle, diet and good health, which explains the emerging demand for products that are able to enhance health beyond providing basic nutrition. Besides the nutritional valaes, ingestion of lactic acid bacteria (LAB) and their fermented foods has been suggested to confer a range of health benefits including immune system modulation, increased resistance to malignancy, and infectious illness (Soccol, et al., 2010). LAB were first isolated from milk. They can be found in fermented products as meat, milk products, vegetables, beverages and bakery products. LAB occur naturally in soil, water, manure, sewage, silage and plants. They are part of the microbiota on mucous membranes, such as the intestines, mouth, skin, urinary and genital organs of both humans and animals, and may have a beneficial influence on these ecosystems. LAB that grow as the adventitious microflora of foods or that are added to foods as cultures are generally considered to be harmless or even an advantage for human health. Since their discovery, LAB has been gained mush interest in various applications, as starter cultures in food and feed fermentations, pharmaceuticals, probiotics and as biological control agents. In food industry, LAB are widely used as starters to achieve favorable changes in texture, aroma, flavor and acidity (Leory and De Vuyst, 2004). However, there has been an important interest in using bacteriocin and/or other inhibitory substance producing LAB for non-fermentative biopreservation applications. Du to their antimicrobial and antioxidant activities some LAB strains are used in food biopreservation. However, LAB are generally regarded as safe (GRAS) to the consumer and during storage, they naturally dominate the microflora of many foods (Osmanağaoğlu and Beyatli, 1999; Parada et al., 2007). Many of the indications for probiotic activity have been obtained from effects observed in various clinical situations. Even, there are few strains that have officially gained the status of pharmaceutical preparation; each of these effects is gradually being supported by a number of clinical studies or human intervention trials, performed in a way that resembles the traditional pharmacological approach (placebo-controlled, double blind, randomized trials) and the strains used in these studies belong to different microbial species, but are mostly lactic acid bacteria (Mercenier et al, 2003).

2. LAB as probiotic agents

2.1. Overview of probiotics

The most tried and tested manner in which the gut microbiota composition may be influenced is through the use of live microbial dietary additions, as probiotics. In fact, the concept dates back as far as prebiblical ages. The first records of ingestion of live bacteria by humans are over 2,000 years old. However, at the beginning of this century probiotics were first put onto a scientific basis by the work of Metchnikoff (1908). He hypothesised that the normal gut microflora could exert adverse effects on the host and that consumption of 'soured milks' reversed this effect. The word "probiotics" was initially used as an anonym of the word "antibiotic". It is derived from Greek words pro and biotos and translated as "for life". The origin of the first use can be traced back to Kollath (1953), who used it to describe the restoration of the health of malnourished patients by different organic and inorganic supplements. Later, Vergin (1954) proposed that the microbial imbalance in the body caused by antibiotic treatment could have been restored by a probiotic rich diet; a suggestion cited by many as the first reference to probiotics as they are defined nowadays. Similarly, Kolb recognized detrimental effects of antibiotic therapy and proposed the prevention by probiotics (Vasiljevic and Shah, 2008) Later on, Lilly and Stillwell (1965) defined probiotics as "...microorganisms promoting the growth of other microorganisms". Following recommendations of a FAO/WHO (2002) working group on the evaluation of probiotics in food, probiotics, are live microorganisms that, when administered in adequate amounts, confer a health benefit on the host (Sanders, 2008; Schrezenmeir and De Vrese, 2001). The idea of health-promoting effects of LAB is by no means new, as Metchnikoff proposed that lactobacilli may fight against intestinal putrefaction and contribute to long life. Such microorganisms may not necessarily be constant inhabitants of the gut, but they should have a "...beneficial effect on the general and health status of man and animal"

Probiotic	Human disease in which benefit is shown	Animal model in which benefit is shown
Yeast		
Saccharomyces boulardii	*Clostridium difficile* infection	*Citrobacter rodentium*-induced colitis
Gram-negative bacteria		
Escherichia coli Nissle 1917	NA	DSS-induced colitis
Gram-positive bacteria		
Bifidobacteria bifidum	NA	Rat model of necrotizing enterocolitis
Bifidobacteria infantis	IBS29	NA
Lactobacillus rhamnosus GG (used with lactoferrin)	Sepsis in very low birth weight infants	NA
Lactococcus lactis (engineered to produce IL-10 or trefoil factors)	Crohn's disease	DSS-induced colitis and IL-10$^{-/-}$ mice (spontaneous IBD)
Lactobacillus plantarum 299v	Antibiotic-associated diarrhea	IL-10$^{-/-}$ mice (spontaneous IBD)
Lactobacillus acidophilus	NA	Visceral hyperalgesia 40 and *C. rodentium*-induced colitis
Lactobacillus rhamnosus	Pediatric antibiotic-associated diarrhea	–
Lactobacillus casei	NA	DNBS-induced colitis
Bacillus polyfermenticus	NA	DSS-induced colitis and TNBS-induced colitis
Combination regimens		
Lactobacillus rhamnosus GG combined with *Bifidobacterium lactis*	Bacterial infections	NA
Lactobacillus rhamnosus combined with *Lactobacillus helveticus*	NA	*C. rodentium*-induced colitis, chronic stress, and early life stress
VSL#3 (*Lactobacillus casei, Lactobacillus plantarum, Lactobacillus acidophilus, Lactobacillus bulgaricus, Bifidobacterium longum, Bifidobacterium breve, Bifidocacterium infantis and Streptococcus thermophilus*)	Pouchitis and pediatric ulcerative colitis	DSS-induced colitis, IL-10$^{-/-}$ mice (spontaneous IBD; DNA only), and SAMP mouse model of spontaneous IBD

Abbreviations: DNBS, dinitrobenzene sulfonic acid; DSS, dextran sodium sulfate; IL-10, interleukin 10; NA, not available; TNBS, trinitrobenzene sulfonic acid.

Table 1. Selected organisms that are used as probiotic agents (Gareau et al., 2010).

(Holzapfel et al., 2001; Belhadj et al., 2010). Other definitions advanced through the years have been restrictive by specification of mechanisms, site of action, delivery format, method, or host. Probiotics have been shown to exert a wide range of effects. The mechanism of action of probiotics (e.g, having an impact on the intestinal microbiota or enhancing immune function) was dropped from the definition to encompass health effects due to novel mechanisms and to allow application of the term before the mechanism is confirmed. Physiologic benefits have been attributed to dead microorganisms. Furthermore, certain mechanisms of action (such as delivery of certain enzymes to the intestine) may not require live cells. However, regardless of functionality, dead microbes are not probiotics (Sanders, 2008). In relation to food, probiotics are considered as "viable preparations in foods or dietary supplements to improve the health of humans and animals". According to these definitions, an impressive number of microbial species are considered as probiotics. (Holzapfel et al., 2001). For gastrointestinal ecosysteme, however, the most important microbial species that are used as probiotics are lactic acid bacteria (LAB) (Table 1).

2.2. Selection of probiotics

Many in vitro tests are performed when screening for potential probiotic strains. The first step in the selection of a probiotic LAB strain is the determination of its taxonomic classification, which may give an indication of the origin, habitat and physiology of the strain. All these characteristics have important consequences on the selection of the novel strains (Morelli, 2007). An FAO/WHO (2002) expert panel suggested that the specificity of probiotic action is more important than the source of microorganism. This conclusion was brought forward due to uncertainty of the origin of the human intestinal microflora since the infants are borne with virtually sterile intestine. However, the panel also underlined a need for improvement of in vitro tests to predict the performance of probiotics in humans. While many probiotics meet criteria such as acid and bile resistance and survival during gastrointestinal transit, an ideal probiotic strain remains to be identified for any given indication. Furthermore, it seems unlikely that a single probiotic will be equally suited to all indications; selection of strains for disease-specific indications will be required (Shanahan, 2003).

The initial screening and selection of probiotics includes testing of the following important criteria: phenotype and genotype stability, including plasmid stability; carbohydrate and protein utilization patterns; acid and bile tolerance and survival and growth; intestinal epithelial adhesion properties; production of antimicrobial substances; antibiotic resistance patterns; ability to inhibit known pathogens, spoilage organisms, or both; and immunogenicity. The ability to adhere to the intestinal mucosa is one of the more important selection criteria for probiotics because adhesion to the intestinal mucosa is considered to be a prerequisite for colonization (Tuomola et al., 2001). The table below (Table 2) indicates key creteria for sellecting probiotic candidat for commercial application, and figure 1 presents major and cardinal steps for sellecting probiotic candidats.

It is of high importance that the probiotic strain can survive the location where it is presumed to be active. For a longer and perhaps higher activity, it is necessary that the strain can proliferate and colonise at this specific location. Probably only host-specific microbial strains are able to compete with the indigenous microflora and to colonise the niches. Besides, the probiotic strain must be tolerated by the immune system and not provoke the formation of antibodies against the probiotic strain. So, the host must be immuno-tolerant to the probiotic. On the other hand, the probiotic strain can act as an adjuvant and stimulate the immune system against pathogenic microorganisms. It goes without saying that a probiotic has to be harmless to the host: there must be no local or general pathogenic, allergic or mutagenic/carcinogenic reactions provoked by the microorganism itself, its fermentation products or its cell components after decrease of the bacteria (Desai, 2008).

General	Property
Safety criteria	Origin
	Pathogenicity and infectivity
	Virulence factors—toxicity, metabolic activity and intrinsic properties, i.e., antibiotic resistance
Technological criteria	Genetically stable strains
	Desired viability during processing and storage
	Good sensory properties
	Phage resistance
	Large-scale production
Functional criteria	Tolerance to gastric acid and juices
	Bile tolerance
	Adhesion to mucosal surface
	Validated and documented health effects
Desirable physiological criteria	Immunomodulation
	Antagonistic activity towards gastrointestinal pathogens, i.e., Helicobacter pylori, Candida albicans
	Cholesterol metabolism
	Lactose metabolism
	Antimutagenic and anticarcinogenic properties

Table 2. Key and desirable criteria for the selection of probiotics in commercial applications (Vasiljevic and Shah, 2008).

When probiotic strains are selected, attributes important for efficacy and technological function must be assessed and a list of characteristics required for all probiotic functions is required. Basic initial characterization of strain identity and taxonomy should be conducted, followed by evaluation with validated assays both in studies of animal models and in controlled studies in the target host. In vitro assays are frequently conducted that have not been proved to be predictive of in vivo function. Technological robustness must also be determined, such as the strain's ability to be grown to high numbers, concentrated, stabilized, and incorporated into a final product with good sensory properties, if applicable, and to be stable, both physiologically and genetically, through the end of the shelf life of the product and at the active site in the host. Assessment of stability can also be a challenge, since factors such as chain length and injury may challenge the typical assessment of colony-forming units, as well as in vivo function (Sanders, 2008). Dose levels of probiotics should be based on levels found to be efficacious in human studies. One dose level cannot be assumed to be effective for all strains. Furthermore, the impact of product format on

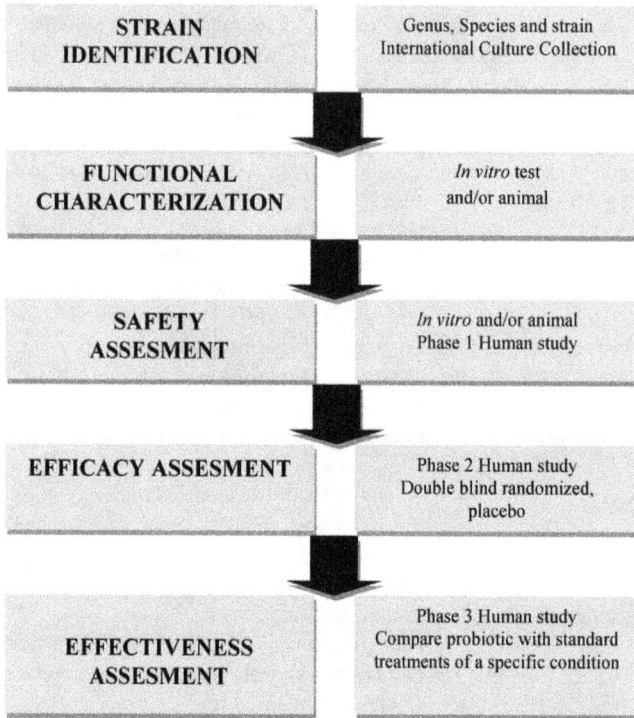

STRAIN IDENTIFICATION	Genus, Species and strain International Culture Collection
FUNCTIONAL CHARACTERIZATION	*In vitro* test and/or animal
SAFETY ASSESMENT	*In vitro* and/or animal Phase 1 Human study
EFFICACY ASSESMENT	Phase 2 Human study Double blind randomized, placebo
EFFECTIVENESS ASSESMENT	Phase 3 Human study Compare probiotic with standard treatments of a specific condition

Figure 1. Scheme of the Guidelines for the Evaluation of Probiotics for Food Use. (Adapted from, Collado et al., 2009).

probiotic function has yet to be explored in depth. The common quality-control parameter of colony-forming units per gram may not be the only parameter indicative of the efficacy of the final product. Other factors, such as probiotic growth during product manufacture, coating, preservation technology, metabolic state of the probiotic, and the presence of other functional ingredients in the final product, may play a role in the effectiveness of a product. More research is needed to understand how much influence such factors have on in vivo efficacy **(Sanders, 2008).**

2.3. Potential mechanisms of action of probiotics

A wide variety of potential beneficial health effects have been attributed to probiotics (Table 3). Claimed effects range from the alleviation of constipation to the prevention of major life-threatening diseases such as inflammatory bowel disease, cancer, and cardiovascular incidents. Some of these claims, such as the effects of probiotics on the shortening of intestinal transit time or the relief from lactose maldigestion, are considered well-established, while others, such as cancer prevention or the effect on blood cholesterol levels, need further scientific backup (Leroy et al., 2008). The mechanisms of action may vary from one probiotic strain to another and are, in most cases, probably a combination of activities, thus making the investigation of the responsible mechanisms a very difficult and complex task. In general, three levels of action can be distinguished: probiotics can influence human

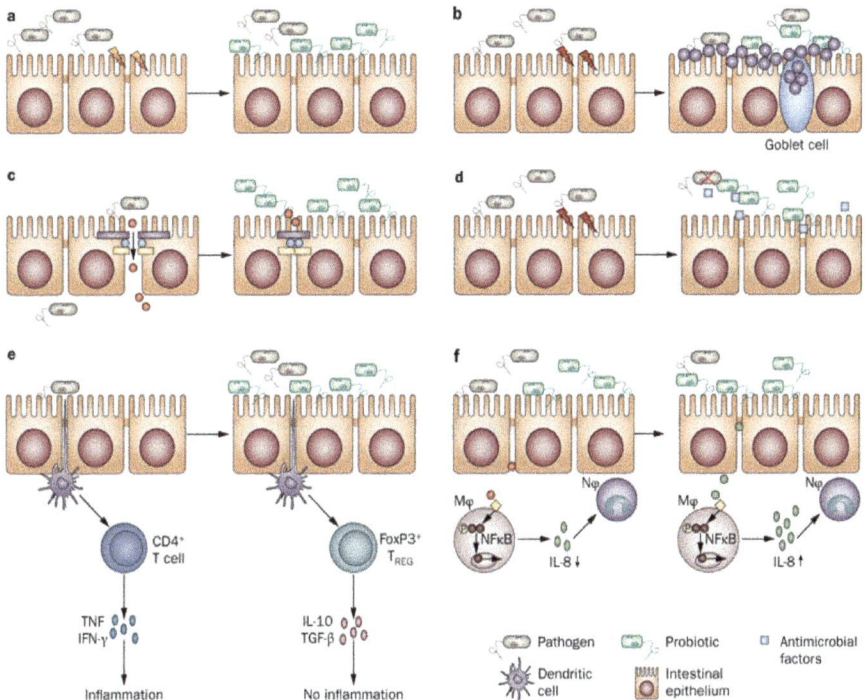

Probiotic organisms can provide a beneficial effect on intestinal epithelial cells in numerous ways. **a**: Some strains can block pathogen entry into the epithelial cell by providing a physical barrier, referred to as colonization resistance or **b**: create a mucus barrier by causing the release of mucus from goblet cells. **c**: Other probiotics maintain intestinal permeability by increasing the intercellular integrity of apical tight junctions, for example, by upregulating the expression of zona-occludens 1 (a tight junction protein), or by preventing tight junction protein redistribution thereby stopping the passage of molecules into the lamina propria. **d**: Some probiotic strains have been shown to produce antimicrobial factors. **e**: Still other strains stimulate the innate immune system by signaling dendritic cells, which then travel to mesenteric lymph nodes and lead to the induction of TREG cells and the production of anti-inflammatory cytokines, including IL-10 and TGF-β. **f**: Some probiotics (or their products) may also prevent (left-hand side) or trigger (right-hand side) an innate immune response by initiating TNF production by epithelial cells and inhibiting (or activitating) NFκB in Mφ and dampening (or priming) the host immune response by influencing the production of IL-8 and subsequent recruitment of Nφ to sites of intestinal injury. Abbreviations: Mφ, macrophage; Nφ, neutrophil; TREG cell, regulatory T cell. Reproduced from, *Gareau M. G., P. M. Sherman & W. A. Walker (2010) Nature Reviews Gastroenterology and Hepatology 7, 503-514.*

Figure 2. Potential mechanisms of action of probiotics.

Health benefit	Proposed mechanism(s)
Cancer prevention	Inhibition of the transformation of pro-carcinogens into active carcinogens, binding/inactivation of mutagenic compounds, production of anti-mutagenic compounds, suppression of growth of pro-carcinogenic bacteria, reduction of the absorption of carcinogens, enhancment of immune function, influence on bile salt concentrations
Control of irritable bowel syndrome	Modulation of gut microbiota, reduction of intestinal gas production
Management and prevention of atopic diseases	Modulation of immune response
Management of inflammatory bowel diseases (Crohn's disease, ulcerative colitis, pouchitis)	Modulation of immune response, modulation of gut microbiota
Prevention of heart diseases/influence on blood cholesterol levels	Assimilation of cholesterol by bacterial cells, deconjugation of bile acids by bacterial acid hydrolases, cholesterol-binding to bacterial cell walls, reduction of hepatic cholesterol synthesis and/or redistribution of cholesterol from plasma to liver through influence of the bacterial production of short-chain fatty acids
Prevention of urogenital tract disorders	Production of antimicrobial substances, competition for adhesion sites, competitive exclusion of pathogens
Prevention/alleviation of diarrhoea caused by bacteria/viruses	Modulation of gut microbiota, production of antimicrobial substances, competition for adhesion sites, stimulation of mucus secretion, modulation of immune response
Prevention/treatment of *Helicobacter pylori* infections	Production of antimicrobial substances, stimulation of the mucus secretion, competition for adhesion sites, stimulation of specific and non-specific immune responses
Relief of lactose indigestion	Action of bacterial β-galactosidase(s) on lactose
Shortening of colonic transit time	Influence on peristalsis through bacterial metabolite production

Table 3. Potential and established health benefits associated with the usage of probiotics (Leroy et al., 2008).

health by interacting with other microorganisms present on the site of action, by strengthening mucosal barriers, and by affecting the immune system of the host (Leroy et al., 2008), and the figure 2 shows the most important mechanisms by whiche probiotics exerce their action inside the gut.

3. Probiotics and gut health

3.1. Gut microbiota

The human gastrointestinal tract is inhabited by a complex and dynamic population of around 500-1000 of different microbial species which remain in a complex equilibrium. It has been estimated that bacteria account for 35–50% of the volume content of the human colon. These include *Bacteroides, Lactobacillus, Clostridium, Fusobacterium, Bifidobacterium, Eubacterium, Peptococcus, Peptostreptococcus, Escherichia and Veillonella*. The bacterial strains with identified beneficial properties include mainly *Bifidobacterium* and *Lactobacillus* species. The dominant microbial composition of the intestine have been shown to be stable over time during adulthood, and the microbial patterns are unique for each individual. However, there are numerous external factors that have potential to influence the microbial composition in the gut as host genetics, birth delivery mode, diet, age, antibiotic treatments and also, other microorganisms as probiotics. (Collado et al., 2009). The intestine is one of the main surfaces of contact with exogenous agents (viruses, bacteria, allergens) in the human body. It has a primary role in the host defense against external aggressions by means of the intestinal mucosa, the local immune system, and the interactions with the intestinal microbiota (resident and in transitbacteria). Gut microbiota influences human health through an impact on the gut defense barrier, immune function, nutrient utilization and potentially by direct signaling with the gastrointestinal epithelium (Collado et al., 2009). Only a limited fraction of bacterial phyla compose the major intestinal microbiota. In healthy adults, 80% of phylotypes belong to four major phylogenetic groups, which are the *Clostiridium leptum, Clostridium coccoides, Bacteroides* and *Bifidobacteria* groups. However, a large fraction of dominant phylotypes is subject specific. Also, studies have found that mucosal microbiota is stable along the distal gastrointestinal tract from ileum to rectum, but mucosa-associated microbiota is different from fecal microbiota. The difference has been estimated to be between 50–90%.

The intestinal microbiota is not homogeneous. The number of bacterial cells present in the mammalian gut shows a continuum that goes from 10^1 to 10^3 bacteria per gram of contents in the stomach and duodenum, progressing to 10^4 to 10^7 bacteria per gram in the jejunum and ileum and culminating in 10^{11} to 10^{12} cells per gram in the colon (Figure 3a). Additionally, the microbial composition varies between these sites. In addition to the longitudinal heterogeneity displayed by the intestinal microbiota, there is also a great deal of latitudinal variation in the microbiota composition (Figure 3b). The intestinal epithelium is separated from the lumen by a thick and physicochemically complex mucus layer. The microbiota present in the intestinal lumen differs significantly from the microbiota attached and embedded in this mucus layer as well as the microbiota present in the immediate

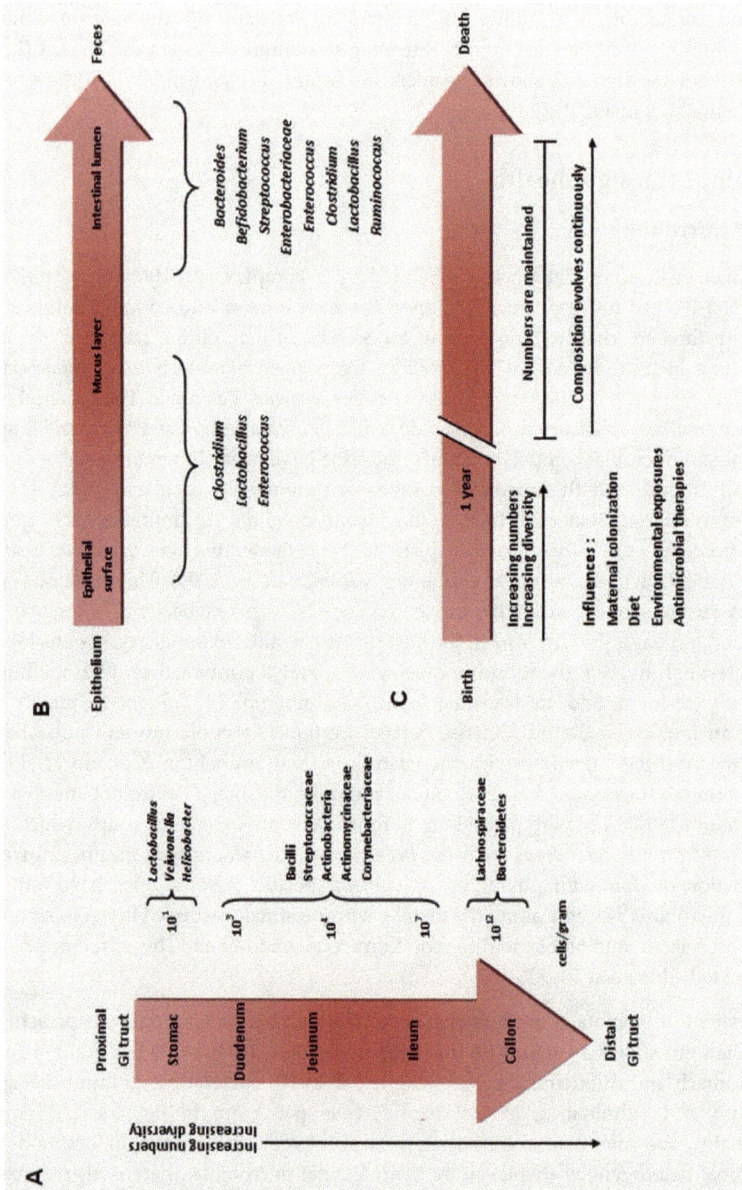

a: variations in microbial numbers and composition across the length of the gastrointestinal tract. *b*: longitudinal variations in microbial composition in the intestine. *c*: temporal aspects of microbiota establishment and maintenance and factors influencing microbial composition. (Sekirov et al., 2010).

Figure 3. Spatial and temporal aspects of intestinal microbiota composition.

proximity of the epithelium. For instance, *Bacteroides*, *Bifidobacterium*, *Streptococcus*, members of *Enterobacteriacea*, *Enterococcus*, *Clostridium*, *Lactobacillus*, and *Ruminococcus* were all found in feces, whereas only *Clostridium*, *Lactobacillus*, and *Enterococcus* were detected in the mucus layer and epithelial crypts of the small intestine (Sekirov et al., 2010). Colonization of the human gut with microbes begins immediately at birth (Figure 3c). Upon passage through the birth canal, infants are exposed to a complex microbial population. After the initial establishment of the intestinal microbiota and during the first year of life, the microbial composition of the mammalian intestine is relatively simple and varies widely between different individuals and also with time. However, after one year of age, the intestinal microbiota of children starts to resemble that of a young adult and stabilizes (Figure 3c) (Sekirov et al., 2010).

3.2. Survival and antagonism effects of probiotics in the gut

The intestinal epithelium is the largest mucosal surface in the human body, provides an interface between the external environment and the host. The gut epithelium is constantly exposed to foreign microbes and antigens derived from digested foods. Thus, the gut epithelium acts as a physical barrier against microbial invaders and is equipped with various elements of the innate defense system. In the gut, two key elements govern the interplay between environmental triggers and the host: intestinal permeability and intestinal mucosal defense. Resident bacteria can interact with pathogenic microorganisms and external antigens to protect the gut using various strategies.

According to the generally accepted definition of a probiotic, the probiotic microorganism should be viable at the time of ingestion to confer a health benefit. Although not explicitly stated, this definition implies that a probiotic should survive GI tract passage and, colonize the host epithelium. A variety of traits are believed to be relevant for surviving GI tract passage, the most important of which is tolerance both to the highly acidic conditions present in the stomach and to concentrations of bile salts found in the small intestine. These properties have consequently become important selection criteria for new probiotic functionality. In addition to tolerating the harsh physical-chemical environment of the GI tract, adherence to intestinal mucosal cells would be necessary for colonization and any direct interactions between the probiotic and host cells leading to the competitive exclusion of pathogens and/or modulation of host cell responses. Moreover, As enteropathogenic Escherichia coli are known to bind to epithelial cells via mannose receptors, probiotic strains with similar adherence capabilities could inhibit pathogen attachment and colonization at these binding sites and thereby protect the host against infection (Marco et al., 2006).

Probiotic bacteria can antagonize pathogenic bacteria by reducing luminal pH, inhibiting bacterial adherence and translocation, or producing antibacterial substances and defensins. One of the mechanisms by which the gut flora resists colonization by pathogenic bacteria is by the production of a physiologically restrictive environment, with respect to pH, redox potential, and hydrogen sulfide production. Probiotic bacteria decrease the luminal pH, as has been demonstrated in patients with ulcerative colitis (UC) following ingestion of the probiotic preparation VSL#3. In a fatal mouse Shiga toxin-producing *E. coli O157:H7* infection model, the probiotic Befidobacterium breve produced a high concentration of acetic acid, consequently lowering the luminal pH. This pH reduction was associated with increased animal survival (Ng et al., 2009).

Production of antimicrobial compounds, termed bacteriocins, by probiotic bacteria is also likely to contribute to their beneficial activity. Several bacteriocins produced by different species from the genus *Lactobacillus* have been described. The inhibitory activity of these bacteriocins varies; some inhibit taxonomically related Gram-positive bacteria, and some are active against a much wider range of Gram-positive and Gram-negative bacteria as well as yeasts and molds. For example, the probiotic *L. salivarius subsp. salivarius* UCC118 produces a peptide that inhibits a broad range of pathogens such as *Bacillus, Staphylococcus, Enterococcus, Listeria,* and *Salmonella* species. Lacticin 3147, a broad-spectrum bacteriocin produced by *Lactococcus lactis,* inhibits a range of genetically distinct *Clostridium difficile* isolates from healthy subjects and patients with IBD. A further example is the antimicrobial effect of *Lactobacillus* species on *Helicobacter pylori* infection of gastric mucosa, achieved by the release of bacteriocins and the ability to decrease adherence of this pathogen to epithelial cells (Gotteland et al., 2006). Probiotics can reduce the epithelial injury that follows exposure to *E. coli O157:H7* and *F. coli O127:H6*. The pretreatment of intestinal (T84) cells with lactic acid-producing bacteria reduced the ability of pathogenic *E. coli* to inject virulence factors into the cells or to breach the intracellular tight junctions. Adhesion and invasion of an intestinal epithelial cell line (Intestine 407) by adherent invasive *E. coli* isolated from patients with Crohn's disease (CD) was substantially diminished by co- or preincubation with the probiotic strain *E. coli* Nissle 1917 (Wehkamp et al., 2004 ; Schlee et al., 2007). These findings demonstrate that probiotics prevent epithelial injury induced by attaching-effacing bacteria and contributes to an improved mucosal barrier and provide a means of limiting access of enteric pathogens (Sherman et al., 2005).

4. Probiotics and the mucous layer

Most mucosal surfaces are covered by a hydrated gel formed by mucins. Mucins are secreted by specialized epithelial cells, such as gastric foveolar mucous cells and intestinal goblet cells, Goblet cells are found along the entire length of the intestinal tract, as well as other mucosal surfaces. Mucins, are abundantly core glycosylated (up to 80% wt/wt) and either localized to the cell membrane or secreted into the lumen to form the mucous layer (Turner, 2009). Of the 18 mucin-type glycoproteins expressed by humans, MUC2 is the predominant glycoprotein found in the small and large bowel mucus. The NH2- and COOH-termini are not glycosylated to the same extent, but are rich in cysteine residues that form intra- and inter-molecular disulfide bonds. These glycan groups confer proteolytic resistance and hydrophilicity to the mucins, whereas the disulfide linkages form a matrix of glycoproteins that is the backbone of the mucous layer (Ohland and MacNaughton, 2010). Although small molecules pass through the heavily glycosylated mucus layer with relative ease, bulk fluid flow is limited and thereby contributes to the development of an unstirred layer of fluid at the epithelial cell surface. As the unstirred layer is protected from convective mixing forces, the diffusion of ions and small solutes is slowed (Turner, 2009). This gel layer provides protection by shielding the epithelium from potentially harmful antigens and molecules including bacteria from directly contacting the epithelial cell layer, while acting as a lubricant for intestinal motility. Mucins can also bind the epithelial cell surface carbohydrates and form the bottom layer, which is firmly attached to the mucosa, whereas the upper layer is loosely adherent. The mucus is the first barrier that intestinal bacteria meet, and pathogens must penetrate it to reach the epithelial cells during infection (Ohland and MacNaughton, 2010).

Probiotics may promote mucus secretion as one mechanism to improve barrier function and exclusion of pathogens. In support of this concept, probiotics have been shown to increase mucin expression in vitro, contributing to barrier function and exclusion of pathogens. Several studies showed that increased mucin expression in the human intestinal cell lines Caco-2 (MUC2) and HT29 (MUC2 and 3), thus blocking pathogenic *E. coli* invasion and adherence. However, this protective effect was dependent on probiotic adhesion to the cell monolayers, which likely does not occur in vivo (Mack et al., 2003; Mattar et al., 2002). Conversely, another study showed that *L. acidophilus A4* cell extract was sufficient to increase MUC2 expression in HT29 cells, independent of attachment (Kim et al., 2008). Additionally, intestinal trefoil factor 3 (TFF3) is coexpressed with MUC2 by colonic goblet cells and is suggested to promote wound repair (Gaudier et al., 2005 ; Kalabis et al., 2006). However, healthy rats did not display increased colonic TFF3 expression after stimulation by VSL#3 probiotics (Caballero-Franco et al., 2007). Furthermore, mice treated with 1% dextran sodium sulfate (DSS) to induce chronic colitis did not exhibit increased TFF3 expression or wound healing when subsequently treated with VSL#3. This observation indicates that probiotics do not enhance barrier function by up-regulation of TFF3, nor are they effective at healing established inflammation. Therefore, use of current probiotics is likely to be effective only in preventing inflammation as shown by studies in animal models (Ohland and MacNaughton, 2010).

5. Interaction of probiotic bacteria with gut epithelium

The composition of the commensal gut microbiota is probably influenced by the combination of food practices and other factors like the geographical localization, various levels of hygiene or various climates. The host-microbe interaction is of primary importance during neonatal period. The establishment of a normal microbiota provides the most substantial antigenic challenge to the immune system, thus helping the gut associated lymphoid tissus (GALT) maturation. The intestinal microbiota contributes to the anti-inflammatory character of the intestinal immune system. Several immunoregulatory mechanisms, including regulatory cells, cytokines, apoptosis among others, participate in the control of immune responses by preventing the pathological processes associated with excessive reactivity. An interesting premise for probiotic physiological action is their capacity to modulate the immune system. Consequently, many studies have focused on the effects of probiotics on diverse aspects of the immune response. Following consumption of probiotic products, the interaction of these bacteria with intestinal enterocytes initiates a host response, since intestinal cells produce various immunomodulatory molecules when stimulated by bacteria (Delcenseri et al., 2009). Furthermore, the indigenous microbiota is a natural resistance factor against potential pathogenic microorganisms and provides colonization resistance, also known as gut barrier, by controlling the growth of opportunistic microorganisms. It has been suggested that commensal bacteria protect their host against microbial pathogens by interfering with their adhesion and toxic effects (Myllyluoma, 2007).

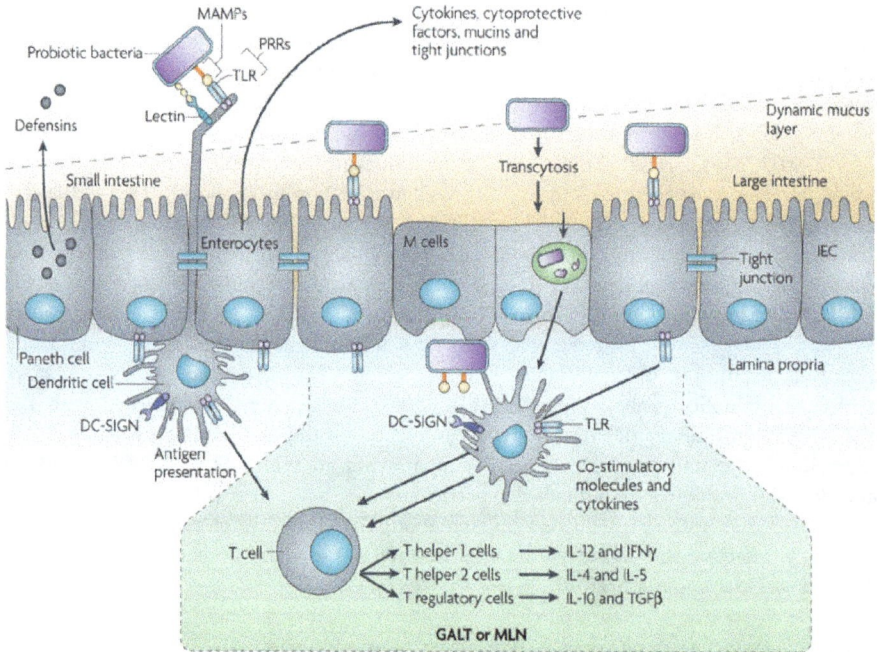

A fraction of ingested probiotics are able to interact with intestinal epithelial cells (IECs) and dendritic cells (DCs), depending on the presence of a dynamic mucus layer. Probiotics can occasionally encounter DCs through two routes: DCs residing in the lamina propria sample luminal bacterial antigens by passing their dendrites between IECs into the gut lumen, and DCs can also interact directly with bacteria that have gained access to the dome region of the gut-associated lymphoid tissue (GALT) through specialized epithelial cells, termed microfold or M cells. The interaction of the host cells with microorganism-associated molecular patterns (MAMPs) that are present on the surface macromolecules of probiotic bacteria will induce a certain molecular response. The host pattern recognition receptors (PRRs) that can perceive probiotic signals include Toll-like receptors (TLRs) and the C type lectin DC-specific intercellular adhesion molecule 3-grabbing non-integrin (DC-SIGN). Some molecular responses of IECs depend on the subtype of cell, for example, Paneth cells produce defensins and goblet cells produce mucus. Important responses of DCs against probiotics include the production of cytokines, major histocompatibility complex molecules for antigen presentation, and co-stimulatory molecules that polarize T cells into T helper or CD4+CD25+ regulatory T cells in the mesenteric lymph nodes (MLNs) or subepithelial dome of the GALT. IFNγ, interferon-γ; IL, interleukin; TGFb; transforming growth factor-β. Reproduced from: *S. Lebeer, J. Vanderleyden & S. C. J. De Keersmaecker (2010). Host interactions of probiotic bacterial surface molecules: comparison with commensals and pathogens. Nature Reviews Microbiology 8, 171-184.*

Figure 4. Interaction of probiotic bacteria with IECs and DCs from the GALT.

The tight epithelial cell barrier forms the another line of defence between the gut luminal contents and the host. Epithelial cells lining the gastrointestinal tract are able to respond to infection by initiating either nonspecific or specific host-defence response (Kagnoff and Eckmann 1997, Strober 1998). Bacterial adhesion to the host cell or recognition by the host cell is often an essential first stage in the disease process. A wide range of gastrointestinal cell surface constituents, such as several glygoconjucates, can serve as receptors for bacterial adherence (Servin and Coconnier 2003, Pretzer et al., 2005). Furthermore, epithelial cells express constitutively host pattern recognition receptors (PRRS), such as Toll-like receptors (TLR). These are a family of transmembrane receptors that recognize repetitive patterns, i.e. the pathogen-associated molecular patterns present in diverse microbes, including gram-positive and gram-negative bacteria (Bäckhed and Hornef 2003, Takeda et al., 2003). TLRs are also found on innate immune cells, such as dendritic cells and macrophages (Vinderola et al., 2005). TLR4 recognizes lipopolysaccharide and gram-negative bacteria, while TLR2 recognizes a variety of microbial components, such as peptidoglycan and lipoteichoic acids, from gram-positive bacteria (Abreu 2003, Matsuguchi et al., 2003, Takeda et al., 2003). Also, several other TLRs with specific actions are known, such as TLR5, which responds to the bacterial flagella (Rhee et al., 2005), and TLR9, which is activated by bacterially derived short DNA fragments containing CpG sequences (Pedersen et al., 2005). Other known recognition receptors are nucleotide-binding oligomerization domain proteins, which recognize both gram-positive and gram-negative bacteria. They are located in cell cytoplasm and are implicated in the induction of defensins. Increased epithelial barrier permeability is frequently associated with gastrointestinal disorders contributing to both disease onset and persistence (Lu and Walker 2001, Berkes 2003). The gatekeeper of the paracellular pathway is the tight junction, which is an apically located cell-cell junction between epithelial cells. The tight junction permits the passage of small molecules, such as ions, while restricting the movement of large molecules, such as antigens and microorganisms, which can cause inflammation. The integral membrane protein family, which are mainly claudins, occluding and zonula occludens 1, are implicated in the formation of the paracellular channels (Berkes et al., 2003).

6. Origine and safety of probiotics

An old dogma of probiotic selection has been that the probiotic strains should be of "human origin". One may argue that from evolutionary point of view, describing bacteria to be of human origin does not make much sense at all. The requirement for probiotics to be of human origin relates actually to the isolation of the strain rather than the "origin" itself. Usually, the strains claimed to be "of human origin" have been isolated from faecal samples of healthy human subjects, and have therefore been considered to be "part of normal healthy human gut microbiota". In reality the recovery of a strain from a faecal sample does not necessarily mean that this strain is part of the normal microbiota of this individual, since microbes passing the GI tract transiently can also be recovered from the faecal samples (Forssten et al., 2011). In practice it is impossible to know the actual origin of the probiotic strains, regardless of whether they have been isolated from faecal samples, fermented dairy products or any other source for that matter. Isolation of a strain from faeces of a healthy individual is also not a guarantee of the safety of the strain—such a sample will also always contain commensal microbes which can act as opportunistic pathogens, or even low levels of true pathogens, which are present in the individual at sub-clinical levels. Therefore, it has been recommend that instead of concentrating on the first point of isolation, the selection processes for new potential probiotic strains should mainly focus on the functional properties of the probiotic strains rather than the "origin" (Forssten et al., 2011; Ouwehand and Lahtinen 2008).

As viable, probiotic bacteria have to be consumed in large quantities, over an extended period of time, to exert beneficial effects; the issue of the safety of these microorganisms is of primary concern (Leroy et al., 2008). Until now, reports of a harmful effect of these microbes to the host are rare. However, many species of the genera Lactobacillus, Leuconostoc, Pediococcus, Enterococcus, and Bifidobacterium were isolated frequently from various types of infective lesions. According to Gasser (1994), L. rhamnosus, L. acidophilus, L. plantarum, L. casei, Lactobacillus paracasei, Lactobacillus salivarius, Lactobacillus lactis, and Leuconostoc mesenteroides are some examples of probiotic bacteria isolated from bacterial endocarditis; L. rhamnosus, L. plantarum, Leuconostoc. mesenteroides, Pediococcus acidilactici, Bifidobacterium eriksonii, and Bifidobacterium adolescentis have been isolated from bloodstream infections and many have been isolated from local infections. Although minor side effects of the use of probiotics have been reported, infections with probiotic bacteria occur and invariably only in immunocompromised patients or those with intestinal bleeding (Leroy et al., 2008).

An issue of concern regarding the use of probiotics is the presence of chromosomal, transposon, or plasmid-located antibiotic resistance genes amongst the probiotic microorganisms. At this moment, insufficient information is available on situations in which these genetic elements could be mobilised, and it is not known if situations could arise where this would become a clinical problem (Leroy et al., 2008). When dealing with the selection of probiotic strains, the FAO/WHO Consultancy recommends that probiotic microorganisms should not harbor transmissible drug resistance genes encoding resistance to clinically used drugs (FAO/WHO, 2002). For the assessment of the safety of probiotic microorganisms and products, FAO/WHO has formulated guidelines, recommending that probiotic strains should be evaluated for a number of parameters, including antibiotic susceptibility patterns, toxin production, metabolic and haemolytic activities, and infectivity in immunocompromised animals (FAO/WHO, 2002). In vitro safety screenings of probiotics may include, among others, antibiotic resistance assays, screenings for virulence factors, resistance to host defence mechanisms and induction of haemolysis. Several different animal models have been utilized in the safety assessment of probiotics. These include models of immunodeficiency, endocarditis, colitis and liver injury. In some cases even acute toxicity of probiotics has been assessed. Last but not least, also clinical intervention trials have yielded evidence on the safety of probiotics for human consumption (Forssten et al., 2011).

7. Conclusion

The individual diversity of the intestinal microflora underscores the difficulty of identifying the entire human microbiota and poses barriers to this field of research. In addition, it is apparent that the actions of probiotics are species and strain specific. It is also apparent that even a single strain of probiotic may exert its actions via multiple, concomitant pathways. Probiotics have long been used as an alternative to traditional medicine with the goal of maintaining enteric homeostasis and preventing disease. However, the actual efficacy of this treatment in still debated. Clinical trials have shown that probiotic treatment can reduce the risk of some diseases, especially antibiotic-associated diarrhea, but conclusive evidence is impeded owing to the wide range of doses and strains of bacteria used. The mechanism of action is also an area of interest (Ohland and MacNaughton, 2010). Many studies, as discussed above, have shown that probiotics increase barrier function in terms of increased mucus, antimicrobial peptides, and sIgA production, competitive adherence for pathogens, and increased TJ integrity of epithelial cells. Current investigation into the mechanism of action of specific probiotics has focused on probiotic-induced changes in the innate immune functions involving TLRs and its downstream systems Like NF-κB, and other pathways (Yoon and Sun, 2011). Although the immunomodulatory effects of probiotics have been demonstrated in experimental animal models of allergy, autoimmunity, and IBD, information from clinical trials in humans is scarce. Furthermore, some studies suggest that probiotics could induce detrimental effects. Therefore, more research, especially in the form of well-designed clinical trials, is needed to evaluate the efficacy and safety of probiotics (Ezendam and Van Loveren, 2008). With evolving knowledge, efective probiotic therapy will be possible in the future.

Author details

Daoud Harzallah and Hani Belhadj
Laboratory of Applied Microbiology, Faculty of Natural and Life Sciences, University Ferhat Abbas, 19000, Sétif, Algeria

8. References

Abreu M.T. (2003). Immunologic regulation of toll-like receptors in gut epithelium. Curr. Opin. Gastroenterol; 19:559-64.

Bäckhed F, and M. Hornef (2003). Toll-like receptor 4-mediated signaling by epithelial surfaces: necessity or threat? Microbes Infect.; 5:951-9.

Belhadj, H., Harzallah, D., Khennouf, S., Dahamna, S., Bouharati, S., and Baghiani, A. (2010). Isolation, identification and antimicrobial activity of lactic acid bacteria from Algerian honeybee collected pollen. Acta Hort. (ISHS) 854: 51 – 58

Berkes J, V.K. Viswanathan, S.D. Savkovic, G. Hecht (2003). Intestinal epithelial responses to enteric pathogens: effects on the tight junction barrier, ion transport, and inflammation. Gut;52:439-51.

Caballero-Franco C., K. Keller, C. De Simone, K. Chadee (2007). The VSL#3 probiotic formula induces mucin gene expression and secretion in colonic epithelial cells. Am J Physiol Gastrointest Liver Physiol 292: G315–G322.

Chow J., (2002). Probiotics and prebiotics: A brief overview, J. Ren. Nutr. 12 76–86.

Collado M. C., E. Isolauri, S. Salminen and Y. Sanz (2009). The Impact of Probiotic on Gut Health. Current Drug Metabolism, , 10, 68-78.

Desai A. (2008) Strain Identification, Viability And Probiotics Properties Of Lactobacillus Casei. PhD Thesis Victoria University, Victoria, Australia 228 pp.

Ezendam J., and H. Van Loveren (2008). Probiotics: Immunomodulation and Evaluation of Safety and Efficacy. Nutrition Reviews, 64, (1) : 1-14.

FAO/WHO. (2002). Guidelines for the evaluation of probiotics in food – Joint Food and Agricultural Organization of the United Nations and World Health Organization Working Group Meeting Report, London Ontario, Canada.

Forssten S. D., S. J. Lahtinen A. C. Ouwehand, and J. J. Malago et al. (eds.), (2011) Chapter 2 :The Intestinal Microbiota and Probiotics 24 pp. In : Probiotic Bacteria and Enteric Infections. Springer Science Business Media B.V.

Gareau M. G., P. M. Sherman and W. A.Walker (2010). Probiotics and the Gut Microbiota in Intestinal Health and Disease. Nature Reviews Gastroenterology and Hepatology 7, 503-514.

Gasser F. (1994) Safety of lactic acid bacteria and their occurrence in human clinical infections. Bull Inst Pasteur; 92:45–67.

Gaudier E., C. Michel, JP. Segain, C. Cherbut, and C. Hoebler (2005). The VSL# 3 probiotic mixture odifies microflora but does not heal chronic dextran- sodium sulfate-induced colitis or reinforce the mucus barrier in mice. J Nutr. 135: 2753–2761.

Gotteland M., O. Brunser, S. Cruchet (2006) Systematic review: are probiotics useful in controlling gastric colonization by Helicobacter pylori? Aliment. Pharmacol. Ther.23:1077–1086.

Holzapfel, W.H., P. Haberer, R. Geisen, J. Björkroth, U. Schillinger (2001) Taxonomy and important features of probiotic microorganisms in food nutrition. Am. J. Clin. Nutr. 73, 365S-373S.

Kagnoff M.F., and L. Eckmann (1997) Epithelial cells as sensors for microbial infection. J Clin Invest;100:6-10.

Kalabis J, I. Rosenberg, and DK. Podolsky (2006). Vangl1 protein acts as a downstream effector of intestinal trefoil factor (ITF)/TFF3 signaling and regulates wound healing of intestinal epithelium. J Biol Chem 281:6434–6441.

Kim Y, SH. Kim, KY. Whang, YJ. Kim, and S. Oh (2008). Inhibition of Escherichia coli O157:H7 attachment by interactions between lactic acid bacteria and intestinal epithelial cells. J Microbiol Biotechnol 18: 1278–1285.

Kollath, W. (1953) Ernährung und Zahnsystem. Deutsche Zahnärztliche Zeitschrift 8: 7-16.

Lebeer S., J. Vanderleyden and S. C. J. De Keersmaecker (2010). Host interactions of probiotic bacterial surface molecules: comparison with commensals and pathogens. Nature Reviews Microbiology 8, 171-184.

Leroy, F. and De Vuyst, L. (2004). Functional lactic acid bacteria starter cultures for the food fermentation industry. Trends in Food Science and Technology 15, 67-78.

Leroy F., G. Falony, and L. de Vuyst (2008) Latest Developments in Probiotics. In Toldrà F. (ed.), Meat Biotechnology, Springer Science+Business Media, LLC. 217-229 pp.

Lilly D.M, and R.H. Stillwell (1965) Probiotics: growth promoting factors produced by microorganisms. Science. 147: 747–8.

Lu L, and W.A. Walker (2001). Pathologic and physiologic interactions of bacteria with the gastrointestinal epithelium. Am. J. Clin. Nutr.;73:1124S-30S.

Mack DR, S. Ahrne, L. Hyde, S. Wei, and MA. Hollingsworth (2003). Extracellular MUC3 mucin secretion follows adherence of Lactobacillus strains to intestinal epithelial cells in vitro. Gut 52: 827–833.

Marco M. L, S. Pavan and M. Kleerebezem (2006). Towards understanding molecular modes of probiotic action . Curr Opin Biotechnol. 17:204–210.

Mattar AF., DH. Teitelbaum, RA. Drongowski, F. Yongyi, CM. Harmon, and AG. Coran (2002). Probiotics up-regulate MUC-2 mucin gene expression in a Caco-2 cell-culture model. Pediatr Surg. Int. 18: 586–590.

Matsuguchi T, A. Takagi, T. Matsuzaki, M. Nagaoka, K. Ishikawa, T. Yokokura, and Y. Yoshikai (2003). Lipoteichoic acids from Lactobacillus strains elicit strong tumor necrosis factor alpha-inducing activities in macrophages through Toll-like receptor 2. Clin Diagn Lab Immunol; 10:259-66.

Mercenier A., S. Pavan and B. Pot (2003) Probiotics as Biotherapeutic Agents: Present Knowledge and Future Prospects. Current Pharmaceutical Design, 8 : 99-110.

Metchnikoff, E. (1908). The Prolongation Of Life, New York, G.P. Putnam's Sons, 376 p.

Morelli, L. (2007). In vitro assessment of probiotic bacteria: From survival to functionality. International Dairy Journal, 17, 1278–1283.

Myllyluoma E. (2007). The Role Of Probiotics In Helicobacter pylori Infection. Dessertation, 86p. Nstitute Of Biomedicine Pharmacology University Of Helsinki, Helsinki, Finland.

Ng S.C., A.L. Hart, M.A. Kamm, A.J. Stagg, and S.C. Knight (2009) Mechanisms of Action of Probiotics: Recent Advances Inflamm Bowel Dis.15:300 –310.

Ohland C. L. and W. K. MacNaughton (2010). Probiotic bacteria and intestinal epithelial barrier function.Am.J. Physiol. Gastrointest. Liver Physiol. 298: G807–G819.

Osmanagaoglu O., and Y. Beyatli (1999) The Use of Bacteriocins Produced by Lactic Acid Bacteria in Food Biopreservation. Türk. Mikrobiyol. Cem. Derg. 32: 295-306.

Ouwehand A.C., Lahtinen S.J. (2008). The (non-) sense of human origin of probiotics. NutraCos, 7:8–10.

Parada J. L., C. R.Caron, A. B. P. Medeiros and C. R. Soccol (2007) Bacteriocins from Lactic Acid Bacteria: Purification, Properties and use as Biopreservatives. Brazilian Archives of Biology And Technology. Vol.50 (3):521-542.

Pedersen G, L. Andresen, M.W. Matthiessen, J. Rask-Madsen, J. Brynskov (2005). Expression of Toll-like receptor 9 and response to bacterial CpG oligodeoxynucleotides in human intestinal epithelium. Clin Exp Immunol;141:298-306.

Pretzer G, J. Snel, D. Molenaar, A. Wiersma, P.A. Bron, J. Lambert, WM de Vos, R. Ven Der Meer, M.A. Smits, and M. Kleerebezem (2005). Biodiversity-based identification and functional characterization of the mannose-specific adhesin of Lactobacillus plantarum. J Bacteriol. 187:6128-36.

Rhee SH, E. Im, M. Riegler, E. Kokkotou, M. O'Brien, C. Pothoulakis. (2005). Pathophysiological role of Toll-like receptor 5 engagement by bacterial flagellin in colonic inflammation. Proc. Natl. Acad .Sci.; 102:13610-5.

Sanders M. E. (2008) Probiotics: Definition, Sources, Selection, and Uses. Clinical Infectious Diseases. 46:S58–61.

Shanahan F. (2003) Probiotics: a perspective on problems and pitfalls, Scand. J. Gastroenterol. 38 (S) :34– 36.

Schlee M., J. Wehkamp, A. Altenhoefer, et al. (2007) Induction of human beta-defensin 2 by the probiotic Escherichia coli Nissle 1917 is mediated through flagellin. Infect Immun.75:2399–2407.

Schrezenmeir J., and M. De Vrese (2001) probiotics, prebiotics, and symbiotics – approaching a definition. Am. J. Clin. Nutr. 73:361S-364S.

Sherman PM, K.C. Johnson-Henry, H.P. Yeung, et al. (2005) Probiotics reduce enterohemorrhagic Escherichia coli O157:H7- and enteropathogenic E. coli O127:H6-induced changes in polarized T84 epithelial cell monolayers by reducing bacterial adhesion and cytoskeletal rearrangements. Infect Immun,73:5183–5188.

Sekirov I., S. L. Russell, L. C. M. Antunes, And B. B. Finlay (2010). Gut Microbiota in Health and Disease. Physiol Rev 90: 859–904.

Servin A.L., M.H. Coconnier (2003). Adhesion of probiotic strains to the intestinal mucosa and interaction with pathogens. Best Pract. Res. Clin. Gastroenterol;17:741-54.

Soccol C. R., L. P. de Souza Vandenberghe, M. R. Spier, A. B. Pedroni Medeiros, C. T. Yamaguishi, J. De D. Lindner, A. Pandey and V. Thomaz-Soccol1(2010). The Potential of Probiotics: A Review. Food Technol. Biotechnol. 48 (4): 413–434.

Strober W. (1998). Interactions between epithelial cells and immune cells in the intestine. Ann N.Y Acad. Sci.; 859:37-45.

Takeda K, T. Kaisho, and S. Akira (2003). Toll-like receptors. Annu. Rev. Immunol.; 21:335-76.

Tuomola, E., R. Crittenden, M. Playne, E. Isolauri, and S. Salminen. (2001). Quality assurance criteria for probiotic bacteria. Am J Clin Nutr.; 73:393S -398S.

Turner J. R. (2009). Intestinal mucosal barrier function in health and disease. Nature Reviews Immunology. 9: 799-809.

Vasiljevic and N.P. Shah, (2008) Review: probiotics — from Metchnikoff to bioactives, International Dairy Journal 18, pp. 714–728.

Vinderola G, C. Matar, and G. Perdigon. (2005). Role of intestinal epithelial cells in immune effects mediated by gram-positive probiotic bacteria: involvement of toll-like receptors. Clin. Diagn. Lab Immunol; 12:1075-84.

Vergin F. (1954) Anti- and probiotika. Hippokrates 25: 16–19.

Yoon S. S., and J. Sun (2011). Probiotics, Nuclear Receptor Signaling, and Anti-Inflammatory Pathways. Gastroenterology Research and Practice Volume 111: 1-16.

Wehkamp J., J. Harder, K. Wehkamp, et al. (2004). NF-kappaB- and AP-1-mediated induction of human beta defensin-2 in intestinal epithelial cells by Escherichia coli Nissle 1917: a novel effect of a probiotic bacterium. Infect Immun.72:5750–5758.

Chapter 13

Dose Effects of LAB on Modulation of Rotavirus Vaccine Induced Immune Responses

Lijuan Yuan, Ke Wen, Fangning Liu and Guohua Li

Additional information is available at the end of the chapter

1. Introduction

Probiotics can influence both mucosal and systemic immune responses and function as adjuvants by promoting proinflammatory cytokine production, enhancing both humoral and cellular immune responses. Adjuvant effects of several probiotic lactic acid bacteria (LAB), mostly *Lactobacillus* strains, including *L. rhamnosus* GG, *L. acidophilus* NCFM, *L. acidophilus* CRL431, *L. acidophilus* La-14, *L. fermentum* CECT5716, *L. casei* DN-114 001, and *Bifidobacterium lactis* Bl-04 have been reported in studies of influenza, polio, rotavirus and cholera vaccines and rotavirus and Salmonella typhi Ty21a infections (Boge et al., 2009; Davidson et al., 2011; Isolauri et al., 1995; Kaila et al., 1992; Mohamadzadeh et al., 2008; Olivares et al., 2007; Paineau et al., 2008; Winkler et al., 2005; Zhang et al., 2008b). The word adjuvant in the phrase "probiotic adjuvant" is not used in its traditional definition in which adjuvant implies a substance included in the vaccine formulation to aid the immune response to the vaccine antigen. Instead, probiotic adjuvants enhance the immunogenicity of vaccines when orally administered repeatedly around the time of vaccination and separately from the vaccine. By skewing the balance of pro- and anti-inflammatory innate immune responses and T helper (Th) 1 and regulatory T (Treg) cell adaptive immune responses in the context of vaccination, probiotic adjuvants act as "signal zero" to reduce Treg cell suppression and unleash effector T cell activation (Rowe et al., 2012).

Although the strain-specific effects of LAB in up- or down-regulating inflammatory immune responses have been well recognized, dose effects of probiotics on innate and adaptive immune responses are not clearly understood. The same *Lactobacillus* strain is oftentimes reported by different research groups to have opposite immune modulating effects. We hypothesized that the dose effect is at least one of the reasons for the conflicting reports. Understanding dose effects of probiotics has significant implications in their use as immunostimulatory (adjuvants) versus immunoregulatory agents.

In this chapter, we discuss findings from our serial studies of gnotobiotic pigs on the dose effects of the *L. acidophilus* NCFM strain (LA) on innate and adaptive immune responses induced by an oral attenuated human rotavirus (HRV) vaccine (AttHRV). We studied the effects of low dose (total 2.11 x 10^6 CFU) and high dose (total 2.22 x 10^9 CFU) LA on the intestinal and systemic (1) rotavirus-specific IFN-γ producing CD4+ and CD8+ T cell responses; (2) CD4+CD25+FoxP3+ and CD4+CD25-FoxP3+ Treg cell responses and the regulatory cytokine TGF-β and IL-10 production; (3) rotavirus-specific antibody-secreting cell (ASC) and serum antibody responses; and (4) plasmacytoid dendritic cell (pDC) and conventional DC (cDC) frequencies, activation status, TLR expression and cytokine production profile. The protective effect of rotavirus vaccine against virus shedding and diarrhea was assessed in AttHRV-vaccinated gnotobiotic pigs fed with high, low, or no LA and challenged with the virulent HRV.

These studies clearly demonstrated differential immune modulating effects of high dose versus low dose LA on DC and T cell responses, and consequently different effects on the protection conferred by the AttHRV vaccine in gnotobiotic pigs challenged with virulent HRV. Low dose LA enhanced the protection against rotavirus diarrhea in AttHRV-vaccinated pigs whereas high dose LA had negative effects on the effectiveness of the vaccine. Thus, the same probiotic strains at different doses can exert qualitatively different modulating effects on immune responses induced by rotavirus vaccines and possibly other vaccines as well.

2. Dose effects of LA on T cell responses

Probiotics have been reported to exert adjuvant properties by inducing pro-Th1 cytokines and promote Th1 type immune responses. For example, *L. lactis* and *L. plantarum* induced production of IL-12 and IFN-γ by splenocytes when the LAB and an allergen were co-administered intranasally to mice (Repa et al., 2003). *L. fermentum* strain CECT5716 enhanced the Th1 responses induced by an influenza vaccine in addition to enhancing virus neutralizing antibody responses (Olivares et al., 2007). Eleven different probiotic strains were tested for cytokine production in human peripheral blood mononuclear cells (MNC) and each tested bacterium was shown to induce production of TNF-α and some strains also induced production of IL-12 and IFN-γ (Kekkonen et al., 2008). Previous studies of gnotobiotic pigs showed that a mixture of LA NCFM strain and *L. reuteri* strain enhanced both Th1 (IL-12, IFN-γ) and Th2 (IL-4 and IL-10) cytokine responses to virulent HRV infection (Azevedo et al., 2012). LA NCFM strain enhanced the HRV-specific IFN-γ producing CD8+ T cell response to a rotavirus vaccine in gnotobiotic pigs, indicating adjuvanticity of the LA strain (Zhang et al., 2008b).

Dose effects of probiotics on modulating T cell immune responses have not been well studied. To address this question, we examined the dose effects of LA NCFM (NCK56) strain on IFN-γ producing CD4+ and CD8+ T cell immune responses induced by an oral rotavirus vaccine in gnotobiotic pigs (Wen et al., 2012). The animal treatment groups included (1) high dose LA plus AttHRV vaccine (HiLA+AttHRV), (2) low dose LA plus

AttHRV (LoLA+AttHRV), (3) AttHRV only, (4) high dose LA only (HiLA), (5) low dose LA only (LoLA), and (6) mock-inoculated control (Mock). Gnotobiotic pigs were orally inoculated at 5 days of age with the AttHRV vaccine at 5 x10^7 fluorescent focus-forming units (FFU) per dose. A booster dose was given 10 days later at the same dose and route. Subsets of the pigs were euthanized at post-inoculation day (PID) 28 to assess immune responses and the rest were challenged with the homotypic virulent Wa (G1,P1A[8]) strain HRV at a dose of 1 x 10^5 FFU to assess protection from post-challenge day (PCD) 1 to 7. The 50% infectious dose and 50% diarrhea dose of the virulent HRV in gnotobiotic pigs are approximately 1 FFU (Ward et al., 1996). The AttHRV inoculation causes virus shedding in about 6% pigs, but it does not cause any illness (Ward et al., 1996). Pigs in the high dose LA groups were fed daily with 10^3 to 10^9 CFU/dose of LA for 14 days with 10-fold incremental dose increases every other day from 3-16 days of age. The accumulative total LA dose was 2.22 x 10^9 CFU. Pigs in the low dose LA groups were fed with 10^3, 10^4, 10^5, 10^6, and 10^6 CFU/dose of LA every other day from 3-11 days of age. The accumulative total LA dose was 2.11 x 10^6 CFU.

2.1. Low dose LA, but not high dose LA, enhanced HRV-specific IFN-γ producing T cell responses

The magnitude of HRV-specific IFN-γ producing T cell responses in pigs was differentially modulated by low versus high dose LA at both prechallenge and postchallenge (PID 28 and PCD 7). AttHRV-vaccinated and low dose LA fed pigs (LoLA+AttHRV) had significantly higher frequencies of HRV-specific IFN-γ+CD8+ T cells in ileum (11- and 5-fold higher pre- and postchallenge, respectively), spleen (3.8- and 2.1-fold higher pre- and postchallenge, respectively) and blood (3- and 20-fold higher pre- and postchallenge, respectively) compared to the AttHRV only pigs (Table 1). The LoLA+AttHRV pigs also had significantly higher frequencies of HRV-specific IFN-γ+CD4+ T cells in blood (3-fold higher for both pre- and postchallenge) compared to the AttHRV only pigs. In contrast, high dose LA did not significantly alter the HRV-specific IFN-γ producing CD4+ and CD8+ T cell responses in the HiLA+HRV pigs compared to AttHRV only pigs.

	Frequencies of IFN-γ+CD8+ T cells among CD3+ cells					
	PID 28			PCD 7		
	Ileum	Spleen	Blood	Ileum	Spleen	Blood
HiLA+AttHRV	0.05	0.34	0.05	0.11	0.16	0.06
LoLA+AttHRV	1.21	0.46	0.24	0.56	0.46	0.98
AttHRV only	0.11	0.12	0.08	0.11	0.22	0.05

(Summarized from Wen et al., 2012)

Table 1. Effect of low dose vs. high dose LA on IFN-γ producing CD8+ T cell responses

2.2. High dose LA significantly increased frequencies of intestinal and systemic CD4+CD25-FoxP3+ Treg cells whereas low dose LA decreased TGF-β and IL-10 producing Treg cell responses

Frequencies and cytokine production of Treg cells in pigs were differentially modulated by low versus high dose LA at both prechallenge and postchallenge. HiLA+AttHRV pigs had significantly higher frequencies of CD4+CD25-FoxP3+ Treg cells (ranging from 6- to 86-fold higher) in all the tissues compared to LoLA+AttHRV and AttHRV only pigs pre- and postchallenge (Table 2).

Because Treg cells exert regulatory functions through mechanisms involving TGF-β and IL-10, we also compared the effects of high and low dose LA on frequencies of the Treg cell subsets that produced TGF-β or IL-10 among the AttHRV-vaccinated pigs.

	Frequencies of CD4+CD25-FoxP3+Treg cells among total MNC					
	PID 28			PCD 7		
	Ileum	Spleen	Blood	Ileum	Spleen	Blood
HiLA+AttHRV	2.96	10.34	9.00	1.56	3.78	3.65
LoLA+AttHRV	0.24	0.54	0.10	0.08	0.22	0.05
AttHRV only	0.23	0.59	0.51	0.27	0.26	0.17

(Summarized from Wen et al., 2012)

Table 2. Effect of low dose vs. high dose LA on frequencies of Treg cells

Low dose LA reduced frequencies of TGF-β producing CD4+CD25+FoxP3+ and CD4+CD25-FoxP3+ Treg cells compared to high dose LA and AttHRV only pigs in all tissues pre- and postchallenge (Table 3; data for CD25+ Treg cells are not shown). Low dose LA also reduced pre- and postchallenge frequencies of IL-10 producing CD4+CD25+FoxP3+ and CD4+CD25-FoxP3+Treg cells compared to high dose LA and AttHRV- only (except for CD4+CD25-FoxP3+ Treg cells in ileum and spleen postchallenge) (Table 3). High dose LA induced 2.6-fold and 20-fold, respectively higher frequencies of IL-10 producing CD4+CD25-FoxP3+ Treg cells in ileum and spleen postchallenge compared to AttHRV only.

	PID 28			PCD 7		
	Ileum	Spleen	Blood	Ileum	Spleen	Blood
	Frequencies of TGF-β+ cells among CD4+CD25-FoxP3+ Treg cells					
HiLA+AttHRV	4.51	1.13	0.92	4.87	10.05	1.39
LoLA+AttHRV	0.32	0.15	0.31	0.00	0.17	0.38
AttHRV only	4.19	1.19	1.55	4.34	0.79	1.53
	Frequencies of IL-10+ cells among CD4+CD25-FoxP3+ Treg cells					
HiLA+AttHRV	4.21	2.97	0.88	10.68	17.70	2.51
LoLA+AttHRV	0.92	0.15	0.00	5.22	6.79	0.00
AttHRV only	4.47	3.01	1.77	4.08	0.90	1.44

(Summarized from Wen et al., 2012)

Table 3. Effect of low dose vs. high dose LA on CD25-FoxP3+ Treg cell cytokine production

These data clearly demonstrated that low dose LA promoted IFN-γ producing T cell and down-regulated Treg cell responses, whereas high dose LA induced a strong Treg cell response and promoted the regulatory cytokine production by tissue-residing Treg cells postchallenge in gnotobiotic pigs. Studies of other lactobacilli strains have reported similar findings. A mixture of *L. plantarum* CEC 7315 and CEC 7316 at high dose (5×10^9 CFU/day) resulted in significant increases in the percentages of activated potential T-suppressor and NK cells, while at low dose (5×10^8 CFU/day) increased activated T-helper cells, B cells and antigen presenting cells (APCs) in institutionalized seniors (Mane et al., 2011). High concentration ($\geq 1\times10^6$ colony forming unit [CFU]/ml) of a combination containing LA and *Bifidobacterium* or *B. infantis* attenuated mitogen-induced immune responses by inhibiting cell proliferation and arresting the cell cycle at the G0/G1 stage in both mitogen-stimulated spleen cells and peripheral blood MNC. However, low concentration ($\leq 1\times10^6$ CFU/ml) promoted a shift in the Th1/Th2 balance toward Th1-skewed immunity by enhancing IFN-γ and inhibiting IL-4 response (Li et al., 2011). The differences between the "low dose" and "high dose" LAB in these studies are small, yet the immunomodulatory effects are qualitatively different.

Dose effects may explain some of the controversies that result from the same probiotic strain used by different research groups in animal studies showing opposite immunomodulatory functions. For example, administration of *L. casei* suppressed pro-inflammatory cytokine expression by CD4+ T cells and up-regulated IL-10 and TGF-β levels in rats (So et al., 2008a; So et al., 2008b). On the contrary, another study found that *L. casei* was a pure Th1 inducer in mice. In addition to the difference in animal species, the *L. casei* doses used by the different studies differed significantly, with much higher doses used in the studies of rats (So et al., 2008a; So et al., 2008b). In the studies of rats, the amount of *L. casei* was 5×10^9 or 2×10^{10} CFU/dose per rat, three times per week for 11-12 weeks. In the study of mice, the amount of *L. casei* was 2×10^8 CFU/dose per mouse, twice per week for 8 weeks (Van Overtvelt et al., 2010). Thus, different dose and frequency of administration of the same LAB strains may result in totally different *in vivo* effects.

The dose effects of LA on immune responses to the AttHRV vaccine in pigs may also partly explain why the efficacies of oral rotavirus vaccines are significantly reduced in low-income countries compared to developed countries. The two licensed rotavirus vaccines, RotaTeq and Rotarix have a protective efficacy of >85% against moderate to severe rotavirus gastroenteritis in middle and high-income countries (O'Ryan et al., 2009). However, the protective efficacy of RotaTeq vaccine is only 39.3% against severe rotavirus gastroenteritis in sub-Saharan Africa (Armah et al., 2010) and 48.3% in developing countries in Asia (Zaman et al., 2010). Rotarix vaccine showed a similar disparity in efficacy in low-income countries in Africa (O'Ryan & Linhares, 2009). In addition to other factors that contribute to the reduction in rotavirus vaccine efficacy (e.g., higher titers of maternal antibodies, malnutrition), during the initial colonization of human infants, exposure to high doses of commensal bacteria (common in countries with lower hygiene standards) would have a suppressive effect on IFN-γ producing T cell responses and promote Treg cell responses, thus leading to the lowered protective immunity after rotavirus vaccination.

3. Dose effects of LA on antibody and B cell responses

Probiotics are known to modulate both humoral and cellular immune responses. Probiotics can induce antigen-specific and non-specific IgA antibody responses at mucosal surfaces (Perdigon et al., 2001; Wells & Mercenier, 2008) to prevent invasion by pathogenic microorganisms. Oral administration of *L. acidophilus* L-92 strain led to a significant increase of IgA production in Peyer's patches in mice (Torii et al., 2007). *L. casei* CRL 431 strain increased induction of intestinal IgA secreting cells in mice (Galdeano & Perdigon, 2006). *L. acidophilus* La1 strain and bifidobacteria enhanced specific serum IgA titers to *S. typhi* strain Ty21a and also total serum IgA in humans (Link-Amster et al., 1994). *L. rhamnosus* GG enhanced rotavirus-specific IgA ASC responses in humans and promoted recovery from rotavirus diarrhea (Kaila et al., 1992). In our earlier study of gnotobiotic pigs, a mixture of LA strain and *L. reuteri* strain did not alter virus-specific intestinal and systemic antibody and ASC responses, but they significantly enhanced total intestinal IgA secreting cell responses and total serum IgM and intestinal IgM and IgG titers in rotavirus infected pigs (Zhang et al., 2008a).

The first reported adjuvant effect of probiotic LAB in vaccination was from a human clinical trial in which *L. rhamnosus* GG was shown to enhance rotavirus-specific IgM secreting cells and rotavirus IgA seroconversion in infants receiving a live oral rhesus-human rotavirus reassortant vaccine (Isolauri et al., 1995). In recent years, an increasing number of human clinical trials have demonstrated adjuvant effects of probiotics in enhancing vaccine-induced antibody responses. In a double-blind randomized controlled trial, *L. rhamnosus* GG or *L. acidophilus* CRL 431 increased serum poliovirus neutralizing antibody titers and poliovirus-specific IgA and IgG titers 2- to 4-fold in adult human volunteers vaccinated with the live oral polio vaccine (de Vrese et al., 2005). In another human clinical trial, six out of the seven probiotic strains tested enhanced cholera-specific IgG antibody concentration in serum; for the *B. lactis* Bl-04 and *L. acidophilus* La-14 strains the increase was more significant (Paineau et al., 2008). Daily consumption of a fermented dairy drink (*L. casei* DN-114 001 and yoghurt ferments, Actimel) was shown to increase virus specific antibody responses to the intramuscular inactivated influenza vaccine in individuals of over 70 years of age (Boge et al., 2009). In a randomized, double-blind placebo-controlled pilot study, *L. rhamnosus* GG significantly improved the development of serum antibody responses to the H3N2 strain influenza virus (84% receiving *L. rhamnosus* GG versus 55% receiving placebo had a protective titer 28 days after vaccination) in healthy adults receiving the live attenuated influenza vaccine (FluMist, Medimmune Vaccines, Gaithersburg, MD, USA) (Davidson et al., 2011). Thus, specific strains of probiotics can act as adjuvants to enhance humoral immune responses following not only mucosal (oral or intranasal) but also parenteral vaccination. Yet, dose effects of probiotics on antibodies responses have not been well studied.

In our studies, we demonstrated that high dose LA did not significantly alter the HRV-specific antibody responses whereas low dose LA had negative effects on the antibody responses. The effect of high and low dose LA NCFM strain on HRV-specific serum IgA and IgG antibody levels and HRV-specific ASC and memory B cell responses in the intestinal and systemic lymphoid tissues of gnotobiotic pigs induced by rotavirus vaccination were examined. The animal treatment groups were the same as listed above in the studies of T cell responses. High dose LA did not significantly alter the HRV-specific antibody responses in serum and ASC responses in any tissue at PID 28 and PCD 7 in the AttHRV-vaccinated pigs (Figs 1 and 2), except to reduce the IgG ASC response in ileum of the mock-vaccinated pigs postchallenge (Fig. 2c). In contrast, low dose LA significantly reduced the HRV-specific IgA antibody titers at PID 7 and 14 (Fig. 1a) and reduced or significantly reduced IgG ASC responses in blood pre- and postchallenge as well as IgA ASC responses in spleen and blood postchallenge (Fig. 2a and 2b). Low dose LA also significantly reduced the IgA and IgG ASC responses in ileum of the mock-vaccinated pigs postchallenge (Fig. 2c).

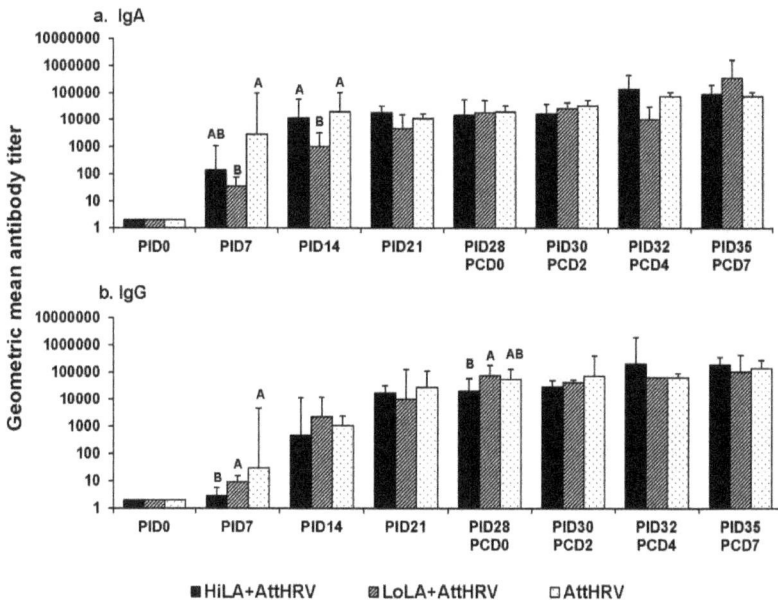

Figure 1. Rotavirus-specific serum IgA and IgG antibody responses in Gn pigs vaccinated with AttHRV with or without high or low dose LA feeding. Rotavirus-specific serum IgA (a) and IgG (b) antibody were measured by an indirect isotype-specific antibody ELISA. Error bars indicate the standard error of the mean. Different capital letters (A, B) indicate significant difference among different pig groups at the same time point (Kruskal Wallis Test, p<0.05, n=3-27), whereas shared letters or no letters on top indicate no significant difference.

The negative effects of low dose LA on the HRV-specific serum antibody responses and ASC responses induced by the AttHRV vaccine were undesirable for the vaccine's immunogenicity; however it is consistent with the strong pro-Th1 effect of the low dose LA. The skewed balance toward a Th1 type immune response in the low dose LA group may have resulted in the weakened antibody responses. In the subsequent studies, we evaluated the effects of a low dose and an intermediate dose of *L. rhamnosus* GG on the effector T cell, antibody and ASC responses induced by the AttHRV vaccine and we found that *L. rhamnosus* GG enhanced the production of a balanced Th1 and Th2 immune responses to the AttHRV vaccine and significantly increased the virus-specific IFN-γ producing T cell responses, the antibody responses and the protection rate of the AttHRV vaccine (manuscripts under preparation).

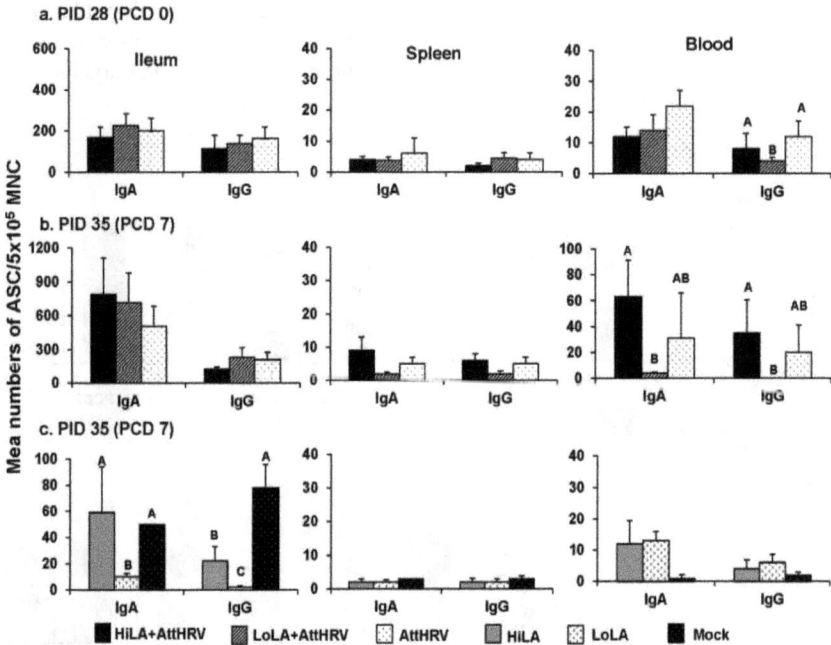

Figure 2. Rotavirus-specific IgA and IgG ASC responses in Gn pigs. Rotavirus-specific IgA and IgG ASC in the MNC isolated from ileum, spleen and blood of AttHRV-vaccinated pigs on PID 28 (PCD 0) (a) and PID 35 (PCD 7) (b) and of mock-vaccinated pigs on PID 35 (PCD 7) (c) were enumerated by using an ELISPOT assay and are presented as the mean numbers of virus-specific IgA and IgG ASC per 5×10⁵ MNC. Error bars indicate the standard error of the mean. Different capital letters (A, B, C) on top of the bars indicate significant difference among the treatment groups for the same isotype in the same tissue (Kruskal Wallis Test, p<0.05, n=3-14), whereas shared letters or no letters on top indicate no significant difference. Note the y axis scale difference (HiLA, high dose LA; LoLA, low dose LA).

4. Dose effects of LA on DC responses

The nature and consequences of a CD4+ T cell response (Th1, Th2, Th17, or Treg type) largely depend on the immune functions of DCs, which are the professional antigen presenting cells that can prime and differentiate naive T cells. Both cDC and pDC are responsible for presenting microbial and dietary antigens to the adaptive immune systems, thereby influencing polarization of the adaptive immune response (Konieczna et al., 2012).

The pDC most effectively sense virus infections and are characterized by their capacity to produce large quantities of IFN-α and the pro-inflammatory cytokines IL-6 and TNF-α. These cytokines promote cDC maturation (Summerfield & McCullough, 2009). MHC II expression in professional APCs is tightly regulated. The MHC II of immature DCs are expressed at low levels at the plasma membrane, but abundantly in endocytic compartments. In the presence of inflammatory cytokines such as IFN-γ, DCs are activated; they stop capturing antigens and markedly increase MHC II expression on their plasma membrane. These MHC II are loaded with peptides derived from antigens captured at the site of inflammation. The mature DCs migrate to lymphoid tissues and up-regulate the co-stimulatory molecules (CD80/86) necessary to activate naïve T cells (Villadangos et al., 2001).

It is known that probiotics can modify the distribution, the phenotype and the function of DC subsets (Grangette, 2012). Both species-specific and strain-specific immunomodulatory effects of different LAB on DCs have been described in a large number of studies and was reviewed previously (Meijerink & Wells, 2010). Among the differential effects, several lactobacilli strains, including *L. acidophilus, L. gasseri, L. fermentum, L. casei, L. plantarum, L. johnsonii,* and *L. rhamnosus* have been reported to stimulate human or murine DCs to produce increased levels of proinflammatory cytokines (IL-2, IL-12, TNF-α) that favored Th1 and cytotoxic T cell polarization, and decreased levels of the regulatory cytokine TGF-β (Chiba et al., 2010; Christensen et al., 2002; Mohamadzadeh et al., 2005; Van Overtvelt et al., 2010; Vitini et al., 2000; Weiss et al., 2010; Yazdi et al., 2010). Such immune stimulating effects are characteristics of adjuvants. However, studies of dose effects of lactobacilli on DC responses are scarce, with most consisting of *in vitro* experiments, and there is a dearth of comparative studies linking *in vitro* and *in vivo* results. *L. rhamnosus* Lcr35 was shown to induce a dose-dependent immunomodulation of human DCs. Lcr35 at 10^7 CFU/ml (10 multiplicity of infection), but not 10^4 CFU/ml induced the semi-maturation of the DCs and a strong pro-inflammatory response (Evrard et al., 2011). LA NCFM induced a concentration dependent production of IL-10, and low IL-12p70 in monocyte derived DCs (Konstantinov et al., 2008). Immature DCs incubated with the LA NCFM at a bacterium to cell ratio of 1000:1 ("high dose") produced significantly higher IL-10 compared with the ratio of 10:1. In contrast, IL-12p70 was up-regulated at a lower concentration of the bacterium (10:1).

In our studies, dose effects of LA on pDC and cDC responses after rotavirus vaccination were examined in gnotobiotic pigs. The animal treatment groups were the same as listed earlier in the studies of T cell and B cell immune responses. Porcine pDC (CD172a+CD4+) and cDC (CD172a+CD11R1+) were defined as previously described (Jamin et al., 2006). The frequencies and tissue distribution, MHC II and costimulatory (CD80/86) molecular, TLR (2, 3, 9) and cytokine (IL-6, IL-10, IFN-α, TNF-α) expression by pDC and cDC in ileum, spleen and blood of gnotobiotic pigs vaccinated with the AttHRV and fed with high dose, low dose or no LA were determined using multi-color flow cytometry.

The low dose LA group had significantly higher frequencies of pDC in ileum and spleen and cDC in spleen and blood compared to the high dose LA and AttHRV only groups (Fig. 3a). The low dose LA group had overall lower MHC II expression on pDC and cDC in all tissues and lower CD80/86 expression in blood, but significantly higher CD80/86 expression on cDC in ileum, compared to the high dose LA and AttHRV only groups (Fig. 3b). High dose LA did not have a significant effect on DC frequencies or activation marker MHC II and CD80/86 expression, except for the significantly increased CD80/86 expression on pDC in ileum compared to the AttHRV only group (Fig. 3b).

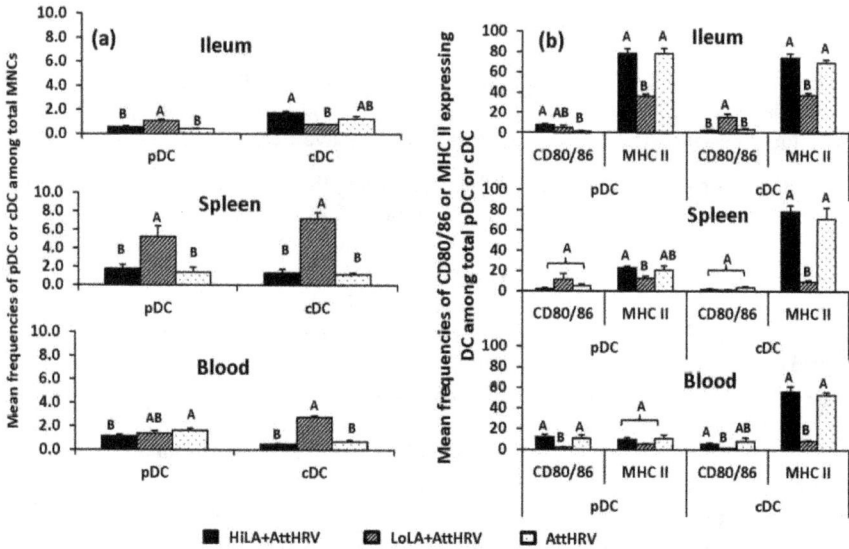

Figure 3. Frequencies of pDC and cDC (a) and the activation marker (CD80/86 and MHC II) expression (b) in intestinal and systemic lymphoid tissues of Gn pigs vaccinated with AttHRV vaccine with high dose, low dose or no LA at PID 28. MNC were stained freshly without *in vitro* stimulation before being subjected to flow cytometry analyses. Data are presented as mean frequency ± standard error of the mean (n = 3-13). Different letters on top of bars indicate significant differences in frequencies among groups for the same cell type and tissue (Kruskal–Wallis test, p < 0.05), while shared letters indicate no significant difference.

The low dose LA group had lower or significantly lower frequencies of TLR3 expression in both pDC and cDC in all tissues and significantly lower TLR2 expression on cDC in spleen compared to the high dose LA and AttHRV only groups (Fig. 4). High dose LA did not have a significant effect on TLR expression in ileum and spleen. In blood, high dose LA group had significantly lower TLR3 expression in cDC (and lower in pDC) compared to the AttHRV only group.

The most striking dose effect of LA on the cytokine production profile in DCs is the significantly increased IL-6 in the low dose LA group (Fig. 5). The low dose LA group had significantly higher frequencies of IL-6 producing pDC and cDC in all tissues compared to the high dose LA and AttHRV only groups. Interestingly, the low dose LA reduced or significantly reduced the other cytokine TNF-α, IL-10 and IFN-α production in pDC in ileum and spleen. In contrast to ileum and spleen, the low dose LA increased or significantly increased TNF-α, IL-10 and IFN-α production in blood compared to the high dose LA or the AttHRV only group. High dose LA did not have a significant effect on IL-6, TNF-α, and IL-10 but lowered or significantly lowered IFN-α production in both pDC and cDC in all tissues compared to the AttHRV only group.

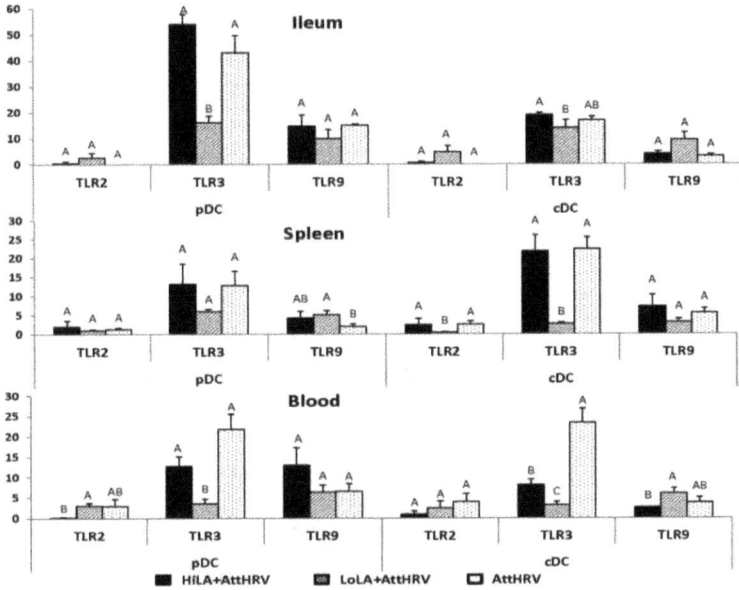

Figure 4. TLR expression patterns of pDC and cDC in intestinal and systemic lymphoid tissues of Gn pigs vaccinated with AttHRV vaccine with high dose, low dose or no LA at PID 28. MNC were stained freshly without *in vitro* stimulation before flow cytometry analyses. Data are presented as mean frequency ± standard error of the mean (n = 3-5). Different letters on top of bars indicate significant differences in frequencies among groups for the same TLR in the same tissue (Kruskal–Wallis test, p < 0.05), while shared letters indicate no significant difference.

Therefore, the effects of high versus low dose LA on the frequencies, maturation status and functions of DCs were strikingly different. Low dose LA significantly increased frequencies of both DC subsets, but these DCs were immature because they expressed lower frequencies of activation makers CD80/86 and MHC II and had reduced TNF-α, IL-10 and IFN-α production compared to the high dose LA and AttHRV only groups. Low dose LA promoted a strong IL-6 response in all tissues and increased all the other cytokine TNF-α, IL-10 and IFN-α production in blood for both pDC and cDC. High dose LA did not have such a significant modulating effect on the DC responses compared to the low dose (with a few exceptions). These findings are consistent with the differential effects of low dose versus high dose LA on the adaptive immune responses. The differential modulating effects of high versus low dose LA are intriguing. The biological and immunological implications of these effects and the underlying mechanisms require further investigation. From these data, it is clear that the same probiotic strain at different doses can exert qualitatively different modulating effects on DCs and consequently on adaptive immune responses induced by rotavirus vaccines. It has been reported that the effect of low dose microbe-associated pattern molecular (MAPM), such as lipopolysaccharide, was strikingly different as compared to that of high dose on macrophage cell functions: low dose lipopolysaccharide induced a strong inflammatory response in macrophages (Maitra et al., 2011). It is plausible that a similar interaction occurs between the MAPM from LA and DCs in the gut. Future studies are needed to identify the molecular mechanisms of the dose responses of different MAPM.

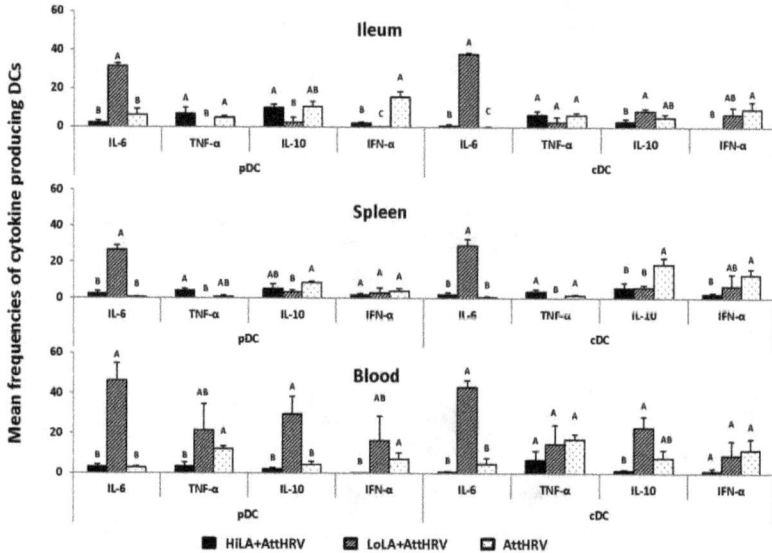

Figure 5. Cytokine production profiles of pDC and cDC in intestinal and systemic lymphoid tissues of Gn pigs vaccinated with AttHRV vaccine with high dose, low dose or no LA at PID 28. MNC were stained freshly without *in vitro* stimulation before flow cytometry analyses. Data are presented as mean frequency ± standard error of the mean (n = 3-8). Different letters on top of bars indicate significant differences in frequencies among groups for the same cytokine in the same tissue (Kruskal–Wallis test, p < 0.05), while shared letters indicate no significant difference.

5. Dose effects of LA on protection conferred by the oral AttHRV vaccine against virulent HRV challenge

To examine the effects of low and high dose LA on improving the protection conferred by the AttHRV vaccine, subsets of gnotobiotic pigs from each treatment group were challenged with the virulent HRV Wa strain at PID 28. Clinical signs and virus shedding were monitored for 7 days postchallenge (Table 4).

After challenge, although the proportion of pigs that developed virus shedding and diarrhea did not differ significantly among the three AttHRV vaccinated pig groups, the LoLA+AttHRV group had the shortest mean durations of fecal virus shedding and diarrhea and the lowest mean cumulative fecal consistency score among all the treatment groups. The durations of diarrhea in the LoLA+AttHRV pigs were significantly shorter compared to the AttHRV only and the mock-vaccinated control pigs. The durations of virus shedding in the LoLA+AttHRV pigs were significantly shorter compared to the HiLA+AttHRV and the mock control pigs. The mean cumulative fecal consistency scores in all the pigs in the LoLA+AttHRV and AttHRV only groups (8.4 and 9.0, respectively) were significantly lower than the control group, indicating significant protection against the severity of diarrhea. Thus, low dose LA slightly, but clearly improved the protection conferred by the AttHRV vaccine against rotavirus diarrhea. In contrast, high dose LA reduced the protection conferred by the AttHRV vaccine as indicated by the significantly longer mean duration of virus shedding (3.8 versus 1.3 days) and higher mean cumulative fecal scores compared to the AttHRV only pigs.

Treatments	n	Clinical signs			Fecal virus shedding (by CCIF and/or ELISA)		
		% with diarrhea [*, a]	Mean duration days [**]	Mean cumulative score [**, c]	% shedding virus [*]	Mean duration days [**]	Mean peak titer (FFU/ml) [**, d]
HiLA+AttHRV	13	92 [A]	4.3 (0.7[b]) [AB]	12.5 (1.4) [AB]	31 [B]	3.8 (0.3) [A]	2.0 [B]
LoLA+AttHRV	8	88 [A]	2.4 (0.7) [B]	8.4 (1.3) [B]	36 [B]	1.0 (0.0) [B]	5.6 [B]
AttHRV only	12	67 [A]	4.6 (0.5) [A]	9.8 (1.4) [B]	50 [B]	1.3 (0.2) [B]	4.9 [B]
Mock control	9	100 [A]	5.6 (0.3) [A]	14.4 (1.0) [A]	100 [A]	4.7 (0.7) [A]	4558 [A]

[A] The data was partially presented previously (Wen et al., 2012).
[a] Pigs with daily fecal scores of ≥2 were considered diarrheic. Fecal consistency was scored as follows: 0, normal; 1, pasty; 2, semiliquid; and 3, liquid.
[b] Standard error of the mean.
[c] Mean cumulative score calculation included all the pigs in each group.
[d] FFU, fluorescent focus forming units. Geometric mean peak titers were calculated among pigs that shed virus.
[*] Proportions in the same column with different superscript letters (A, B) differ significantly (Fisher's exact test, p≤0.05).
[**] Means in the same column with different superscript letters (A, B, C) differ significantly (Kruskal Wallis Test, p≤0.05).

Table 4. Clinical signs and rotavirus fecal shedding in Gn pigs after virulent HRV challenge[Δ]

We reported previously that protection rates against rotavirus diarrhea are correlated with virus-specific intestinal IgA ASC and IFN-γ producing T cell responses at PID 28 in Gn pigs (Yuan et al., 1996; Yuan et al., 2008). A balanced Th1 and Th2 type response is needed for the optimal protective immunity against rotavirus. Although low dose LA further reduced the duration of diarrhea in the AttHRV-vaccinated pigs postchallenge, neither low nor high dose LA significantly altered protection rate against rotavirus challenge (proportions of pigs that were infected and developed diarrhea after challenge). Because virus-specific intestinal

IgA ASC responses probably play a more important role in rotavirus protective immunity than the IFN-γ producing CD8+ T cell responses (Yuan et al., 1996; Yuan et al., 2008), the effect of LA on virus-specific ASC responses also need to be taken into consideration regarding the differences in the protection conferred by the AttHRV vaccine with high or low dose LA. Although the low dose LA enhanced IFN-γ producing CD8+ T cell responses, it had negative effects on the serum antibody and ASC responses induced by the AttHRV vaccine. To improve the AttHRV vaccine efficacy, a different dose of LA (possible an intermediate dose) or a different probiotic strain (i.e. LGG) may be optimal to promote a balanced Th1 and Th2 response without increasing Treg cell responses.

6. Conclusion

Differential modulating effects on innate and adaptive immune responses by low dose versus high dose of the same LA NCFM strain were clearly demonstrated in gnotobiotic pigs. Low dose LA significantly enhanced the Th1 type effector T cell responses and decreased Treg cell functions in AttHRV-vaccinated pigs. Meanwhile, low dose LA resulted in a suppressed Th2 response, as evidenced by significantly reduced virus-specific ASC responses and serum antibody titers compared to the AttHRV only group. The dose effects of LA on IFN-γ producing T cell and CD4+CD25-FoxP3+ Treg cell immune responses were similar between the intestinal and systemic lymphoid tissues. Thus the same probiotic strain used in different doses can either increase or reduce mucosal and systemic immune responses induced by vaccines. These findings have significant implications in the use of probiotic lactobacilli as immunostimulatory versus immunoregulatory agents. Probiotic products are increasingly used to improve health, alleviate disease symptoms, and enhance vaccine efficacy. Our findings suggest that probiotics can be ineffective or even detrimental if not used at the optimal dosage for the appropriate purposes, highlighting the importance of not only strain but also dose selection in probiotic studies.

The gnotobiotic pig model is a valuable animal model for study of probiotic-virus-host interaction because of the many similarities between human and porcine intestinal physiology and mucosal immune system (Meurens et al., 2012). The gnotobiotic status prevents confounding factors from commensal microflora that are present in conventionally reared animals or in humans. Unlike gnotobiotic mice, gnotobiotic pigs are devoid of maternal antibodies, thus providing an immunologically naïve background that allows clear identification of the immune responses to a single vaccine in hosts colonized with a qualitatively and quantitatively defined probiotic bacterial strain (Butler, 2009; Yuan & Saif, 2002). Although data from studies of gnotobiotic animal models may not be generalized directly to normal animals or humans, gnotobiotic animals provide a medium in which investigating the complex interrelationships of the host and its associated microbes become possible (Coates, 1975). Our findings provide a good starting point for identification of the optimal dosage of a probiotic strain. But nonetheless, the optimal dosage needs to be confirmed in conventionalized gnotobiotic pigs and in human clinical trials in order to achieve the appropriate adjuvant effect for rotavirus and other vaccines.

Author details

Lijuan Yuan, Ke Wen, Fangning Liu and Guohua Li
Department of Biomedical Sciences and Pathobiology,
Virginia Polytechnic Institute and State University, USA

Acknowledgement

Studies of probiotics' immune modulating effects on the rotavirus vaccine in gnotobiotic pigs were supported by a grant (R01AT004789) from the National Center of Complementary and Alternative Medicine (NCCAM), National Institutes of Health, Bethesda, MD.

7. References

Armah, G.E., Sow, S.O., Breiman, R.F., Dallas, M.J., Tapia, M.D., Feikin, D.R., Binka, F.N., Steele, A.D., Laserson, K.F., Ansah, N.A., Levine, M.M., Lewis, K., Coia, M.L., Attah-Poku, M., Ojwando, J., Rivers, S.B., Victor, J.C., Nyambane, G., Hodgson, A., Schodel, F., Ciarlet, M. & Neuzil, K.M. (2010). Efficacy of pentavalent rotavirus vaccine against severe rotavirus gastroenteritis in infants in developing countries in sub-Saharan Africa: a randomised, double-blind, placebo-controlled trial. *Lancet.* 376, 9741, (Aug 21, 2010), 606-614.

Azevedo, M.S., Zhang, W., Wen, K., Gonzalez, A.M., Saif, L.J., Yousef, A.E. & Yuan, L. (2012). Lactobacillus acidophilus and Lactobacillus reuteri modulate cytokine responses in gnotobiotic pigs infected with human rotavirus. *Benef Microbes.* 3, 1, (Mar 1, 2012), 33-42.

Boge, T., Remigy, M., Vaudaine, S., Tanguy, J., Bourdet-Sicard, R. & van der Werf, S. (2009). A probiotic fermented dairy drink improves antibody response to influenza vaccination in the elderly in two randomised controlled trials. *Vaccine.* 27, 41, (Sep 18, 2009), 5677-5684.

Butler, J.E. (2009). Isolator and other neonatal piglet models in developmental immunology and identification of virulence factors. *Anim Health Res Rev.* 10, 1, (Jun, 2009), 35-52.

Chiba, Y., Shida, K., Nagata, S., Wada, M., Bian, L., Wang, C., Shimizu, T., Yamashiro, Y., Kiyoshima-Shibata, J., Nanno, M. & Nomoto, K. (2010). Well-controlled proinflammatory cytokine responses of Peyer's patch cells to probiotic Lactobacillus casei. *Immunology.* 130, 3, (Jul, 2010), 352-362.

Christensen, H.R., Frokiaer, H. & Pestka, J.J. (2002). Lactobacilli differentially modulate expression of cytokines and maturation surface markers in murine dendritic cells. *J Immunol.* 168, 1, (Jan 1, 2002), 171-178.

Coates, M.E. (1975). Gnotobiotic animals in research: their uses and limitations. *Lab Anim.* 9, 4, (Oct, 1975), 275-282.

Davidson, L.E., Fiorino, A.M., Snydman, D.R. & Hibberd, P.L. (2011). Lactobacillus GG as an immune adjuvant for live-attenuated influenza vaccine in healthy adults: a randomized double-blind placebo-controlled trial. *Eur J Clin Nutr.* 65, 4, (Apr, 2011), 501-507.

de Vrese, M., Rautenberg, P., Laue, C., Koopmans, M., Herremans, T. & Schrezenmeir, J. (2005). Probiotic bacteria stimulate virus-specific neutralizing antibodies following a booster polio vaccination. *Eur J Nutr.* 44, 7, (Oct, 2005), 406-413.

Evrard, B., Coudeyras, S., Dosgilbert, A., Charbonnel, N., Alame, J., Tridon, A. & Forestier, C. (2011). Dose-dependent immunomodulation of human dendritic cells by the probiotic Lactobacillus rhamnosus Lcr35. *PLoS One*. 6, 42011), e18735.

Galdeano, C.M. & Perdigon, G. (2006). The probiotic bacterium Lactobacillus casei induces activation of the gut mucosal immune system through innate immunity. *Clin Vaccine Immunol*. 13, 2, (Feb, 2006), 219-226.

Grangette, C. (2012). Bifidobacteria and subsets of dendritic cells: friendly players in immune regulation! *Gut*. 61, 3, (Mar, 2012), 331-332.

Isolauri, E., Joensuu, J., Suomalainen, H., Luomala, M. & Vesikari, T. (1995). Improved immunogenicity of oral D x RRV reassortant rotavirus vaccine by Lactobacillus casei GG. *Vaccine*. 13, 3, (Feb, 1995), 310-312.

Jamin, A., Gorin, S., Le Potier, M.F. & Kuntz-Simon, G. (2006). Characterization of conventional and plasmacytoid dendritic cells in swine secondary lymphoid organs and blood. *Vet Immunol Immunopathol*. 114, 3-4, (Dec 15, 2006), 224-237.

Kaila, M., Isolauri, E., Soppi, E., Virtanen, E., Laine, S. & Arvilommi, H. (1992). Enhancement of the circulating antibody secreting cell response in human diarrhea by a human Lactobacillus strain. *Pediatr Res*. 32, 2, (Aug, 1992), 141-144.

Kekkonen, R.A., Kajasto, E., Miettinen, M., Veckman, V., Korpela, R. & Julkunen, I. (2008). Probiotic Leuconostoc mesenteroides ssp. cremoris and Streptococcus thermophilus induce IL-12 and IFN-gamma production. *World J Gastroenterol*. 14, 8, (Feb 28, 2008), 1192-1203.

Konieczna, P., Groeger, D., Ziegler, M., Frei, R., Ferstl, R., Shanahan, F., Quigley, E.M., Kiely, B., Akdis, C.A. & O'Mahony, L. (2012). Bifidobacterium infantis 35624 administration induces Foxp3 T regulatory cells in human peripheral blood: potential role for myeloid and plasmacytoid dendritic cells. *Gut*. 61, 3, (Mar, 2012), 354-366.

Konstantinov, S.R., Smidt, H., de Vos, W.M., Bruijns, S.C., Singh, S.K., Valence, F., Molle, D., Lortal, S., Altermann, E., Klaenhammer, T.R. & van Kooyk, Y. (2008). S layer protein A of Lactobacillus acidophilus NCFM regulates immature dendritic cell and T cell functions. *Proc Natl Acad Sci U S A*. 105, 49, (Dec 9, 2008), 19474-19479.

Li, C.Y., Lin, H.C., Lai, C.H., Lu, J.J., Wu, S.F. & Fang, S.H. (2011). Immunomodulatory Effects of Lactobacillus and Bifidobacterium on Both Murine and Human Mitogen-Activated T Cells. *Int Arch Allergy Immunol*. 156, 2, (May 16, 2011), 128-136.

Link-Amster, H., Rochat, F., Saudan, K.Y., Mignot, O. & Aeschlimann, J.M. (1994). Modulation of a specific humoral immune response and changes in intestinal flora mediated through fermented milk intake. *FEMS Immunol Med Microbiol*. 10, 1, (Nov, 1994), 55-63.

Maitra, U., Gan, L., Chang, S. & Li, L. (2011). Low-dose endotoxin induces inflammation by selectively removing nuclear receptors and activating CCAAT/enhancer-binding protein delta. *J Immunol*. 186, 7, (Apr 1, 2011), 4467-4473.

Mane, J., Pedrosa, E., Loren, V., Gassull, M.A., Espadaler, J., Cune, J., Audivert, S., Bonachera, M.A. & Cabre, E. (2011). A mixture of Lactobacillus plantarum CECT 7315 and CECT 7316 enhances systemic immunity in elderly subjects. A dose-response, double-blind, placebo-controlled, randomized pilot trial. *Nutr Hosp*. 26, 1, (Jan-Feb, 2011), 228-235.

Meijerink, M. & Wells, J.M. (2010). Probiotic modulation of dendritic cells and T cell responses in the intestine. *Benef Microbes*. 1, 4, (Nov, 2010), 317-326.

Meurens, F., Summerfield, A., Nauwynck, H., Saif, L. & Gerdts, V. (2012). The pig: a model for human infectious diseases. *Trends Microbiol*. 20, 1, (Jan, 2012), 50-57.

Mohamadzadeh, M., Duong, T., Hoover, T. & Klaenhammer, T.R. (2008). Targeting mucosal dendritic cells with microbial antigens from probiotic lactic acid bacteria. *Expert Rev Vaccines*. 7, 2, (Mar, 2008), 163-174.

Mohamadzadeh, M., Olson, S., Kalina, W.V., Ruthel, G., Demmin, G.L., Warfield, K.L., Bavari, S. & Klaenhammer, T.R. (2005). Lactobacilli activate human dendritic cells that skew T cells toward T helper 1 polarization. *Proc Natl Acad Sci U S A*. 102, 8, (Feb 22, 2005), 2880-2885.

O'Ryan, M. & Linhares, A.C. (2009). Update on Rotarix: an oral human rotavirus vaccine. *Expert Rev Vaccines*. 8, 12, (Dec, 2009), 1627-1641.

O'Ryan, M.L., Hermosilla, G. & Osorio, G. (2009). Rotavirus vaccines for the developing world. *Curr Opin Infect Dis*. 22, 5, (Oct, 2009), 483-489.

Olivares, M., Diaz-Ropero, M.P., Sierra, S., Lara-Villoslada, F., Fonolla, J., Navas, M., Rodriguez, J.M. & Xaus, J. (2007). Oral intake of Lactobacillus fermentum CECT5716 enhances the effects of influenza vaccination. *Nutrition*. 23, 3, (Mar, 2007), 254-260.

Paineau, D., Carcano, D., Leyer, G., Darquy, S., Alyanakian, M.A., Simoneau, G., Bergmann, J.F., Brassart, D., Bornet, F. & Ouwehand, A.C. (2008). Effects of seven potential probiotic strains on specific immune responses in healthy adults: a double-blind, randomized, controlled trial. *FEMS Immunol Med Microbiol*. 53, 1, (Jun, 2008), 107-113.

Perdigon, G., Fuller, R. & Raya, R. (2001). Lactic acid bacteria and their effect on the immune system. *Curr Issues Intest Microbiol*. 2, 1, (Mar, 2001), 27-42.

Repa, A., Grangette, C., Daniel, C., Hochreiter, R., Hoffmann-Sommergruber, K., Thalhamer, J., Kraft, D., Breiteneder, H., Mercenier, A. & Wiedermann, U. (2003). Mucosal co-application of lactic acid bacteria and allergen induces counter-regulatory immune responses in a murine model of birch pollen allergy. *Vaccine*. 22, 1, (Dec 8, 2003), 87-95.

Rowe, J.H., Ertelt, J.M. & Way, S.S. (2012). Foxp3(+) regulatory T cells, immune stimulation and host defence against infection. *Immunology*. 136, 1, (May, 2012), 1-10.

So, J.S., Kwon, H.K., Lee, C.G., Yi, H.J., Park, J.A., Lim, S.Y., Hwang, K.C., Jeon, Y.H. & Im, S.H. (2008a). Lactobacillus casei suppresses experimental arthritis by down-regulating T helper 1 effector functions. *Mol Immunol*. 45, 9, (May, 2008a), 2690-2699.

So, J.S., Lee, C.G., Kwon, H.K., Yi, H.J., Chae, C.S., Park, J.A., Hwang, K.C. & Im, S.H. (2008b). Lactobacillus casei potentiates induction of oral tolerance in experimental arthritis. *Mol Immunol*. 46, 1, (Nov, 2008b), 172-180.

Summerfield, A. & McCullough, K.C. (2009). The porcine dendritic cell family. *Dev Comp Immunol*. 33, 3, (Mar, 2009), 299-309.

Torii, A., Torii, S., Fujiwara, S., Tanaka, H., Inagaki, N. & Nagai, H. (2007). Lactobacillus Acidophilus strain L-92 regulates the production of Th1 cytokine as well as Th2 cytokines. *Allergol Int*. 56, 3, (Sep, 2007), 293-301.

Van Overtvelt, L., Moussu, H., Horiot, S., Samson, S., Lombardi, V., Mascarell, L., van de Moer, A., Bourdet-Sicard, R. & Moingeon, P. (2010). Lactic acid bacteria as adjuvants for sublingual allergy vaccines. *Vaccine*. 28, 17, (Apr 9, 2010), 2986-2992.

Villadangos, J.A., Cardoso, M., Steptoe, R.J., van Berkel, D., Pooley, J., Carbone, F.R. & Shortman, K. (2001). MHC class II expression is regulated in dendritic cells independently of invariant chain degradation. *Immunity*. 14, 6, (Jun, 2001), 739-749.

Vitini, E., Alvarez, S., Medina, M., Medici, M., de Budeguer, M.V. & Perdigon, G. (2000). Gut mucosal immunostimulation by lactic acid bacteria. *Biocell*. 24, 3, (Dec, 2000), 223-232.

Ward, L.A., Rosen, B.I., Yuan, L. & Saif, L.J. (1996). Pathogenesis of an attenuated and a virulent strain of group A human rotavirus in neonatal gnotobiotic pigs. *J Gen Virol*. 77 (Pt 7), (Jul, 1996), 1431-1441.

Weiss, G., Rasmussen, S., Zeuthen, L.H., Nielsen, B.N., Jarmer, H., Jespersen, L. & Frokiaer, H. (2010). Lactobacillus acidophilus induces virus immune defence genes in murine dendritic cells by a Toll-like receptor-2-dependent mechanism. *Immunology*. 131, 2, (Oct, 2010), 268-281.

Wells, J.M. & Mercenier, A. (2008). Mucosal delivery of therapeutic and prophylactic molecules using lactic acid bacteria. *Nat Rev Microbiol*. 6, 5, (May, 2008), 349-362.

Wen, K., Li, G., Bui, T., Liu, F., Li, Y., Kocher, J., Lin, L., Yang, X. & Yuan, L. (2012). High dose and low dose Lactobacillus acidophilus exerted differential immune modulating effects on T cell immune responses induced by an oral human rotavirus vaccine in gnotobiotic pigs. *Vaccine*. 30, 6, (Feb 1, 2012), 1198-1207.

Winkler, P., de Vrese, M., Laue, C. & Schrezenmeir, J. (2005). Effect of a dietary supplement containing probiotic bacteria plus vitamins and minerals on common cold infections and cellular immune parameters. *International journal of clinical pharmacology and therapeutics*. 43, 7, (Jul, 2005), 318-326.

Yazdi, M.H., Soltan Dallal, M.M., Hassan, Z.M., Holakuyee, M., Agha Amiri, S., Abolhassani, M. & Mahdavi, M. (2010). Oral administration of Lactobacillus acidophilus induces IL-12 production in spleen cell culture of BALB/c mice bearing transplanted breast tumour. *Br J Nutr*. 104, 2, (Jul, 2010), 227-232.

Yuan, L. & Saif, L.J. (2002). Induction of mucosal immune responses and protection against enteric viruses: rotavirus infection of gnotobiotic pigs as a model. *Vet Immunol Immunopathol*. 87, 3-4, (Sep 10, 2002), 147-160.

Yuan, L., Ward, L.A., Rosen, B.I., To, T.L. & Saif, L.J. (1996). Systematic and intestinal antibody-secreting cell responses and correlates of protective immunity to human rotavirus in a gnotobiotic pig model of disease. *J Virol*. 70, 5, (May, 1996), 3075-3083.

Yuan, L., Wen, K., Azevedo, M.S., Gonzalez, A.M., Zhang, W. & Saif, L.J. (2008). Virus-specific intestinal IFN-gamma producing T cell responses induced by human rotavirus infection and vaccines are correlated with protection against rotavirus diarrhea in gnotobiotic pigs. *Vaccine*. 26, 26, (Jun 19, 2008), 3322-3331.

Zaman, K., Dang, D.A., Victor, J.C., Shin, S., Yunus, M., Dallas, M.J., Podder, G., Vu, D.T., Le, T.P., Luby, S.P., Le, H.T., Coia, M.L., Lewis, K., Rivers, S.B., Sack, D.A., Schodel, F., Steele, A.D., Neuzil, K.M. & Ciarlet, M. (2010). Efficacy of pentavalent rotavirus vaccine against severe rotavirus gastroenteritis in infants in developing countries in Asia: a randomised, double-blind, placebo-controlled trial. *Lancet*. 376, 9741, (Aug 21, 2010), 615-623.

Zhang, W., Azevedo, M.S., Gonzalez, A.M., Saif, L.J., Van Nguyen, T., Wen, K., Yousef, A.E. & Yuan, L. (2008a). Influence of probiotic Lactobacilli colonization on neonatal B cell responses in a gnotobiotic pig model of human rotavirus infection and disease. *Vet Immunol Immunopathol*. 122, 1-2, (Mar 15, 2008a), 175-181.

Zhang, W., Azevedo, M.S., Wen, K., Gonzalez, A., Saif, L.J., Li, G., Yousef, A.E. & Yuan, L. (2008b). Probiotic Lactobacillus acidophilus enhances the immunogenicity of an oral rotavirus vaccine in gnotobiotic pigs. *Vaccine*. 26, 29-30, (Jul 4, 2008b), 3655-3661.

Permissions

The contributors of this book come from diverse backgrounds, making this book a truly international effort. This book will bring forth new frontiers with its revolutionizing research information and detailed analysis of the nascent developments around the world.

We would like to thank J. Marcelino Kongo, for lending his expertise to make the book truly unique. He has played a crucial role in the development of this book. Without his invaluable contribution this book wouldn't have been possible. He has made vital efforts to compile up to date information on the varied aspects of this subject to make this book a valuable addition to the collection of many professionals and students.

This book was conceptualized with the vision of imparting up-to-date information and advanced data in this field. To ensure the same, a matchless editorial board was set up. Every individual on the board went through rigorous rounds of assessment to prove their worth. After which they invested a large part of their time researching and compiling the most relevant data for our readers. Conferences and sessions were held from time to time between the editorial board and the contributing authors to present the data in the most comprehensible form. The editorial team has worked tirelessly to provide valuable and valid information to help people across the globe.

Every chapter published in this book has been scrutinized by our experts. Their significance has been extensively debated. The topics covered herein carry significant findings which will fuel the growth of the discipline. They may even be implemented as practical applications or may be referred to as a beginning point for another development. Chapters in this book were first published by InTech; hereby published with permission under the Creative Commons Attribution License or equivalent.

The editorial board has been involved in producing this book since its inception. They have spent rigorous hours researching and exploring the diverse topics which have resulted in the successful publishing of this book. They have passed on their knowledge of decades through this book. To expedite this challenging task, the publisher supported the team at every step. A small team of assistant editors was also appointed to further simplify the editing procedure and attain best results for the readers.

Our editorial team has been hand-picked from every corner of the world. Their multi-ethnicity adds dynamic inputs to the discussions which result in innovative outcomes. These outcomes are then further discussed with the researchers and contributors who give their valuable feedback and opinion regarding the same. The feedback is then collaborated with the researches and they are edited in a comprehensive manner to aid the understanding of the subject.

Apart from the editorial board, the designing team has also invested a significant amount of their time in understanding the subject and creating the most relevant covers. They scrutinized every image to scout for the most suitable representation of the subject and create an appropriate cover for the book.

The publishing team has been involved in this book since its early stages. They were actively engaged in every process, be it collecting the data, connecting with the contributors or procuring relevant information. The team has been an ardent support to the editorial, designing and production team. Their endless efforts to recruit the best for this project, has resulted in the accomplishment of this book. They are a veteran in the field of academics and their pool of knowledge is as vast as their experience in printing. Their expertise and guidance has proved useful at every step. Their uncompromising quality standards have made this book an exceptional effort. Their encouragement from time to time has been an inspiration for everyone.

The publisher and the editorial board hope that this book will prove to be a valuable piece of knowledge for researchers, students, practitioners and scholars across the globe.

List of Contributors

A.K. Szczepankowska, R.K. Górecki and J.K. Bardowski
Institute of Biochemistry and Biophysics of Polish Academy of Sciences, Warsaw, Poland

P. Kołakowski
Danisco Biolacta, Innovation, Olsztyn, Poland

J. Marcelino Kongo
Instituto de Inovação Tecnológica dos Açores (INOVA), Currently at Canadian Research Institute of Food Safety

F. Martin, B. Ebel, C. Rojas, P. Gervais, N. Cayot and R. Cachon
Unité Procédés Alimentaires et Microbiologiques, UMR A 02.102, AgroSup Dijon/Université deourgogne, 1 esplanade Erasme, Dijon, France

Lothar Kröckel
Max Rubner-Institute Location Kulmbach, Department of Safety and Quality of Meat, Kulmbach, Germany

Jirasak Kongkiattikajorn
School of Bioresources and Technology, King Mongkut's University of Technology Thonburi, Thailand

Lavinia Claudia Buruleanu, Magda Gabriela Bratu, Iuliana Manea, Daniela Avram and Carmen Leane Nicolescu
Department of Food Engineering, Faculty of Environmental Engineering and Biotechnology, Valahia University of Targoviste, Romania

Belal J. Muhialdin and Nazamid Saari
Universiti Putra Malaysia (UPM), Malaysia

Zaiton Hassan
Universiti Sains Islam Malaysia (USIM), Malaysia

Gülden Başyiğit Kiliç
Mehmet Akif Ersoy University, Department of Food Engineering, Faculty of Engineering-Architecture, Burdur, Turkey

Katia Sivieri,
Department of Food & Nutrition, Faculty of Pharmaceutical Sciences, São Paulo State University, Araraquara, SP, Brazil

Raquel Bedani
University of São Paulo, Faculty of Pharmaceutical Sciences, São Paulo, SP, Brazil

Daniela Cardoso Umbelino Cavallini
Department of Food & Nutrition, Faculty of Pharmaceutical Sciences, São Paulo State University,
Araraquara, SP, Brazil

Elizeu A. Rossi
Department of Food & Nutrition, Faculty of Pharmaceutical Sciences, São Paulo State University, Araraquara, SP, Brazil

Anne Pihlanto
MTT Agrifood Finland, Biotechnology and Food Research, Jokioinen, Finland

Nditange Shigwedha and Li Jia
University of Namibia & Meat Corporation of Namibia, Namibia

Daoud Harzallah and Hani Belhadj
Laboratory of Applied Microbiology, Faculty of Natural and Life Sciences, University Ferhat Abbas, 19000, Sétif, Algeria

Lijuan Yuan, Ke Wen, Fangning Liu and Guohua Li
Department of Biomedical Sciences and Pathobiology, Virginia Polytechnic Institute and State University, USA